EQUATIONS WITH TRANSFORMED ARGUMENT

AN ALGEBRAIC APPROACH

Modern Analytic *and* Computational Methods *in* Science *and* Mathematics

A GROUP OF MONOGRAPHS AND ADVANCED TEXTBOOKS

Richard Bellman, EDITOR
University of Southern California

Published

1. R. E. Bellman, R. E. Kalaba, and Marcia C. Prestrud, Invariant Imbedding and Radiative Transfer in Slabs of Finite Thickness, 1963
2. R. E. Bellman, Harriet H. Kagiwada, R. E. Kabala, and Marcia C. Prestrud, Invariant Imbedding and Time-Dependent Transport Processes, 1964
3. R. E. Bellman and R. E. Kalaba, Quasilinearization and Nonlinear Boundary-Value Problems, 1965
4. R. E. Bellman, R. E. Kalaba, and Jo Ann Lockett, Numerical Inversion of the Laplace Transform: Applications to Biology, Economics, Engineering, and Physics, 1966
5. S. G. Mikhlin and K. L. Smolitskiy, Approximate Methods for Solution of Differential and Integral Equations, 1967
6. R. N. Adams and E. D. Denman, Wave Propagation and Turbulent Media, 1966
7. R. L. Stratonovich, Conditional Markov Processes and Their Application to the Theory of Optimal Control, 1968
8. A. G. Ivakhnenko and V. G. Lapa, Cybernetics and Forecasting Techniques, 1967
9. G. A. Chebotarev, Analytical and Numerical Methods of Celestial Mechanics, 1967
10. S. F. Feshchenko, N. I. Shkil', and L. D. Nikolenko, Asymptotic Methods in the Theory of Linear Differential Equations, the Theory of Linear Differential Equations, 1967
11. A. G. Butkovskiy, Distributed Control Systems, 1969
12. R. E. Larson, State Increment Dynamic Programming, 1968
13. J. Kowalik and M. R. Osborne, Methods for Unconstrained Optimization Problems, 1968
14. S. J. Yakowitz, Mathematics of Adaptive Control Processes, 1969
15. S. K. Srinivasan, Stochastic Theory and Cascade Processes, 1969
16. D. U. von Rosenberg, Methods for the Numerical Solution of Partial Differential Equations, 1969
17. R. B. Banerji, Theory of Problem Solving: An Approach to Artificial Intelligence, 1969
18. R. Lattès and J.-L. Lions, The Method of Quasi-Reversibility: Applications to Partial Differential Equations. Translated from the French edition and edited by Richard Bellman, 1969
19. D. G. B. Edelen, Nonlocal Variations and Local Invariance of Fields, 1969
20. J. R. Radbill and G. A. McCue, Quasilinearization and Nonlinear Problems in Fluid and Orbital Mechanics, 1970
21. W. Squire, Integration for Engineers and Scientists, 1970
22. T. Parthasarathy and T. E. S. Raghavan, Some Topics in Two-Person Games, 1971
23. T. Hacker, Flight Stability and Control, 1970

24. D. H. Jacobson and D. Q. Mayne, Differential Dynamic Programming, 1970
25. H. Mine and S. Osaki, Markovian Decision Processes, 1970
26. W. Sierpiński, 250 Problems in Elementary Number Theory, 1970
27. E. D. Denman, Coupled Modes in Plasmas, Elastic Media, and Parametric Amplifiers, 1970
28. F. A. Northover, Applied Diffraction Theory, 1971
29. G. A. Phillipson, Identification of Distributed Systems, 1971
30. D. H. Moore, Heaviside Operational Calculus: An Elementary Foundation, 1971
31. S. M. Roberts and J. S. Shipman, Two-Point Boundary Value Problems: Shooting Methods, 1971
32. V. F. Demyanov and A. M. Rubinov, Approximate Methods in Optimization Problems, 1970
33. S. K. Srinivasan and R. Vasudevan, Introduction to Random Differential Equations and Their Applications, 1971
34. C. J. Mode, Multitype Branching Processes: Theory and Applications, 1971
35. R. Tompvić and M. Vukobratović, General Sensitivity Theory, 1971
36. J. G. Krzyż, Problems in Complex Variable Theory, 1971
37. W. T. Tutte, Introduction to the Theory of Matroids, 1971
38. B. W. Rust and W. R. Burrus, Mathematical Programming and the Numerical Solution of Linear Equations, 1972
39. J. O. Mingle, The Invariant Imbedding Theory of Nuclear Transport, 1973
40. H. M. Lieberstein, Mathematical Physiology, 1973

EQUATIONS WITH TRANSFORMED ARGUMENT
AN ALGEBRAIC APPROACH

by

DANUTA PRZEWORSKA-ROLEWICZ

Institute of Mathematics of the Polish Academy of Sciences

ELSEVIER SCIENTIFIC PUBLISHING COMPANY
AMSTERDAM

PWN—POLISH SCIENTIFIC PUBLISHERS
WARSZAWA
1973

Distribution of this book is being handled by the following publishers:

for the U.S.A. and Canada

American Elsevier Publishing Company, Inc.
52 Vanderbilt Avenue
New York, New York 10017

for Albania, Bulgaria, Chinese People's Republic, Czechoslovakia, Cuba, German Democratic Republic, Hungary, Korean People's Democratic Republic, Mongolia, Poland, Rumania, Democratic Republic of Vietnam, U.S.S.R. and Yugoslavia

PWN—Polish Scientific Publishers
00-251 Warszawa (Poland), Miodowa 10

for all remaining areas

Elsevier Scientific Publishing Company
355 Jan van Galenstraat
P.O. Box 1270, Amsterdam, The Netherlands

ISBN 0-444-41078-3
Library of Congress Catalog Card Number 72-87962

Printed in Poland (DRP)

To the Memory
of my Professor
WITOLD POGORZELSKI

Contents

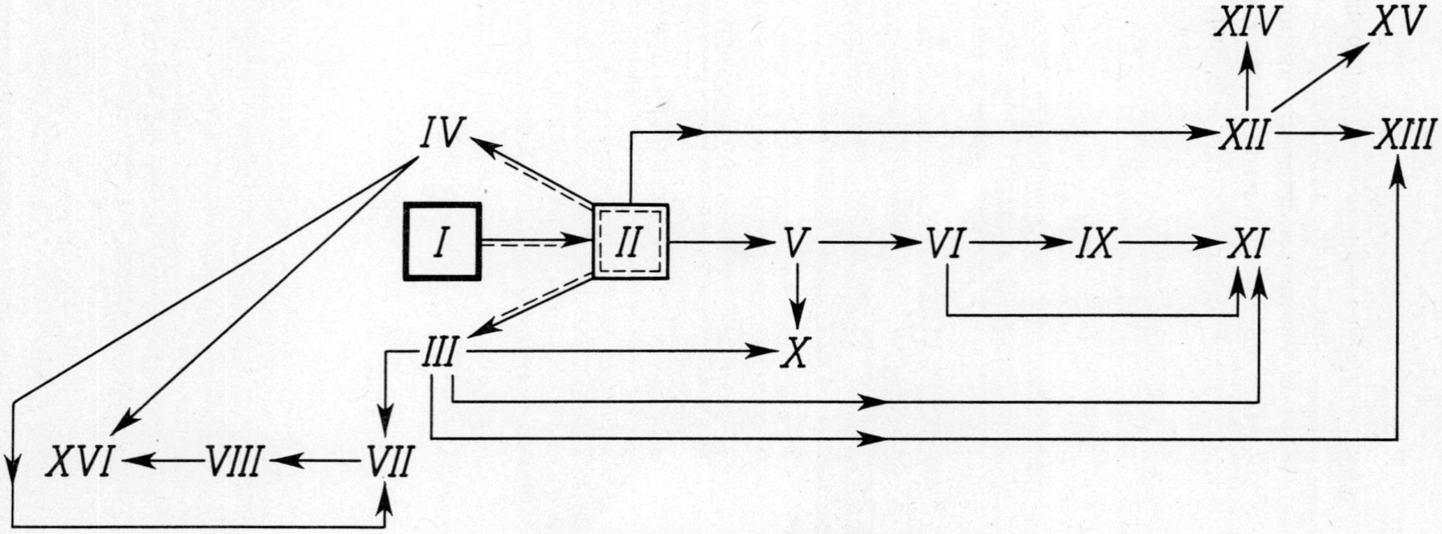

Fig. 0. Internal order of the book

"Alias aequationes differentio differentiales nisi huius modi, nemo adhuc, quantum scio, ad differentiales primi gradusi unquam reduxit, nisi forte in promptu fuerit eas prorsus integrare. Hic autem methodus exponam, qua non quidem omnes sed tamen innumerabiles aequationes differentio-differentiales utut ab utraque indeterminata affectae ad simpliciter differentiales reduci poterant."

L. Euler, 1728, [1]*

---

* Indeed, in the cited paper Euler did not consider differential-difference equations in the present sense, as has been wrongly suggested in many books concerning these equations. As regards J. Bernoulli's papers (printed in the same volume of Comentarii) we can say exactly the same. Euler examined equations which can be written as a difference of differentials, i.e. differential equations with separated variables, called by him "difference differential equations".

Preface

By an equation with a transformed argument we shall mean an equation, in which together with the unknown function $x(t)$ and possibly its derivatives, integrals, etc., the values $x(a(t))$, $x(b(t))$, ... (and the respective derivatives, integrals, ...) appear. We shall not enumerate or define here all the different types of equations with transformed argument and their applications in engineering, physics and economics, but remark only that the following types of equations with transformed argument play an important role in many applications: differential-difference equations, functional-differential equations (especially those with a retarded argument and of the neutral type), singular integral equations of Carleman type and the Wiener–Hopf equations.

The study of these equations is rather difficult and leads more often to existence theorems and qualitative results than to effective formulae giving solutions.

The present book consist largely of an attempt to fill this gap at least partially. This attempt, however, is limited by the method used: that is, all questions will be tackled by means of different applications of the method of algebraic operators developed by the author from 1960 onwards (see: Przeworska–Rolewicz [1]–[15], Przeworska–Rolewicz and Rolewicz [6]–[8]). Since this method is purely algebraic, it can only be applied to particular classes of transformations of argument in various types of equations. Thus the only feature common to all parts of the book, which in fact deals with a wide variety of types of equations, is the method of algebraic operators.

The book includes some results which, although not concerned with equations with transformed argument, are also obtained by the method of algebraic operators.

The list of problems which can be solved by the method of algebraic operators increased rapidly while this book was being written, and for this reason we here indicate only some of the possibilities of new applications.

The book is not self-contained. To keep it in a reasonable size we generally omit the proofs of the auxiliary theorems and of the existence and qualitative theorems concerning the types of equations that are considered. The extensive bibliography should enable the reader to find the relevant information on many topics. Since the number of types of equations considered is quite large, this bibliography cannot be complete. Since most of the results given in this book are the work of the author, no name is indicated for these theorems by way of attribution.

The book can be read by engineers, physicists and economists as well as by mathematicians, for such applied mathematical tools as are used are fairly elementary.

I wish to express my thanks to Professor J. Kudrewicz of the Technical University of Warsaw and to Dr. K. Malanowski of the Institute of Applied Cybernetics of the Polish Academy of Sciences for their valuable remarks and suggestions.

I am especially indebted to Dr. A. Manitius of the Technical University of Warsaw, who read the manuscript, for his much appreciated corrections and suggestions.

A debt of gratitude is also due to my husband Professor S. Rolewicz for his patience and help while this book was being written.

Finally, I wish to thank Mr. J. Panz, editor, whose initiative and advice was a great help in the preparation of this book.

Warszawa, July 1, 1971

DANUTA PRZEWORSKA-ROLEWICZ

CHAPTER I

Preliminaries

This chapter contains auxiliary notions and theorems. The proofs of the theorems from Section 1 can be found in any book on groups and algebras, those from Sections 2, 3, 4 — in the book of the present author and Rolewicz [6].

1. LINEAR SPACES. LINEAR RINGS

We assume the reader to be acquainted with the notions of a set, a subset and a family of sets, and with the standard operations on sets. We denote by $A \cup B$ the union (sum), by $A \setminus B$ the difference, and by $A \cap B$ the intersection (product) of the sets A and B. Let a family $\mathfrak{A}$ of sets be given. We denote by $\bigcup_{A \in \mathfrak{A}} A$ the union and by $\bigcap_{A \in \mathfrak{A}} A$ the intersection of the sets of the family $\mathfrak{A}$. In the sequel we shall often speak of the "points" rather than the "elements" of a set A. A set containing only one element x will be denoted by $\{x\}$.

We also assume that the reader is acquainted with the notions of a relation, a function and a of one-to-one function. In the sequel, the terms "map" and "transformation" will have the same meaning as "function".

Sets A and B are said to be *of the same power* if there exists a one-to-one map of A onto B. Sets of the same power as the set of all positive integers are called *countable*.

Let a set X be given. Suppose further that to every two elements x and y of the set X there corresponds a unique third element $z \in X$, which will be denoted by $z = x \odot y$. If the operation $\odot$ is *associative*, i.e.

$$(x \odot y) \odot z = x \odot (y \odot z) \quad \text{for all } x, y, z \in X$$

and if for any two elements $x, y \in X$ there exist elements $z_1, z_2 \in X$ such that

$$x \odot z_1 = y \quad \text{and} \quad z_2 \odot x = y,$$

then the set X is called a *group*.

If X is a group and the equality $x \odot y = y \odot x$ holds for any two elements

$x, y \in X$, then X is called a *commutative group* or an *Abelian group*. The operation $\odot$ in an Abelian group is denoted traditionally by "+" and called *addition*. The element z satisfying the equation $x+z = y$ is called the *difference* of the elements and is denoted by $z = y-x$. The operation "$-$" is called *subtraction*. The element $x-x$ is denoted by 0 and is called the *zero element* or the *neutral element* of the group X. Note that $x+0 = x$ for all $x \in X$. Evidently, the zero element 0 is unique. Indeed, if 0_1 and 0_2 are two zeros in the group X, then $0_1 = 0_1+0_2 = 0_2$. A subset M of a group X is called a *subgroup* of the group X if M is a group under the same group operation as that of X. If M, N are subsets of a group X, we write

$$M \odot N = \{x \odot y\colon\ x \in M,\ y \in N\}.$$

If X is an Abelian group, the set $M+N$ is called the *algebraic sum* of the sets M and N. The sum $\{x\}+M$ is denoted briefly by $x+M$.

A commutative group X endowed with a further associative operation $a \cdot b$ besides $a+b$ is called a *ring* if the following *distributivity* conditions are satisfied:

$$(1.1) \qquad z(x+y) = zx+zy; \quad (x+y)z = xz+yz \quad \text{for all } x, y, z \in X.$$

The operation $a \cdot b$ is called *multiplication*.

If there exists an element $e_r \in X$ (resp. $e_l \in X$) satisfying the condition $x \cdot e_r = x$ (resp. $e_l \cdot x = x$) for all $x \in X$, then this element is called the *right unit* (resp. the *left unit*). If the units e_r and e_l both exist, then they are equal, for $e_l = e_l \cdot e_r = e_r$. Such an element is called the *unit* and it will be denoted by e. A ring possessing a unit is known as a *ring with unit*. Every ring X can be embedded in a ring X_1 with unit. Indeed, let us define the product of an element $x \in X$ with an integer p as follows:

if $p = 0$, then $p \cdot x = 0$;

if $p = n$, then $p \cdot x = \underbrace{x+x+\ldots+x}_{n\text{-fold}}$, where n is a positive integer;

if $p = -n$, then $(-n)x = n(-x)$, where $-x$ is the element $0-x$.

Consider now the set

$$X_1 = \{(x, p)\colon\ x \in X;\ p \text{ is an integer}\}.$$

The operations of multiplication by a scalar, addition of elements and multiplication of elements can be defined in X_1 in the following manner:

$$q(x, p) = (qx, qp),$$

$$(x, p)+(y, q) = (x+y, p+q), \quad \text{where } x, y \in X;\ p, q \text{ are integers},$$

$$(x, p)(y, q) = (xy+qx+py, qp).$$

It is easy to check that the set X_1 is a ring under the operations defined above and that the pair $(0, 1)$ is the unit of this ring. Moreover, the set X is identifiable with the subset of X_1 consisting of all pairs of the form $(x, 0)$.

Let x be an element of a ring X with unit. If there is an element x_r (resp. x_l) such that $xx_r = e$ (resp. $x_l x = e$), the element x is said to be *right* (*left*) *invertible*. If x is both right invertible and left invertible, it is called an *invertible element*. In this case we have $x_l = x_l x x_r = x_r$. The element x_r is called the *inverse* of x and is denoted by x^{-1}.

A subset M of a ring X is called a *subring* of the ring X if it is itself a ring under the same operations. If M and N are subsets of a ring X, we write

$$M \cdot N = \{xy: x \in M;\ y \in N\}.$$

The set $M \cdot N$ is called the *algebraic product* of the sets M and N. For brevity we write $xM = \{x\} \cdot M$, $Mx = \{M\}x$, where $M \subset X$, $x \in X$.

We say that the elements $x, y \in X$ are *commutative* if $xy = yx$. We sometimes say alternatively that x *commutes with* y. If every pair of elements of a ring X is commutative, X is called a *commutative ring*.

A ring X such that every non-zero element of X has an inverse is called a *field*. For instance the set $\boldsymbol{R}$ of real numbers and the set $\boldsymbol{C}$ of complex numbers are fields with the usual definitions of addition and multiplication as ring operations.

A *right ideal* (*left ideal*) in a ring X is a non-empty set $M \subset X$ such that

(i) if $x, y \in M$, then $x - y \in M$,

(ii) if $x \in X$, then $Mx \subset M$ (resp. $xM \subset M$).

If M is both a right ideal and a left ideal, M is called a *two-sided ideal*. It is obvious that the whole ring X and the set $\{0\}$ are ideals. These ideals are called *trivial ideals*. All other ideals are said to be *non-trivial* or *proper ideals*. It is easily verified that the intersection of an arbitrary number of left (right) ideals is a left (right) ideal, which is not necessarily proper.

A proper ideal cannot contain the unit of the ring. A proper right (left) ideal cannot contain a right (left) invertible element. Conversely if an element x is not right (left) invertible, then there exists a proper right (left) ideal containing x, namely the set xX (resp. Xx). A proper right (left, two-sided) ideal is called *maximal* if every proper right (left, two-sided) ideal M_1 containing M is equal to M, i.e. $M_1 \supset M$ implies $M_1 = M$. Every proper right (left, two-sided) ideal is contained in a maximal right (left, two-sided) ideal. Note that if M is a two-sided ideal, the maximal right- and the maximal left-ideals containing M need not to be the same.

A *radical* of a ring X with unit is the set $\boldsymbol{R}(X)$ of all elements z such that the element $e + xzy$ is invertible for all $x, y \in X$. $\boldsymbol{R}(X)$ is a two-sided ideal in X.

Indeed, since x and y are arbitrary elements of the ring, $z \in \boldsymbol{R}(X)$ implies $x_1 z$ and $zy_1 \in \boldsymbol{R}(X)$ for arbitrary $x_1, y_1 \in X$. Now suppose that $z, u \in \boldsymbol{R}(X)$ and let $x, y \in X$. Then the element

$$e+x(z-u)y = e+xzy-xuy = (e+xzy)\,[e-(e+xzy)^{-1}xuy]$$

is invertible.

The radical may also be defined as the intersection of all right (left) ideals (see Jacobson [1]).

Suppose that a group X and a subgroup M of X are given such that $y\odot M = M\odot y$ for every $y \in X$. To each element $x \in X$ there correspons its *coset*, i.e. the set of elements of the form $[x] = xM = \{x\odot z\colon z \in M\}$. We define: $[x]\odot[y] = [x\odot y]$. Let us observe that $[x]\odot[y] \supset \{z = x\odot y\colon x \in [x],\ y \in [y]\}$. The set of all cosets together with the group operation defined above is denoted by X/M and is called the *quotient group* or the *factor group* of X by M.

Let X be a ring, and let $M \subset X$ be a two-sided ideal. The operation of multiplication in the quotient group X/M is defined as follows: $[xy] \supset [x]\,[y] = \{z = xy\colon x \in [x],\ y \in [y]\}$. This operation is associative and distributive. Hence the set X/M is a ring, known as the *quotient ring*. We can plainly write $[x] = x+ +M$. If X is a commutative ring, then X/M is a field if and only if M is a maximal ideal.

A *linear space over the field* $\mathscr{F}$ *of scalars* is a commutative group X such that multiplication of elements of X by scalars is defined and satisfies the following conditions:

$$t(x+y) = tx+ty; \quad (t+s)x = tx+sx; \quad (ts)x = t(sx); \quad 1\cdot x = x; \quad 0\cdot x = 0$$

for all $x, y \in X$ and $t, s \in \mathscr{F}$. It follows from the above conditions that if $tx = 0$ and $x \neq 0$, then $t = 0$.

Since most results for linear spaces over the field $\boldsymbol{R}$ of reals and over the field $\boldsymbol{C}$ of complexes are the same, the term "linear space" in the sequel will include both real and complex linear spaces, provided there is no risk of a confusion. The word "number" or "scalar" will obviously mean an element from the particular field under consideration. The same should be said with regard to functions assuming numerical values; considering real-valued and complex valued functions, we obtain examples of linear spaces over the field of real numbers and the field of complex numbers respectively.

Let a linear space X and a subset $Y \subset X$ be given, and suppose that the sum of two elements of Y and the product of an element of Y with a scalar again belong to Y. Such a subset $Y \subset X$ is called a *linear subset*, a *linear manifold* or a *subspace* of X.

Let E be an arbitrary subset of a linear space X. The least linear subset containing the set E is called the *space spanned by the set E*, or the *linear span* of E and it is denoted by $\operatorname{lin} E$. It is easy to show that

$$\operatorname{lin} E = \left\{x \in X\colon x = \sum_{j=1}^{n} t_j x_j, \text{ where the } t_j \text{ are scalars and } x_j \in E\right\}.$$

We say that an *element* $x \in X$ is *linearly dependent on a set* E (or *on elements of the set* E) if $x \in \operatorname{lin} E$. If x is not linearly dependent on the set E, we say that x is *linearly independent of the set E*. A set E is called *linearly independent* if there is no element $x \in E$ linearly dependent on the set consisting of the remaining elements of E, i.e. if for all $x \in E$, $x \notin \operatorname{lin}(E \setminus \{x\})$.

It follows from the form of the set $\operatorname{lin} E$ that the elements x_j are linearly independent if the equality $t_1 x_1 + \ldots + t_n x_n = 0$ implies $t_1 = t_2 = \ldots = t_n = 0$. If the maximal number of linearly independent elements belonging to a linear space X is finite, we call this number the *dimension of the space* X and denote it by $\dim X$. Otherwise we say that the dimension of the space X is *infinite* and we write $\dim X = +\infty$. If $\dim X < +\infty$, we call X *finite dimensional*, and if $\dim X = +\infty$, X is said to be *infinite dimensional*. Observe that if the same finite dimensional space X is considered both as a space over the field of complex numbers and as a space over the field of reals, its dimension in the second case is twice its value in the first case. When discussing linear spaces over the field of complex numbers, in what follows, we shall always have in mind the dimension over this field.

The linear span of n linearly independent elements x_j is of dimension n:

$$\dim \operatorname{lin}(x_1, \ldots, x_n) = n.$$

The *Cartesian product* $X \times Y$ of two linear spaces X and Y is the space of all ordered pairs (x, y) with addition of elements and multiplication of an element by a scalar defined by the formulae:

$$(x_1, y_1) + (x_2, y_2) = (x_1 + x_2, y_1 + y_2); \qquad t(x, y) = (tx, ty)$$

for all $x, x_1, x_2 \in X$; $y, y_1, y_2 \in Y$; $t \in \mathscr{F}$.

If Y and Z are subspaces of a linear space X and if $Y \cap Z = \{0\}$, i.e. if the intersection of Y and Z consists solely of the element 0, then the set $Y+Z$ is called the *direct sum* of the spaces Y and Z and is denoted by $Y \oplus Z$. Note that the condition $Y \cap Z = \{0\}$ implies that every element $x \in Y \oplus Z$ can be written in the form $x = y + z$, where $y \in Y$ and $z \in Z$, in a unique way. If $X = Y \oplus Z$, we say that *X can be decomposed into a direct sum*. If X is a linear space and Y is a subspace of the space X, then there always exists a subspace $Z \subset X$ such that X can be decomposed into a direct sum $X = Y \oplus Z$ (provided that the axiom of choice is assumed to hold).

Let X be a linear space over the field of real numbers. Then X can be embedded in a linear space over the field of complexes in a natural way. Indeed, let us consider the space of all ordered pairs (x, y) with operations defined as follows:

$$(x_1, y_1)+(x_2, y_2) = (x_1+x_2, y_1+y_2); \qquad (a+ib)(x, y) = (ax-by, ay+bx).$$

We denote this space by $X+iX$. The rules of distributivity for the operations defined above are easily verified. We shall only prove here that the multiplication is associative. Indeed,

$$\begin{aligned}[(a+ib)(c+id)](x, y) &= (ac-bd, bc+ad)(x, y)\\ &= ((ac-bd)x-(bc+ad)y, (ac-bd)y+(bc+ad)x);\\ (a+ib)[(c+id)(x, y)] &= (a+ib)(cx-dy, cy+dx)\\ &= ((ac-bd)x-(bc+ad)y, (ac-bd)y+(bc+ad)x).\end{aligned}$$

On the other hand, if X is a linear space over the field of complex numbers, there always exists a linear space Y over the reals such that $Y+iY = X$.

Let X_0 be a subspace of a linear space X. The quotient group X/X_0 is also a linear space, known as the *quotient space*.

The *defect* (or *codimension*) of a subspace X_0 of a linear space X is the dimension of the quotient space X/X_0: $\operatorname{codim} X_0 = \dim X/X_0$. If X_0 is of finite codimension, then there exists a subspace X_1 such that $X = X_0 \oplus X_1$ and $\dim X_1 = \operatorname{codim} X_0$.

If a linear space X is a ring (under the same definition of addition), then X is called a *linear ring* or an *algebra*. A subset X_0 of an algebra X is called a *subalgebra* if it is an algebra with respect to the same operations as those of X. If a subalgebra is also a two-sided ideal, we can form the quotient ring X/X_0 which is clearly also an algebra, since $a[x] = [ax]$ for an arbitrary scalar a. It can be proved that every algebra can be embedded in an algebra with a unit, it suffices to consider in the corresponding extension of a ring an arbitrary scalar t in place of an integer P. We give here some examples of linear spaces.

EXAMPLE 1.1. The n-dimensional vector space $\boldsymbol{R}^n$ with addition and multiplication by scalars defined in the usual way is a linear space.

EXAMPLE 1.2. The space $C[0, 1]$ of continuous functions on the closed interval $[0, 1]$ is a linear space, for the sum of two continuous functions is again a continuous function.

EXAMPLE 1.3. The space $C^n[0, 1]$ of functions defined on the closed interval $[0, 1]$ and possessing an nth continuous derivative is a linear space, for the derivative of a linear combination of two such functions is (the same) linear combination of their derivatives.

EXAMPLE 1.4. The space $C^\infty[0, 1]$ of functions infinitely differentiable on the interval $[0, 1]$ is a linear space.

EXAMPLE 1.5. The space $S[0, 1]$ of all measurable functions defined on the interval $[0, 1]$ is a linear space, for a linear combination of measurable functions is itself a measurable function.

2. LINEAR OPERATORS. PERTURBATIONS

A *linear operator* is a map A of a linear subset $\boldsymbol{D}_A$ of a linear space X into a linear space Y, over the same field of scalars as X, such that

$$A(x+y) = Ax+Ay, \quad A(tx) = t(Ax) \quad \text{for all } x, y \in \boldsymbol{D}_A \text{ and all scalars } t.$$

The set $\boldsymbol{D}_A$ is called the *domain of the operator A*. To be precise, a linear operator is a pair $(\boldsymbol{D}_A, A)$, since it is defined both by the domain $\boldsymbol{D}_A$ and by the form of the map A. However, to be brief, we shall use the traditional notation A rather than the pair $(\boldsymbol{D}_A, A)$.

Let G be a subset of $\boldsymbol{D}_A$. We write

$$AG = \{y \in Y\colon\ y = Ax,\ x \in G\}.$$

The set $\boldsymbol{E}_A = A\boldsymbol{D}_A$ is called the *range of the operator A* or the *set of its values*. The *graph of an operator* is the set

$$W_A = \{(x, y) \in X \times Y\colon\ x \in \boldsymbol{D}_A;\ y = Ax\}.$$

The set of all linear operators defined in the space X and taking values in Y will be denoted by $L(X \to Y)$.

A linear operator $A \in L(X \to Y)$ is called an *isomorphism* if $\boldsymbol{D}_A = X$, $\boldsymbol{E}_A = Y$ and A is one-to-one. Two linear spaces X and Y are said to be *isomorphic* if there is an isomorphism mapping the space X onto the space Y.

If an operator $A \in L(X \to Y)$ is one-to-one, we can define the *inverse operator* A^{-1} in such a way that every element $x \in \boldsymbol{D}_A$ corresponds to an element $y \in \boldsymbol{E}_A$ satisfying the condition $y = Ax$. It is easily verified that the operator A^{-1} is linear and $\boldsymbol{D}_{A^{-1}} = \boldsymbol{E}_A$, $\boldsymbol{E}_{A^{-1}} = \boldsymbol{D}_A$. Hence, if A is an isomorphism, then A^{-1} exists and is also an isomorphism.

We define the *addition of operators* $A, B \in L(X \to Y)$ and the *multiplication of an operator* by a scalar as follows:

$$(A+B)x = Ax+Bx; \quad (tA)x = t(Ax) \quad \text{for} \quad x \in \boldsymbol{D}_{A+B} = \boldsymbol{D}_A \cap \boldsymbol{D}_B.$$

Evidently, an operator C such that $A+C = B$ does not exist for any pair of operators $A, B \in L(X \to Y)$. This follows from the fact that the domains $\boldsymbol{D}_A$ and $\boldsymbol{D}_B$ may be different. If the operator C does exist, we write $C = B-A$

and C is then called the *difference of the operators B and A*; the operation "$-$" is called a *subtraction.*

We denote by $L_0(X \to Y)$ the set of all operators $A \in L(X \to Y)$ such that $\boldsymbol{D}_A = X$. The operations of addition, subtraction and multiplication by a scalar are well-defined in the set $L_0(X \to Y)$. Hence $L_0(X \to Y)$ is a linear space. The zero element is the *zero operator* $\theta : \theta x = 0$ for all $x \in X$. In the sequel we will denote this operator by 0, the same symbol denotes the zero element of the space X; as this should not lead to any misunderstanding.

Let $A \in L(Y \to Z)$ and $B \in L(X \to Y)$, and suppose that $\boldsymbol{D}_A \supset \boldsymbol{E}_B$. The *superposition* (or the *product*) *of the operators A and B* is defined as an operator AB by means of the equality

$$(AB)x = A(Bx) \quad \text{for all } x \in \boldsymbol{D}_B.$$

We have $AB \in L(X \to Z)$. Provided all the operators appearing below exist, the following rules of distributivity hold:

$$\text{(2.1)} \quad \begin{aligned} &A(B_1+B_2) = AB_1+AB_2, &&A \in L(Y \to Z),\ B_1, B_2 \in L(X \to Y), \\ &(A_1+A_2)B = A_1B+A_2B, &&A_1, A_2 \in L(Y \to Z),\ B \in L(X \to Y). \end{aligned}$$

We shall write $L_0(X)$ instead of $L_0(X \to X)$ and $L(X)$ instead of $L(X \to X)$ for short. Formulae (2.1) show that $L_0(X)$ is not only a linear space but also a linear ring. The algebra $L_0(X)$ contains a unit, namely the *identity operator* $I : Ix = x$ for all $x \in X$.

An operator $P \in L_0(X)$ is called a *projector* (*projection operator*) if $P^2 = P$ (where $P^2 = P \cdot P$). Each projector induces a decomposition of the space X onto a direct sum $X = Y \oplus Z$, where

$$Y = \{x \in X : Px = x\}, \qquad Z = \{x \in X : Px = 0\}.$$

Indeed, if $x \in Y \cap Z$, then $x = 0$ for $x = Px = 0$. If $x \in X$, then $z = x - Px \in Z$, because $P(x-Px) = Px - P^2x = Px - Px = 0$ and so $x = y+z$, where $y = Px \in Y$, $z = x - Px = (I-P)x \in Z$. On the other hand, if $X = Y \oplus Z$ is a decomposition of the space X into a direct sum, then the operator $Px = y$, where $x = y+z$, $y \in Y$, $z \in Z$, is a projector on the space Y "in the direction" Z.

Note that the operator $I-P$ is also a projector on the space Z "in the direction" Y.

Consider now a subspace X_0 of a linear space X. Every operator $A \in L_0(X \to Y)$ induces an operator $[A] \in L_0([X] \to [Y])$, where $[X] = X/X_0$, $[Y] = Y/AX_0$, defined as follows:

$$[A][x] = [Ax] \quad \text{for } x \in [x],$$

where $[x]$ is the coset defined by the element x.

Let X_0 be a subspace of a linear space X, and let $A \in L(X \to Y)$. The operator $A_0 \in L_0(X_0 \to Y)$ defined by means of the formula

$$A_0 x = Ax \quad \text{for } x \in \boldsymbol{D}_A \cap X_0$$

is called the *restriction of the operator* A to the subspace X_0. The operator $A_1 \in L_0(X \to Y)$ is called an *extension of the operator* $A \in L_0(X_0 \to Y)$ to the space X if $A_1 x = Ax$ for $x \in X_0$. Every operator $A \in L(X \to Y)$ defined on a subspace X_0 of a linear space X can be extended to an operator $\tilde{A}$ such that $\boldsymbol{D}_{\tilde{A}} = X$, $\boldsymbol{E}_{\tilde{A}} = \boldsymbol{E}_A$. Indeed, let us write X as a direct sum $X_0 \oplus Z$. Then there exists a projector $P \in L_0(X \to Y)$ projecting the space X onto the subspace X_0. The operator $\tilde{A} = AP$ possesses all the required properties.

A linear operator whose domain is $\boldsymbol{D}_A = X$ and which takes values in the field of scalars (here we consider only the field $\boldsymbol{R}$ of reals and the field $\boldsymbol{C}$ of complexes) is called a *linear functional* defined on the space X. We denote by X' the set of all linear functionals defined on the space X.

If X is an n-dimensional space spanned by the elements $(x_1, \ldots, x_n)$, every linear functional f is of the form

$$f(x) = \sum_{j=1}^{n} t_j a_j, \quad \text{where } x = \sum_{j=1}^{n} t_j x_j,$$

i.e. it is determined by the values $a_j = f(x_j)$, $j = 1, \ldots, n$.

Linear functionals defined on the whole space X clearly form a linear space $L_0(X \to \boldsymbol{C})$ or $L_0(X \to \boldsymbol{R})$, depending on the field of scalars under consideration. We can therefore say, what is meant by the "linear independence" of a set of functionals. Noting the form of functionals in finite dimensional spaces, we conclude that $\dim X' = \dim X$. We have the following condition for the linear dependence of functionals: If $g, f_1, \ldots, f_n$ are linear functionals defined on a linear space X and if $f_j(x) = 0$ for $j = 1, 2, \ldots, n$ implies $g(x) = 0$, then the functional g is linearly dependent on the functionals f_j.

Let X and Y be linear spaces and let $A \in L(X \to Y)$. We denote by Z_A the set of zeros of the operator A:

$$Z_A = \{x \in \boldsymbol{D}_A \colon Ax = 0\}.$$

The set Z_A is clearly a linear space: it is called the *kernel of the operator* A, and the quotient space $Y/\boldsymbol{E}_A$ is called the *cokernel of the operator* A. The number $\alpha_A = \dim Z_A$ is called the *nullity of the operator* A, and the number $\beta_A = \dim Y/\boldsymbol{E}_A = \operatorname{codim} \boldsymbol{E}_A$ is called its *deficiency*. Hence, by definition, the deficiency of an operator A is equal to the defect of its range (Fig. 1).

The ordered pair (α_A, β_A) is called the *dimensional characteristic* of the oper-

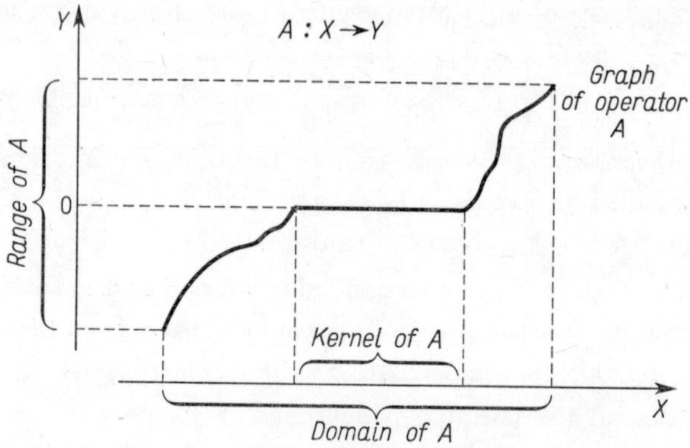

Fig. 1. Graph of a linear operator

ator A, or briefly its *d-characteristic.* We say that the d-characteristic of an operator A is *finite* if the nullity α_A and the deficiency β_A are both finite. If at least one of these numbers is finite, we say that the operator A has *semifinite d-characteristic.*

The *index of an operator* $A \in L(X \to Y)$ is defined in the following manner:

$$\varkappa_A = \begin{cases} \beta_A - \alpha_A & \text{if } A \text{ has finite } d\text{-characteristic,} \\ +\infty & \text{if } A \text{ has finite nullity and infinite deficiency,} \\ -\infty & \text{if } A \text{ has finite deficiency and infinite nullity.} \end{cases}$$

The notions introduced above are of use in solving linear equations. Indeed, if A is a linear operator, then a solution of the equation $Ax = y$ exists if and only if $y \in \boldsymbol{E}_A$. On the other hand, if $Ax_1 = y$ for a certain x_1, the general solution of the equation $Ax = y$ is of the form $x = x_0 + x_1$, where x_0 is an arbitrary element of the space Z_A. Thus it is important to have some information about the sets $\boldsymbol{E}_A$ and Z_A. The nullity and the deficiency characterize these sets in a certain way (although they do not describe them). As we shall see later, knowledge of these numbers can be very useful when investigating the solvability of the equation $Ax = y$. Very often the given equation can be reduced to another in such a way that the nullity and the deficiency of the considered operator are easy to determine. The following theorems therefore play a fundamental role in our subsequent studies:

THEOREM 2.1. *If* $A \in L(Y \to Z)$, $B \in L(X \to Y)$, $\boldsymbol{D}_A \supset \boldsymbol{E}_B$ *and* A, B *have both finite nullity (deficiency, d-characteristic), then the superposition* AB *exists and has the finite nullity (deficiency, d-characteristic, respectively) and*

$$\varkappa_{AB} = \varkappa_A + \varkappa_B.$$

THEOREM 2.2. *Let $A \in L(Y \to Z)$, $B \in L(X \to Y)$. By AB we mean the superposition of operators A and B whenever this superposition is well-defined. If AB has finite nullity (deficiency), B has finite nullity (respectively: A has finite deficiency).*

This immediately implies

COROLLARY 2.3. *If $A \in L_0(X \to Y)$, $B \in L_0(Y \to X)$ and if the superpositions $AB \in L_0(Y)$ and $BA \in L_0(X)$ both have finite d-characteristic, then both A and B have finite d-characteristic.*

It follows from the definition, that A is an isomorphism if and only if $\alpha_A = \beta_A = 0$.

We say that an operator $A \in L_0(X \to Y)$ is *right invertible* (*left invertible*) if there is an operator $B \in L_0(Y \to X)$ such that $AB = I_Y$ (resp. $BA = I_X$), where I_X and I_Y are the identity operators in the spaces X and Y respectively. If there is no danger of misunderstanding, we shall denote all operators such as I_X and I_Y by the same letter I.

THEOREM 2.4. *An operator $A \in L_0(X \to Y)$ is right (left) invertible if and only if $\beta_A = 0$ $(\alpha_A = 0)$.*

EXAMPLE 2.1. The operator A defined on the space $C[0, 1]$ of all continuous functions on the interval $[0, 1]$ by the formula $(Ax)(t) = \int_0^t x(s)ds$ is left invertible because $\frac{d}{dt}\int_0^t x(s)\,ds = x(t)$ for all $x \in C[0, 1]$ and so $\frac{d}{dt}A = I$, hence $\alpha_A = 0$. It follows from these formulae that the operator d/dt is right invertible. Hence $\beta_{d/dt} = 0$.

EXAMPLE 2.2. The operator p of multiplication by a function $p(t) \in C[0, 1]$, where $p(t)$ is different from zero for $0 \leqslant t \leqslant 1$, considered as an operator acting in $C[0, 1]$, is invertible. Hence $\alpha_p = \beta_p = \varkappa_p = 0$.

An operator $K \in L(X \to Y)$ such that $\dim E_K = n < +\infty$ is called a *finite dimensional operator*. The number n is called the *dimension of the operator* K. Each n-dimensional operator is of the form $Kx = \sum_{j=1}^{n} f_j(x)y_j$, where the elements $y_j \in Y$ are linearly independent, and the f_j are linear functionals defined on the space X. Note that the sum of two finite dimensional operators and the superpositions AK and KA of an arbitrary linear operator and a finite dimensional operator K (if they exist) are finite dimensional operators.

THEOREM 2.5. *If the operator $K \in L(X)$ is finite dimensional, then the operator $I+K$ has finite d-characteristic and $\varkappa_{I+K} = 0$.*

Equations with finite dimensional operators are known in the theory of integral equations as *equations with degenerate kernels* (see, for instance, Pogorzelski [1]).

Let a class $\mathfrak{A}$ of operators be given. An operator B is called the *$\mathfrak{A}$-perturbation of an operator $A \in \mathfrak{A}$* if $A+B \in \mathfrak{A}$. If an operator B is the $\mathfrak{A}$-perturbation of all operators $A \in \mathfrak{A}$, we call B an *$\mathfrak{A}$-perturbation* (*perturbation of the class* $\mathfrak{A}$). The set of all $\mathfrak{A}$-perturbations is *additive*, i.e. if the operators T_1, T_2 are $\mathfrak{A}$-perturbations, then the operator T_1+T_2 is also an $\mathfrak{A}$-perturbation. However, this set not always is linear. The following theorem is a consequence of Theorems 2.5 and 2.1:

THEOREM 2.6. *If $A \in L(X \to Y)$ has finite nullity (deficiency, d-characteristic) and $K \in L(X \to Y)$ is finite dimensional, then the operator $A+K$ has finite nullity (deficiency, d-characteristic, resp.) and $\varkappa_{A+K} = \varkappa_A$.*

This means that every finite dimensional operator is an index preserving perturbation of the class of all operators with finite nullity (deficiency, *d*-characteristic). On the other hand we have the following

THEOREM 2.7. *If the class of all operators with finite nullity in $L(X \to Y)$ is non-empty and if an operator $K \in L(X \to Y)$ is a perturbation of this class, then K is a finite dimensional operator.*

Consider the operator $A = T-\lambda I$, where $T \in L_0(X)$ and λ is a scalar. If

$$\alpha_{T-\lambda I} = \beta_{T-\lambda I} = 0,$$

the number λ is called a *regular value of the operator T*. The set of all numbers λ which are not regular values is called the *spectrum of the operator T*. If $\alpha_{T-\lambda I} > 0$, then such an element λ of the spectrum is called an *eigenvalue of the operator T*. If λ_0 is an eigenvalue of the operator T, then there is an $x \neq 0$ such that $Tx = \lambda_0 x$. All elements x possessing this property are called *eigenvectors of the operator T corresponding to the eigenvalue λ_0*. The linear span of all eigenvectors is called the *eigenspace of the operator T corresponding to the eigenvalue λ_0*. A *principal vector corresponding to the number λ* is an element x such that $(T-\lambda I)^n x = 0$ for some positive integer n. The linear span of all principal vectors is called the *principal space corresponding to the value λ*.

The dimension of the principal space corresponding to the eigenvalue λ_0 is called the *multiplicity of the eigenvalue λ_0*. If there principal vectors exist, then clearly so do eigenvectors. Indeed, if n is the least positive integer such that $(T-\lambda_0 I)^n x = 0$, then $x_0 = (T-\lambda_0 I)^{n-1}x$ is an eigenvector. Every eigenvector is a principal vector. Hence the dimension of the principal space is not less than the dimension of the eigenspace.

A number λ is a regular value of the operator T^n if and only if the nth roots of the number λ are regular values of the operator T.

If there is a positive integer N such that for all $n > N$ the operator T^n has a discrete spectrum and the operator $I-T^n$ has finite d-characteristic and index equal to zero, then the operator $I-T$ has also finite d-characteristic and index equal to zero.

EXAMPLE 2.3. Every real number is an eigenvalue of the operator $d/dt \in L_0(C^\infty [0,1])$. Indeed, the equation $\frac{d}{dt}x(t)-\lambda x(t) = 0$ has a solution for every value λ. To the value λ there corresponds an eigenvector $x_\lambda(t) = e^{\lambda t}$. It follows that the eigenspace is one-dimensional.

EXAMPLE 2.4. Every complex number is an eigenvalue of the operator $d^2/dt^2 \in L_0(C^\infty [0, 1])$. As is well known, the following linearly independent eigenvectors correspond to the eigenvalue λ:

$e^{\lambda_1 t}$, $e^{\lambda_2 t}$, where λ_1, λ_2 are the square roots of λ, if $\lambda \neq 0$;

$1, t$ if $\lambda = 0$.

Hence the eigenspace is two-dimensional.

3. ALGEBRAS OF LINEAR OPERATORS. CONJUGATE OPERATORS AND Φ_Ξ-OPERATORS

Let $\mathscr{X}(X) \subset L_0(X)$ be an *algebra of operators*, i.e. suppose that linear combinations and superpositions of two operators belonging to $\mathscr{X}(X)$ also belong to $\mathscr{X}(X)$. We always assume further that the identity operator I belongs to $\mathscr{X}(X)$.

An example of such an algebra is the whole algebra $L_0(X)$. It is easy to verify that all the finite dimensional operators belonging to an algebra $\mathscr{X}(X)$ form a two-sided ideal in that algebra. Note that every algebra can be represented as an algebra of operators over a certain linear space.

If for a given operator $A \in \mathscr{X}(X) \subset L_0(X)$ there exists an operator $R_A \in \mathscr{X}(X)$ such that $R_A A = I+T$ (resp. $AR_A = I+T$), where $T \in J$ and J is a two-sided ideal in the algebra $\mathscr{X}(X)$, then the operator R_A is called a *left* (*right*) *regularizer* of the operator A to the ideal J. If a regularizer is simultaneously left and right, it is called a *simple regularizer*.

A left (right) regularizer to the zero ideal is a left (right) inverse of the operator A. If there exists a regularizer to a proper ideal J, it does not belong to J. Indeed, if we assume that $R_A \in J$, we conclude that $I = -T+R_A A \in J$, which contradicts the assumption that J is proper.

If an operator A has a left (right) regularizer to an ideal J, then the coset $[R_A]$ is a left (right) inverse of the coset $[A]$ in the quotient algebra $\mathscr{X}(X)/J$. Hence, if the operator A has a simple regularizer, then the coset $[A]$ is invertible and $[A]^{-1} = [R_A]$. Note that this equality does not imply that A itself is invertible. In the sequel we will make use of the following properties of regularizers:

PROPERTY 3.1. *If an operator* $A \in \mathfrak{X}(X) \subset L_0(X)$ *has a left regularizer* R_1 *and a right regularizer* R_2 *to an ideal* $J \subset \mathfrak{X}(X)$, *then each of them is simple and* $R_1 - R_2 \in J$.

PROPERTY 3.2. *A simple regularizer to an ideal J is unique in the sense that two simple regularizers differ only by an element of the ideal J.*

PROPERTY 3.3. *If an operator* $A \in \mathfrak{X}(X)$ *is of the form* $A = B + T$, *where B has a left (right) inverse* $B_1 \in \mathfrak{X}(X)$ *and* $T \in J \subset \mathfrak{X}(X)$, *then A possesses a left (right) regularizer* $R_A = B_1$ *to the ideal J. Conversely, if a left (right) regularizer of an operator A has a right (left) inverse* $B \in \mathfrak{X}(X)$, *then* $A = B + T$, *where* $T \in J$.

PROPERTY 3.4. *If an operator* $A \in \mathfrak{X}(X)$ *has a left (right simple), regularizer* R_A *to an ideal* $J \subset \mathfrak{X}(X)$, *then for every* $T \in J$ *there exists a left (right, simple) regularizer* R_{A+T} *of the operator* $A+T$, *namely* $R_{A+T} = R_A$.

Note that the following equality holds for the superposition of two operators: $R_{AB} = R_B R_A$ (provided these regularizers exist).

Let $A, B \in L_0(X)$, then obviously

(3.1) $$Z_{AB} \supset Z_A \quad \text{and} \quad E_{AB} \subset E_A.$$

Suppose now that the operator A has a left (right) regularizer to an ideal $J \subset \mathfrak{X}(X)$. Then

$$R_A A = I + T_1 \quad (AR_A = I + T_2), \quad \text{where } T_1, T_2 \in J.$$

Formulae (3.1) show that to study the kernel of the operator A it suffices to study the restriction of the operator A to the space $Z_{I+T_1} = Z_{R_A A} \supset Z_A$. Similarly, to investigate the cokernel of the operator A it suffices to consider the operator $\hat{A}$ induced by the operator A in the quotient space X/E_{I+T_2}, for $E_{I+T_2} = E_{AR_A} \subset E_A$. In the case, when $\alpha_{I+T_1} < +\infty$ ($\beta_{I+T_2} < +\infty$), this is an essential simplification, since the study of infinite dimensional spaces reduces to that of certain finite dimensional spaces. In the sequel we will therefore consider only those ideals of operators for which either the nullity of the operator $I+T_1$ or the deficiency of the operator $I+T_2$ or the d-characteristic of both operators is finite, for all $T_1, T_2 \in J$.

Moreover, if the operator A has a left regularizer R_A which is left invertible, then the first formula of (3.1) implies $Z_A = Z_{I+T_1}$. Hence $\alpha_A = \alpha_{I+T_1} < +\infty$. Analogously, if the operator A has a right regularizer R_A which is right invertible, then $E_A = E_{I+T_2}$. Hence $\beta_A = \beta_{I+T_2} < +\infty$.

A proper two-sided ideal J of operators is said to be a *quasi-Fredholm ideal*, if the operator $I+T$ has finite d-characteristic for every $T \in J$. Moreover, if $\varkappa_{I+T} = 0$ for every $T \in J$, then J is called a *Fredholm ideal.*

The set of all finite dimensional operators belonging to an algebra of operators is a Fredholm ideal in this algebra (Theorem 2.5). There exists a quasi-Fredholm ideal which is not Fredholm: this was proved by means of an example of G. Neubauer (see the present author and Rolewicz [6], p. 42).

An immediate consequence of Theorem 2.2 and Corollary 2.2 is the following: Let $\mathscr{X}(X)$ be an algebra of operators. If an operator $A \in \mathscr{X}(X)$ has a simple regularizer to a quasi-Fredholm ideal $J \subset \mathscr{X}(X)$, then A has finite d-characteristic. If J is Fredholm, then we also have $\varkappa_A = -\varkappa_{R_A}$.

An algebra of operators $\mathscr{X}(X) \subset L_0(X)$ is said to be *regularizable* to a two-sided ideal $J \subset \mathscr{X}(X)$ if each operator with finite d-characteristic which belongs to the algebra has a simple regularizer to the ideal J. For instance the algebra $L_0(X)$ is regularizable to the ideal of all finite dimensional operators belonging to $L_0(X)$.

THEOREM 3.1. *If J is a quasi-Fredholm ideal in an algebra of operators $\mathscr{X}(X)$ regularizable to a quasi-Fredholm ideal $J_1 \subset \mathscr{X}(X)$, then every operator $T \in J$ is a perturbation of the class of all operators with finite d-characteristic belonging to $\mathscr{X}(X)$. Moreover, if J and J_1 are Fredholm ideals, then this perturbation does not change the index, i.e. $\varkappa_{A+T} = \varkappa_A$ for all $T \in J$ and all $A \in \mathscr{X}(X)$ with finite d-characteristic.*

We do not assume here that $J_1 \neq J$. The same is true for operators with semifinite d-characteristic. Namely, let $\mathscr{X}(X)$ be an algebra such that every operator $A \in \mathscr{X}(X)$ with finite d-characteristic possesses a left (right) regularizer to the ideal of all finite dimensional operators belonging to $\mathscr{X}(X)$. If for every $T \in J$ the operator $I+T$ has finite nullity (deficiency), then all the operators $T \in J$ are perturbations of the class of all operators with finite nullity (deficiency) belonging $\mathscr{X}(X)$.

Let $\mathscr{X}(X) \subset L_0(X)$ be an algebra of operators, and let $J \subset \mathscr{X}(X)$ be a quasi-Fredholm ideal in this algebra. By definition J is assumed to be a two-sided ideal. Hence we can consider the quotient algebra $\mathscr{X}_0 = \mathscr{X}(X)/J$. The coset $[I]$ corresponding to the identity operator is the unit of $\mathscr{X}_0$. The radical $\boldsymbol{R}(\mathscr{X}_0)$ of the algebra $\mathscr{X}_0$ is the set of all $x \in \mathscr{X}_0$ such that the element $[I]+axb$ is invertible for all $a, b \in \mathscr{X}_0$. We write

$$J_0 = \{U \in \mathscr{X}(X)\colon\ [U] \in \boldsymbol{R}(\mathscr{X}_0)\}.$$

The set J_0 is a quasi-Fredholm ideal in $\mathscr{X}(X)$ and it is the *maximal quasi-Fredholm ideal*, i.e. every quasi-Fredholm ideal $\tilde{J} \subset \mathscr{X}(X)$ is contained in J_0: $\tilde{J} \subset J_0$ (provided that the algebra $\mathscr{X}(X)$ is regularizable to a quasi-Fredholm ideal $J_1 \subset \mathscr{X}(X)$).

In the algebra $L_0(X)$ the ideal of all finite dimensional operators belonging to $L_0(X)$ is the maximal quasi-Fredholm ideal. However, this ideal is also Fredholm by virtue of Theorem 2.5.

Suppose we are given an algebra $\mathscr{X}(X)$ regularizable to a Fredholm ideal J. Observe that the index $\varkappa_A$ satisfies the following conditions:

(i) *$\varkappa_A$ is an integer-valued function defined on the set of all operators with finite d-characteristic belonging to $\mathscr{X}(X)$.*

(ii) $\varkappa_{A+T} = \varkappa_A$ *for all* $T \subset J$ (Theorem 3.1).

(iii) $\varkappa_{AB} = \varkappa_A + \varkappa_B$ (Theorem 2.1).

Properties (i), (ii), (iii) characterize the index of an operator in a certain sense. Namely, if $\mathscr{X}(X)$ is an algebra regularizable to a Fredholm ideal J, and if a function $\nu(A)$ satisfies conditions (i), (ii), (iii), and $\varkappa_A = 0$ implies $\nu(A) = 0$, then there exists an integer p such that $\nu(A) = p\varkappa_A$.

If $A \in L_0(X \to Y)$ has finite nullity (deficiency, d-characteristic), then $\alpha_A \leqslant \beta_A$ ($\beta_A \leqslant \alpha_A$, $\varkappa_A = 0$, resp.) if and only if the operator A can be written in the form $A = S + K$, where the operator K is finite dimensional and the operator S is left invertible (right invertible, invertible, resp.).

Results similar to those given above can be obtained for operators mapping one linear space into another (see the present author and Rolewicz [6], p. 51–59).

We denote by X' the space of all linear functionals defined on the space X [1]. A space $\Xi \subset X'$ is *total* if the condition $\xi(x) = 0$ for all $\xi \in \Xi$ implies $x = 0$ ($x \in X$). Clearly the whole space X' is total.

Note that the elements $x \in X$ can be regarded as functionals defined on a total space Ξ using the formula $F_x(\xi) = \xi(x)$.

Thus if we denote by Ξ' the space of all linear functionals defined on the space Ξ, the space X can be mapped into the space Ξ' in a one-to-one manner. This map is called the *natural embedding* and will be denoted by $\varkappa$. The image $\varkappa X$ of the space X under this embedding is a total space of functionals on the space Ξ, since $\xi(x) = 0$ for all x implies $\xi = 0$.

Each total subspace of the space X' is said to be a *space conjugate* to X (*adjoint space, dual space*).

Let $H \subset Y'$ be a conjugate space. To any operator $A \in L(X \to Y)$ there corresponds an operator ηA defined on the space H with values in the space X':

$$(\eta A)x = \eta(Ax) \quad (\text{for all } x \in \boldsymbol{D}_A \text{ and for all } \eta \in H).$$

The operator ηA is called the *conjugate operator* (*adjoint operator*) to the operator A. We shall denote it by A', i.e. $A'\eta = \eta A$. Let us observe that $(A+B)' = A' + B'$ and $I' = I$.

[1] This section may be omitted at a first reading.

Let $\Xi \subset X'$ be an arbitrary conjugate space. We consider the operator A' restricted to those functionals $\eta \in H$ for which $A'\eta = \eta A \in \Xi$. In this manner to every operator $A \in L(X \to Y)$ there corresponds an operator $A' \in L(H \to \Xi)$. With this degree of generality it may happen that the operator A' is only defined at 0. In the sequel we shall consider only operators $A \in L_0(X \to Y)$ for which $A' \in L_0(H \to \Xi)$, i.e. operators $A \in L_0(X \to Y)$ such that $A'\eta \in \Xi$ for every $\eta \in H$. We denote the set of all such operators by $L_0(X \to Y, H \to \Xi)$. Clearly it is a linear space. If we write $A \in L_0(X \to Y, H \to \Xi)$ we tacitly assume that we regard the conjugate operator A' as an operator mapping the space H into the space Ξ. The space $L_0(X \to X, \Xi \to \Xi)$ will be denoted for short by $L_0(X, \Xi)$. The space $L_0(X, \Xi)$ is an algebra for $A'B' = (BA)'$. If $A \in L_0(X \to Y, H \to \Xi)$, then $\alpha_{A'} \leqslant \beta_A$.

If $A \in L_0(X \to Y, H \to \Xi)$, then $A' \in L_0(H \to \Xi, X \to Y)$.

If E is a subset of the space Y, then the set

$$E^{\perp} = \{\eta \in H\colon \eta(y) = 0 \text{ for all } y \in E\}$$

is called the *H-orthogonal complement of the set E*. We say that a subspace $E \subset Y$ is *H-describable* if $(E^{\perp})^{\perp} = E$, where $(E^{\perp})^{\perp} = \{y \in Y\colon \eta(y) = 0 \text{ for } \eta \in E^{\perp}\})$

An operator $A \in L_0(X \to Y, H \to \Xi)$ is said to be *H-resolvable* (*H-solvable.* if the set E_A is H-describable.

If an operator A is H-solvable, then $\dim E_A^{\perp} = \beta_A$. Hence if $A \in L_0(X \to Y, H \to \Xi)$ is H-solvable, then $\alpha_{A'} = \beta_A$. This last equality does not hold in general as the following example shows:

EXAMPLE 3.1. Let $X = Y = C^{\infty}[0, 1]$ be the space of functions infinitely differentiable on the interval $[0, 1]$. Let $\Xi = H$ be the space of all functionals ξ of the form $\xi(x) = \int_0^1 x(t)\xi(t)dt$, where $\xi(t) \in C^{\infty}[0, 1]$ and $\xi^{(n)}(0) = 0$ $(n = 0, 1, \ldots)$, where $\xi^{(n)}(t)$ denotes the nth derivative of the function $\xi(t)$. Finally let the operator A be defined by means of the equality: $(Ax)(t) = y(t) = \int_t^1 x(s)ds$. It is easy to check that $\alpha_A = 0$ and $\beta_A = 1$. We calculate $\eta(Ax)$, where the functional η is defined by means of the function $\eta(t)$:

$$\eta(Ax) = \int_0^1 \Big[\int_t^1 x(s)ds\Big]\,\eta(t)dt$$

$$= \Big[\int_0^1 x(s)ds \cdot \int_0^t \eta(s)ds\Big]_0^1 + \int_0^1 x(t)\Big[\int_0^t \eta(s)ds\Big]\,dt = \int_0^1 x(t)\Big[\int_0^t \eta(s)ds\Big]\,dt.$$

Hence the conjugate operator A' maps the functional η of the form $\eta(x) = \int_0^1 \eta(t)x(t)dt$ to the functional ξ of the form

$$\xi(x) = \int_0^1 \xi(t)x(t)dt, \quad \text{where } \xi(t) = \int_0^t \eta(s)ds.$$

This operator is a one-to-one map of the space H onto itself. Hence $\alpha_{A'} = \beta_{A'} = 0$ and $\alpha_{A'} < \beta_A$.

On the other hand, if $A \in L_0(X \to Y, H \to \Xi)$, then $\alpha_A \leqslant \beta_{A'}$, and if $E_{A'} = Z_A^\perp$, then $\alpha_A = \beta_{A'}$. In particular, if $A \in L_0(X \to Y, Y' \to X')$, then $\alpha_A = \beta_{A'}$. It follows from these considerations that we do not always have $\varkappa_{A'} = -\varkappa_A$. We therefore introduce the *H-index* as the difference of the numbers $\beta_H^A = \alpha_{A'}$ and α_A:

$$\varkappa_A^H = \beta_A^H - \alpha_A \qquad \text{for } A \in L_0(X \to Y, H \to \Xi).$$

Since $\beta_A^H \leqslant \beta_A$ we have $\varkappa_A^H \leqslant \varkappa_A$. The pair of numbers (α_A, β_A^H) is called the *d_H-characteristic of the operator A*. By definition, the d_X-characteristic of the operator A' is clearly $(\alpha_{A'}, \alpha_A)$. Hence $\varkappa_A^H = -\varkappa_{A'}^X$.

As is shown by Example 3.1, the d_H-characteristic is not always equal to the d-characteristic, even if $X = Y$, $\Xi = H$. In a similar manner it can be shown that a theorem about the H-index analogous to Theorem 2.1 does not hold. Indeed, if we put $B = d/dt$ in Example 3.1, we get $BA = -I$. Hence $\varkappa_{BA}^H = \varkappa_{BA} = 0$. On the other hand, $\varkappa_B^H = \varkappa_B = -1$ and $\varkappa_A^H = 0$. Thus $\varkappa_{BA}^H \neq \varkappa_A^H + \varkappa_B^H$.

An operator $A \in L_0(X \to Y)$ is said to be a *Φ_H-operator* if its d-characteristic and d_H-characteristic are equal. For instance, any finite dimensional operator belonging to $L_0(X, \Xi)$ is a Φ_Ξ-operator (Fig. 2).

We say that a *subspace X_0 is described* by a family Ξ_0 (not necessarily linear) of linear functionals defined on the space X if $x \in X_0$ if and only if $\xi(x) = 0$ for all functionals $\xi_0 \in \Xi_0$. In other words, a subspace X_0 is described by a family Ξ_0 if and only if it is Ξ_0-describable and $X_0^\perp = \operatorname{lin} \Xi_0$. An operator $A \in L_0(X \to Y, H \to \Xi)$ is H-solvable if and only if the set E_A of its values is described by a family $H_0 \subset H$. It is easy to verify that an operator $A \in L_0(X \to Y, H \to \Xi)$ with finite d-characteristic is a Φ_H-operator if and only if the set E_A can be described by a finite system of functionals. If a subspace $X_0 \subset X$ can be described by a finite system of functionals $\Xi_0 \subset \Xi$, it follows that every space $X_1 \subset X$ containing X_0 can also be described by a finite system of functionals. If we apply these consideration to a linear equation

$$Ax = y \quad (y \in E_A), \tag{3.2}$$

where A is a Φ_H-operator, we conclude that:

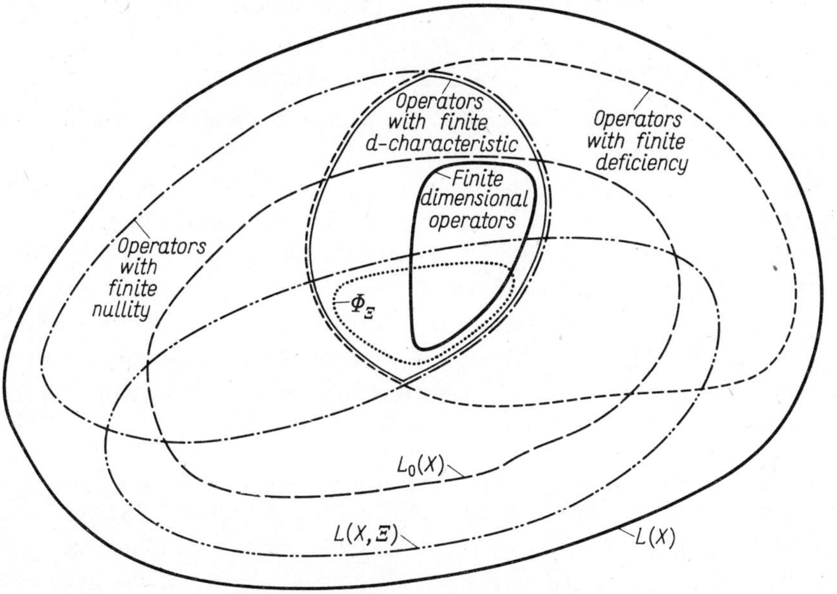

Fig. 2. Classification of linear operators

(i) *the homogeneous equations* $Ax = 0$ *and* $A'\xi = 0$ *both have only a finite number of linearly independent solutions (these numbers are of course* α_A *and* $\alpha_{A'} = \beta_A$ *resp.),*

(ii) *equation* (3.2) *has a solution if and only if the set* $\mathbf{E}_A$ *is described by the solutions of the conjugate homogeneous equation* $A'\xi = 0$, *i.e. if and only if*

$$\xi_i(x) = 0 \quad \textit{for } i = 1, 2, \ldots, \alpha_{A'}, \quad \textit{where } A\xi_i = 0.$$

Theorems of type (i)–(ii) are a generalization of the classical Fredholm alternative concerning the case $\alpha_{A'} = \alpha_A$. They are often called *Noether theorems* (see Noether [1]).

For Φ_{Ξ}-operators we have theorems analogous to Theorems 2.1 and 2.2:

THEOREM 3.2. *If a* Φ_H*-operator* B *belongs to* $L_0(X \to Y)$ *and if a* Φ_{Ξ}*-operator* A *belong to* $L_0(Y \to Z)$, *where* $HA \subset \Xi$, *then the superposition* AB *is also a* Φ_{Ξ}*-operator and* $\varkappa_{AB}^{\Xi} = \varkappa_A^{\Xi} + \varkappa_B^{H}$.

THEOREM 3.3. *If* $A \in L_0(X \to Y)$ *and* $B \in L_0(Y \to X)$ *and if* AB *is a* Φ_{Ξ}*-operator and* BA *is a* Φ_H*-operator, then* A *is a* Φ_{Ξ}*-operator and* B *a* Φ_H*-operator respectively.*

For instance, if $T \in L_0(X)$ and if there exists a positive integer m such that the operator $I - T^m$ is a Φ_{Ξ}-operator, then $I - T$ is also a Φ_{Ξ}-operator.

We can similarly define a *Φ_{Ξ}^{+}-operator* (*Φ_{Ξ}^{-}-operator*) as an operator A such that $\alpha_{A'} = \beta_A = +\infty$, $\alpha_A = \beta_{A'} < +\infty$ ($\alpha_{A'} = \beta_A < +\infty$, $\alpha_{A'} = \beta_{A'} = +\infty$ respectively), and Theorems 3.2 and 3.3 both remain true.

We say that an ideal $J \subset L_0(X, \Xi)$ is a *Ξ-quasi-Fredholm ideal* if $I+T$ is a Φ_{Ξ}-operator for all $T \in J$.

If an operator $A \in L_0(X, \Xi)$ has a simple regularizer R_A to a Ξ-quasi-Fredholm ideal $J \subset L_0(X, \Xi)$, then A is a Φ_{Ξ}-operator. Every quasi-Fredholm ideal contained in the algebra $L_0(X, \Xi)$ is a Ξ-quasi-Fredholm ideal. If J is a Ξ-quasi-Fredholm ideal in the algebra $L_0(X, \Xi)$, then all the operators belonging to J are perturbations of the class of all Φ_{Ξ}-operators belonging to this algebra.

As is shown by Example 3.1, not all d_{Ξ}-characteristics are equal if we vary the conjugate space Ξ. The following theorem shows that certain changes of the space X and Ξ leave the d_{Ξ}-characteristic unchanged.

THEOREM 3.4 (FIRST THEOREM ON REDUCTION). *Let X be a linear space and Ξ a conjugate space. Suppose that an operator $T \in L(X)$ is such that the operator $A = I+T$ has finite d_{Ξ}-characteristic. Let X_0 be an arbitrary subspace of the space X, containing TX, and let Ξ_0 be an arbitrary subspace of the space Ξ containing ΞT. Then the operator A restricted to the space X_0 has a finite d_{Ξ_0}-characteristic equal to the d_{Ξ}-characteristic of the operator A on the space X.*

THEOREM 3.5 (SECOND THEOREM ON REDUCTION). *Let X_0 be a subspace of a linear space X and let Ξ_0 be a subspace of a space Ξ conjugate to X. If an operator $A \in L(X_0, \Xi_0)$ has a simple regularizer R_A such that*

$$AR_A = I+T, \qquad R_A A = I+T_1,$$

where the operators T, T_1 can be extended to operators $\tilde{T}, \tilde{T}_1 \in L_0(X, \Xi)$ such that $I+\tilde{T}$ and $I+\tilde{T}_1$ are Φ_{Ξ}-operators, then the operator A is a Φ_{Ξ_0}-operator.

THEOREM 3.6. *If $A, T \in L(X, \Xi)$, A is a Φ_{Ξ}-operator, T is an isomorphism and $\Xi T \subset \Xi, \Xi A \subset \Xi$, then the operator TAT^{-1} is a Φ_{Ξ}-operator and $\varkappa_{TAT^{-1}}^{\Xi} = \varkappa_{A}^{\Xi}$ (i.e. $\varkappa_{TAT^{-1}} = \varkappa_A$).*

4. BANACH SPACES

A linear space X is said to be *normed* if there is a non-negative function, denoted by $\|\ \|$, defined for all $x \in X$ which satisfies the following conditions:

(a) $\|x\| = 0$ if and only if $x = 0$,

(b) $\|ax\| = |a| \cdot \|x\|$ for all scalars a (*homogeneity*),

(c) $||x+y|| \leqslant ||x||+||y||$ (*triangle inequality*).

The number $||x||$ is called the *norm of* x.

EXAMPLE 4.1. We say that an *inner product* (scalar product) is defined in a linear space X if there exists a function defined for all pairs (x, y), where $x, y \in X$, with values in the complex field, such that

(1) $$(x_1+x_2, y) = (x_1, y)+(x_2, y),$$

(2) $$(x, y) = \overline{(y, x)} \quad \text{(where } \bar{a} \text{ is the complex number conjugate to } a\text{)},$$

(3) $$(ax, y) = a(x, y),$$

(4) $$(x, x) > 0 \quad \text{for } x \neq 0.$$

A linear space with an inner product is called a *unitary space*. Such a space is a normed space when we define the norm as follows:

$$||x|| = \sqrt{(x, x)}.$$

Indeed, condition (1) implies $||0|| = \sqrt{(0, 0)} = 0$. Condition (4) implies $||x|| > 0$ for $x \neq 0$. In order to prove the triangle inequality, we first prove the following *Schwarz inequality*:

$$|(x, y)| \leqslant ||x|| \cdot ||y||.$$

For an arbitrary complex number a we have

$$0 \leqslant (x+ay, x+ay) = (x, x)+a[(x, y)+(y, x)]+a^2(y, y)$$
$$= ||x||^2+a[(x, y)+(y, x)]+a^2||y||^2.$$

Hence the discriminant of the last trinomial satisfies the inequality

$$\frac{[(x, y)+(y, x)]^2}{4} - ||x||^2||y||^2 \leqslant 0.$$

There is a number b such that $|b| = 1$ and the product (x, by) is a real number. If $y_0 = by$, then

$$|(x, y)| = \left|\frac{1}{b}(x, y_0)\right| = \left|\frac{(x, y_0)+(y_0, x)}{2}\right| \leqslant ||x|| \cdot ||y_0|| = ||x|| \cdot ||y||.$$

The Schwarz inequality implies that

$$||x+y||^2 = |(x+y, x+y)| = |(x, x)+(y, y)+(y, x)+(x, y)|$$
$$\leqslant ||x||^2+||y||^2+2||x|| \cdot ||y|| = (||x||+||y||)^2$$

which proves the triangle inequality.

In a normed linear space we can easily define the *distance* between two points, namely we set

$$\varrho(x, y) = ||x-y|| \quad \text{for } x, y \in X.$$

Let X be a normed linear space. A sequence $\{x_n\}$ of elements of X is *convergent to an element* $x \in X$ if and only if

$$\lim_{n \to \infty} ||x_n - x|| = 0,$$

i.e. if for every $\varepsilon > 0$ there is a positive integer N such that $||x_n - x|| < \varepsilon$ for all $n > N$. When this holds we write $x = \lim x_n$.

A sequence $\{x_n\} \subset X$ is *fundamental* (or a *Cauchy sequence*) if for every $\varepsilon > 0$ there is an N such that

$$\|x_n - x_m\| < \varepsilon \quad \text{whenever } m, n > N.$$

It is easy to check that each convergent sequence is fundamental. The converse statement is not true in general.

A normed linear space is *complete* if every fundamental sequence is convergent. A complete normed linear space is called a *Banach space.* From the definition it follows that a necessary and sufficient condition for a sequence $\{x_n\}$ of elements of a Banach space X to be convergent is that $\{x_n\}$ be a fundamental sequence (the Cauchy condition).

Let X be a Banach space. A set $E \subset X$ is said to be *compact* if every sequence $\{x_n\}$ of elements of E contains a subsequence $\{x_{n_k}\}$ convergent to an element $x \in E$.

The *closure* of a set $E \subset X$ is the set

$$\bar{E} = \{x \in X\colon\ x = \lim_{n\to\infty} x_n,\ x_n \in E\}.$$

A set $E \subset X$ is *closed* if $\bar{E} = E$. A set $E \subset X$ is called *precompact* if the set $\bar{E}$ is compact. The following criterion for the precompactness is very useful:

(PC) A set $E \subset X$ is precompact if and only if for every $\varepsilon > 0$ there exists a *finite* ε*-net* of the set E, i.e. a system of points $x_1, \ldots, x_n$ such that for every $x \in E$ there is an index i such that $\|x - x_i\| < \varepsilon$.

EXAMPLE 4.2. Let Ω be a compact set in n-dimensional Euclidean space $\boldsymbol{R}^n$ (for instance take Ω to be a closed interval of the real line). We denote by $C(\Omega)$ the set of all functions $x(t)$ defined and continuous on the set Ω (real or complex valued) endowed with the norm

$$\|x\| = \sup_{t\in\Omega} |x(t)|. \tag{4.1}$$

Since any linear combination of continuous functions is a continuous function, $C(\Omega)$ is a linear space. Moreover, it is a normed space with respect to the norm (4.1). We will show that $C(\Omega)$ is complete. Indeed, let $\{x_n(t)\}$ be a fundamental sequence in X. This sequence is convergent at each point $t \in \Omega$. Hence it is convergent to a function $x(t)$. The function $x(t)$ is continuous on Ω as it is the limit of a uniformly convergent sequence of continuous functions. Let ε be an arbitrary positive number. Since $\{x_n\}$ is fundamental, there exists an index k such that $|x_k(t) - x_{k'}(t)| \leqslant \varepsilon$ for $k' > k$ and for every $t \in \Omega$. Letting $k' \to \infty$ we obtain $|x_k(t) - x(t)| \leqslant \varepsilon$ for all $t \in \Omega$. Hence $\|x_k - x\| \leqslant \varepsilon$, which was to be proved. $C(\Omega)$ is thus a Banach space (with respect to the norm (4.1)).

EXAMPLE 4.3. Let Ω be a compact set in n-dimensional Euclidean space $\boldsymbol{R}^n$ and let $|t-s|$ denote the Euclidean distance of points $t = (t_1, \ldots, t_n)$ and $s = (s_1, \ldots, s_n)$, i.e.

$$|t-s| = \Big(\sum_{j=1}^{n} (t_j - s_j)^2\Big)^{1/2}.$$

By $H^\mu(\Omega)$ we denote the set of all bounded functions on Ω satysfying the *Hölder condition* with exponent μ, $0 < \mu \leqslant 1$, i.e. those functions $x(t)$ defined on Ω for which there is a constant $C > 0$ (dependent on x) such that

$$|x(t)-x(t')| \leqslant C|t-t'|^\mu \quad \text{for all } t, t' \in \Omega.$$

(If $\mu = 1$ this inequality is called the *Lipschitz condition.*) It is easily verified that $H^\mu(\Omega)$ is a linear space and, moreover, it is a linear subset of $C(\Omega)$. We define

$$||x|| = \sup_{t\in\Omega} |x(t)| + \sup_{t,t'\in\Omega} \frac{|x(t)-x(t')|}{|t-t'|^\mu}.$$

It is clear that

(a) if $||x|| = 0$, then $x(t) \equiv 0$ and $x = 0$. On the other hand, $x = 0$ implies $||x|| = 0$;

(b) if a is an arbitrary scalar, then $||ax|| = |a| \cdot ||x||$;

(c) $||x+y|| \leqslant ||x|| + ||y||$.

Hence $H^\mu(\Omega)$ is a normed space. We now show that this space is complete.

Indeed, if $\{x_n(t)\}$ is a fundamental sequence in the space $H^\mu(\Omega)$, then it is also a fundamental sequence in the space $C(\Omega)$. Hence the sequence $\{x_n(t)\}$ is uniformly convergent to a continuous function $x(t)$. Since $\{x_n(t)\}$ is fundamental, for every $\varepsilon > 0$ there is an index n such that $||x_n - x_{n'}|| < \varepsilon$ whenever $n' > n$. Hence

$$|[x_n(t)-x_{n'}(t)]-[x_n(t')-x_{n'}(t')]| \leqslant \varepsilon|t-t'|^\mu \quad \text{for arbitrary } t, t' \in \Omega.$$

Letting $n' \to \infty$ we obtain

$$|[x_n(t)-x(t)]-[x_n(t')-x(t')]| \leqslant \varepsilon|t-t'|^\mu \quad \text{for } t, t' \in \Omega.$$

Thus $x \in H^\mu(\Omega)$, as it is the sum of the two functions $x(t)-x_n(t)$ and $x_n(t)$ belonging to $H^\mu(\Omega)$. On the other hand, $|x_n(t)-x(t)| \leqslant \varepsilon$. Hence $||x_n-x|| \leqslant 2\varepsilon$ which proves that the space $H^\mu(\Omega)$ is complete. Thus we have shown that $H^\mu(\Omega)$ is a Banach space.

EXAMPLE 4.4. Let Ω be a set, Σ a countably additive algebra of subsets of Ω and μ a measure defined on Σ. We consider all μ-measurable complex valued functions $x(t)$ such that

$$||x(t)||_p = \Big(\int_\Omega |x(t)|^p d\mu\Big)^{1/p} < +\infty, \quad \text{where } p \geqslant 1.$$

(The assumption that $x(t)$ is a complex valued function is not essential, the same results are true for real valued functions.) We identify the functions $x(t)$ and $y(t)$ whenever the set $\{t\colon x(t) \neq y(t)\}$ is of μ-measure zero, and denote the set of all classes of identifiable functions by $L^p(\Omega, \Sigma, \mu)$. Note that for all $x, y \in L^p(\Omega, \Sigma, \mu)$ the identification defined above implies that $||x||_p = 0$ if and only if $x = 0$. Moreover, $||ax|| = |a| \cdot ||x||$ for all numbers a. The set $L^p(\Omega, \Sigma, \mu)$ is a linear space because a linear combination of measurable functions is again a measurable function. The proof of the triangle inequality closely resembles the proof of the so-called *Minkowski inequality* which in turn results from the following

LEMMA. *If* $x \in L^p(\Omega, \Sigma, \mu)$, $y \in L^q(\Omega, \Sigma, \mu)$ *and* $1/p+1/q = 1$, *then* $xy \in L^1(\Omega, \Sigma, \mu)$ *and*

$$||xy||_1 \leqslant ||x||_p ||y||_q. \tag{4.2}$$

Inequality (4.2) is called the *Hölder inequality*. In particular, if $p = q = 2$ this inequality is called the *Buniakovski–Schwarz inequality*.

PROOF OF INEQUALITY (4.2). Let $s = f(t) = t^a$, where $a > 0$. Since $f'(t) = at^{a-1} > 0$ for $t > 0$ the function $f(t)$ is increasing for $t > 0$. Hence the inverse function $g(s) = s^{1/a} = t$ is well-defined. Let ξ and η be arbitrary positive numbers. We draw two segments parallel to the axes Os and Ot starting at the points $(\xi, 0)$ and $(0, \eta)$ respectively and extending as far as their point of intersection with the graph of the function $f(t)$. We obtain two curvilinear triangles S_1, S_2 with areas

$$|S_1| = \frac{\xi^{a+1}}{a+1} \quad \text{and} \quad |S_2| = \frac{\eta^{1/a+1}}{1/a+1}$$

respectively (Fig. 3).

On the other hand, the sum of the areas of the triangles S_1 and S_2 is not less than the area of the rectangle bounded by parallels to the coordinate axes passing through the points $(\xi, 0)$ and $(0, \eta)$ respectively: $|S_1|+|S_2| \geqslant \xi\eta$. The equality holds only in the case $\eta = \xi^a$. Hence

$$\xi\eta \leqslant \frac{\xi^a}{a+1} + \frac{\eta^{1/a+1}}{1/a+1}.$$

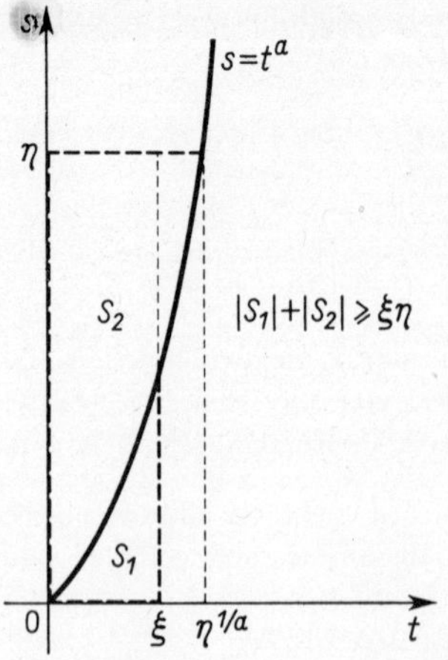

Fig. 3. Proof of the Hölder inequality

Substituting in the last equality

$$\xi = |x(t)|/||x||_p, \quad \eta = |y(t)|/||y||_q,$$

where $x \in L^p(\Omega, \Sigma, \mu)$, $y \in L^q(\Omega, \Sigma, \mu)$, $p = a+1$, $q = 1/a+1$, we have $1/p+1/q = 1$ and

$$\frac{|x(t)y(t)|}{||x||_p||y||_q} = \xi\eta \leqslant \frac{\xi^p}{p} + \frac{\eta^q}{q} = \frac{|x(t)|^p}{p||x||_p^p} + \frac{|y(t)|^q}{q||y||_q^q}.$$

Thus $xy \in L^1(\Omega, \Sigma, \mu)$. Moreover,

$$||\xi\eta||_1 \leqslant \frac{||x||_p^p}{p||x||_p^p} + \frac{||y||_q^q}{q||y||_q^q} = \frac{1}{p} + \frac{1}{q} = 1,$$

whence

$$||xy||_1 = ||x||_p||y||_q||\xi\eta||_1 \leqslant ||x||_p||y||_q$$

which was to be proved.

We shall now show that

(4.3) $$||x+y||_p \leqslant ||x||_p+||y||_p \quad \text{for all } x, y \in L^p(\Omega, \Sigma, \mu),\ p \geqslant 1.$$

This is the so-called Minkowski inequality.

PROOF OF THE MINKOWSKI INEQUALITY (4.3). If $z \in L^p(\Omega, \Sigma, \mu), p \geqslant 1$, then $|z^{p-1}| \in L^q(\Omega, \Sigma, \mu)$, where $1/p+1/q = 1$. Indeed, $(|z|^{p-1})^q = (|z|^{p-1})^{p/(p-1)} = |z|^p$. Hence $z|z|^{p-1} \in L^1(\Omega, \Sigma, \mu)$. We apply the Hölder inequality twice to the functions $x, y \in L^p(\Omega, \Sigma, \mu)$ and to the function $|x+y|^{p-1} \in L^q(\Omega, \Sigma, \mu)$, to obtain

$$||x+y||_p^p \leqslant ||\, |x+y|^{p-1}|x|\, ||_1+||\, |x+y|^{p-1}|y|\, ||_1 \leqslant ||\, |x+y|^{p-1}||_q||x||_p+||\, |x+y|^{p-1}||_q||y||_p$$
$$\leqslant (||x+y||_p)^{p/q}(||x||_p+||y||_p) \leqslant ||x+y||_p^{p-1}(||x||_p+||y||_p).$$

Dividing both sides of this inequality by $||x+y||_p^{p-1}$ we obtain the required inequality (4.3).

We now show that the space $L^p(\Omega, \Sigma, \mu)$, $p \geqslant 1$, is complete. Consider a sequence $\{x_n\} \subset L^p(\Omega, \Sigma, \mu)$ such that $||x_n||_p < 1/4^n$. Let $A_n = \{t\colon |x_n(t)| > 1/2^{n/p}\}$. Evidently, $||x_n||_p < 1/4^n$ implies $\mu(A_n) < 1/2^n$. Let $B_k = \bigcup_{i=k}^{\infty} A_i$. We have $|x_k(t)| < 1/2^{k/p}$ in the complement of the set B_k. Hence the sum of the series $\sum_{n=1}^{\infty} x_n(t)$ exists in the complement of the set $B = \bigcap_{k=1}^{\infty} B_k$. Moreover, this series is uniformly convergent on each of the sets $\Omega\setminus B_k$. If we set $x(t) = \sum_{n=1}^{\infty} x_n(t)$, the function $x(t)$ is measurable on the set $\Omega\setminus B$. Note that $\mu(B_k) \leqslant \sum_{i=k}^{\infty} \mu(A_i) \leqslant 1/2^{k-1}$, and so $\mu(B) = 0$.

The function $x(t)$ is thus measurable on the whole set Ω and is uniquely determined everywhere with the exception of a set of μ-measure equal to zero. Since the series $\sum_{n=1}^{\infty} x_n(t)$ is uniformly convergent on the sets $\Omega\setminus B_k$, the function $x|_{\Omega\setminus B_k} = x(t)$ for $t \notin B_k$ belongs to the space $L^p(\Omega\setminus B_k, \Sigma, \mu)$ and the sequence $\{\sum_{i=1}^{n} x_i - x|_{\Omega\setminus B_k}\}$ tends to zero in the norm, for arbitrary k. Hence the series $\sum_{n=1}^{\infty} x_n$ is convergent to the function $x \in L^p(\Omega, \Sigma, \mu)$.

Let $\{y_n\}$ be an arbitrary fundamental sequence. We extract a subsequence $\{y_{n_k}\}$ such that $||y_{n_{k+1}}-y_{n_k}||_p < \varepsilon_k$, where the series $\sum_{k=1}^{\infty} \varepsilon_k$ is convergent, and so the series $\sum_{k=1}^{\infty} x_k$, where $x_k = y_{n_{k+1}}-y_{n_k}$, is convergent. Denote its sum by x. So that that the sequence $\{y_{n_k}\}$ is convergent to x. We show that $y_n \to x$. For any positive number ε there is a number N such that $||y_n-y_m||_p < \varepsilon/2$ for $n, m > N$. Let $n_k > N$ be an index such that $||y_{n_k}-x||_p < \varepsilon/2$. Then

$$||y_n-x||_p \leqslant ||y_n-y_{n_k}||_p+||y_{n_k}-x||_p < \varepsilon \quad \text{for } n > N.$$

Hence the space $L^p(\Omega, \Sigma, \mu)$ is complete.

If Ω is either an interval or $\boldsymbol{R}^n$ and μ is Lebesgue measure on these sets, $L^p(\Omega, \Sigma, \mu)$ is denoted briefly by $L^p(a, b)$ and $L^p(\boldsymbol{R}^n)$ respectively. If Ω is the set of all integers and Σ is the family consisting of all of its subsets and if $\mu(n) = 1$ for $n = 0, \pm 1, \pm 2, \ldots$, then $L^p(\Omega, \Sigma, \mu)$ is the space of all sequences which are summable with power p and we denote it by l^p_Z. If Ω is the set of all positive integers and $\mu(n) = 1$ for $n = 1, 2, \ldots$, then we denote $L^p(\Omega, \Sigma, \mu)$ by l^p. The space $L^2(\Omega, \Sigma, \mu)$ can be regarded as a *Hilbert space*, i.e. a complete unitary space, if we define the inner product by the formula

$$(x, y) = \int_\Omega x(t)y(t)d\mu, \quad \text{where } x, y \in L^p(\Omega, \Sigma, \mu).$$

If in the space $L^2(\boldsymbol{R}^n)$ a sequence $\{x_n\}$ is convergent in the norm $\|\ \|_2$ to an element $x \in L^2(\boldsymbol{R}^n)$, i.e. if $\|x_n - x\|_2 \to 0$ as $n \to \infty$, then this is traditionally denoted in the following manner

$$\operatorname*{l.i.m.}_{n\to\infty} x_n = x,$$

where "l.i.m." is to be read "limes inter medias" which means "limit in mean".

We recall that the functions of a set $E \subset C(\Omega)$ are *uniformly bounded* if there is a constant $C > 0$ such that $\|x\| \leqslant C$ for all $x \in E$.

The functions of a set $E \subset C(\Omega)$ are *equicontinuous* if for every $\varepsilon > 0$ there is a $\delta > 0$ such that for all $x \in E$ and for all $t, t' \in \Omega$

$$|x(t) - x(t')| < \varepsilon \quad \text{whenever } |t - t'| < \delta.$$

We use the following criterion for precompactness in the space $C(\Omega)$:

THEOREM 4.1 (Arzelà [1]). *A set $E \subset C(\Omega)$ is precompact if and only if it consists of uniformly bounded and equicontinuous functions.*

PROOF. *Necessity.* Since the set is precompact, for every $\varepsilon > 0$ there is a finite ε-net of E, i.e. these is a system of functions $x_1, \ldots, x_n \in C(\Omega)$ such that for every function $x \in E$ one can choose a function x_i satisfying the inequality $\|x - x_i\| = \sup_{t\in\Omega} |x_i(t) - x(t)| < -\frac{1}{3}\varepsilon$. This implies that

$$\|x\| < \|x_i\| + \tfrac{1}{3}\varepsilon \leqslant \sup_{1\leqslant i\leqslant n} \|x_i\| + \tfrac{1}{3}\varepsilon.$$

Since ε is arbitrary it follows that all the functions $x \in E$ are uniformly bounded. Moreover, since the functions $x_i(t)$ are continuous on the compact set Ω, they are uniformly continuous on Ω. Hence for every i there is a number $\delta_i > 0$ such that the condition $|t - t'| < \delta_i$ implies $|x_i(t) - x_i(t')| < \frac{1}{3}\varepsilon$ $(i = 1, 2, \ldots, n)$. Let $\delta = \min_{1\leqslant i\leqslant n} \delta_i$ and let $|t - t'| < \delta$. Then

$$|x(t) - x(t')| \leqslant |x(t) - x_i(t)| + |x_i(t) - x_i(t')| + |x_i(t') - x(t')| < \varepsilon.$$

Hence all the functions $x(t) \in E$ are equicontinuous.

Sufficiency. The assumptions imply that for every $\varepsilon > 0$ there is a $\delta > 0$ such that for all $t, t' \in \Omega$,

$$|t-t'| < \delta \text{ implies that } |x(t)-x(t')| < \varepsilon/4 \text{ for all } x \in E.$$

Moreover, there is a constant $M > 0$ such that $|x(t)| < M$ for all $t \in \Omega$, $x \in E$. Since the set Ω is compact in $\boldsymbol{R}^n$, there exists a finite $\delta/2$-net of Ω, a system of points $t_1, \ldots, t_m$ say. We now decompose the set Ω into a sum of m disjoint subsets Ω_j such that the diameter of Ω_j is less than $\delta/2$ and $t_j \in \Omega_j$. Let χ_A denote the characteristic function of the set A, i.e. $\chi_A(t) = 1$ for $t \in A$ and 0 otherwise.

Let E_M denote the set of all linear combinations of functions $\chi_{\Omega_1}, \ldots, \chi_{\Omega_m}$ which are bounded by M, i.e.

$$E_M = \Big\{x(t) = \sum_{k=1}^{m} a_j \chi_{\Omega_j} : |x(t)| < M, a_j \text{ are arbitrary reals}\Big\}.$$

Since the set is finite dimensional and bounded, it is compact. Hence there is an $\varepsilon/3$-net of the set E_M which we denote by $x_1, \ldots, x_N$. If for some fixed $\boldsymbol{i}$ there is an $y \in E$ such that for all $t \in \Omega$ we have $|x_i(t)-y(t)| < \varepsilon/3$, then we denote it by y_i. When such a y does not exist we omit this index. The set $E_0 \subset E$ of all elements y_i constitutes a finite ε-net for the set E. Indeed, let x be an arbitrary function of the set E and let

$$\tilde{x}(t) = \sum_{j=1}^{m} x(t_j)\chi_{\Omega_j}(t).$$

By definition $\tilde{x} \in E_M$ and $x(t) = x(t_i)$ for $t \in \Omega_i$. Hence, there is an index k such that for $i = 1, 2, \ldots, m$ and for all $t \in \Omega_i$

$$|x(t)-x_k(t)| \leqslant |x(t)-\tilde{x}(t)|+|\tilde{x}(t)-x_k(t)| = |x(t)-x(t_i)|+|\tilde{x}(t)-x_k(t)|$$
$$\leqslant \frac{\varepsilon}{3}+\frac{\varepsilon}{3} = \frac{2}{3}\varepsilon.$$

Hence also

$$|x(t)-x_k(t)| < \frac{2\varepsilon}{3} \qquad \text{for all } t \in \Omega.$$

This implies that there is an $y_k \in E_0 \subset E$ such that $|x_k(t)-y_k(t)| < \varepsilon/3$ for all $t \in \Omega$. Therefore, for all $t \in \Omega$,

$$|x(t)-y_k(t)| \leqslant |x(t)-x_k(t)|+|x_k(t)-y_k(t)| < \frac{2\varepsilon}{3}+\frac{\varepsilon}{3} = \varepsilon.$$

Finally we find that for every $x \in E$ there is an $y_k \in E_0 \subset E$ such that $||x-y_k|| < \varepsilon$, which means that the set E_0 constitutes a finite ε-net for the set E. This implies that the set E is precompact, which was to be proved.

COROLLARY 4.1. *The unit ball in the space $H^\mu(\Omega)$ is a compact set with respect to the norm of the space $C(\Omega)$.*

PROOF. The unit ball of the space $H^\mu(\Omega)$ is the set

$$K = \{x \in H^\mu(\Omega): ||x||_{H^\mu(\Omega)} \leqslant 1\}$$

$$= \left\{x \in H^\mu(\Omega): \sup_{t\in\Omega}|x(t)| + \sup_{t,t'\in\Omega} \frac{|x(t)-x(t')|}{|t-t'|^\mu} \leqslant 1\right\}.$$

Since $\sup_{t\in\Omega}|x(t)| \leqslant 1$, all the functions belonging to K are uniformly bounded. Since

$$\sup_{t,t'\in\Omega} \frac{|x(t)-x(t')|}{|t-t'|^\mu} \leqslant 1,$$

we infer that, for every $\varepsilon > 0$, if $|t-t'| < \varepsilon^{1/\mu}$, then $|x(t)-x(t')| \leqslant |t-t'|^\mu < (\varepsilon^{1/\mu})^\mu = \varepsilon$ for all $x \in K$. Therefore all functions belonging to K are equicontinuous. This and the Arzelà Theorem imply that K, being a closed set, is compact.

Let X and Y be Banach spaces. An operator $A \in L_0(X \to Y)$ is said to be *continuous* (or: *bounded*) if there exists a constant $M > 0$ such that

$$||Ax||_Y \leqslant M||x||_X \quad \text{for all } x \in X.$$

If $A \in L_0(X \to Y)$ is bounded, the infimum

$$(4.4) \qquad ||A|| = \inf\{M: ||Ax||_Y \leqslant M||x||_X \text{ for } x \in X\} = \sup_{||x||_X\leqslant 1} ||Ax||_Y$$

exists.

The number $||A||$ is called the *norm of the bounded operator A*. We denote by $B(X \to Y)$ the set of all continuous operators belonging to $L_0(X \to Y)$, where X and Y are Banach spaces. $B(X \to Y)$ is a Banach space with the norm defined by Formula (4.4).

Indeed,

(a) $||A|| = 0$ if and only if $A = 0$.

(b) For an arbitrary scalar a

$$||aA|| = \sup_{||x||_X\leqslant 1} ||aAx||_Y = \sup_{||x||_X\leqslant 1} |a|\cdot||Ax||_Y = |a| \sup_{||x||_X\leqslant 1} ||Ax||_Y = |a|\cdot||A||.$$

(c) Since $||u+v||_Y \leqslant ||u||_Y+||v||_Y$ for all $u, v \in Y$, we find

$$||A+B|| = \sup_{||x||_X \leqslant 1} ||(A+B)x||_Y \leqslant \sup_{||x||_X \leqslant 1} [||Ax||_Y+||Bx||_Y]$$

$$\leqslant \sup_{||x||_X \leqslant 1} ||Ax||_Y + \sup_{||x||_X \leqslant 1} ||Bx||_Y = ||A||+||B||.$$

Let $A \in B(X \to Y)$ and $B \in B(Y \to Z)$. Then the superposition BA is well defined and $BA \in B(X \to Z)$. Moreover,

$$||BA|| \leqslant ||B|| \cdot ||A||. \tag{4.5}$$

Let $Y = X$. We denote $B(X \to X)$ briefly by $B(X)$. It is easy to show that $B(X)$ is a *Banach algebra* (i.e. a linear ring which is simultaneously a Banach space with respect to the same addition and multiplication by scalars, and is such that the multiplication is continuous with respect to the norm), since Formula (4.5) holds.

An operator $T \in B(X \to Y)$ is *compact* (or *completely continuous*) if there is an $\varepsilon > 0$ such that the set TU, where $U = \{x \in X: ||x|| < \varepsilon\}$ is a precompact set in the space Y. There exist continuous operators which are not compact, although every compact operator is continuous.

It can easily be shown that the sum, the superposition of two compact operators, the superposition of a compact operator and a continuous operator, and the limit of a sequence of compact operators are all compact, provided that they exist. Hence the set $T(X)$ of all compact operators belonging to $B(X)$ is a proper closed two-sided ideal in the algebra $B(X)$. Moreover, it is a Fredholm ideal, for $\varkappa_{I+T} = 0$ for all $T \in T(X)$. The algebra $B(X)$ is regularizable to the ideal $T(X)$. It is also regularizable to the ideal of all finite dimensional operators contained in $B(X)$. We will make use of the following

THEOREM 4.2 (F. Riesz [1]). *If the operator $T \in B(X)$ is compact and if $\alpha_{I+T} = 0$, then the operator $I+T$ has an inverse and $(I+T)^{-1} \in B(X)$.*

In the sequel we will apply the following theorem (see the author [7]):

THEOREM 4.3. *Let X be a Banach space under the norm $|| \; ||_X$ and let Y be a linear subset of X which is a Banach space under the norm $|| \; ||_Y$ such that $||x||_Y \geqslant ||x||_X$ for all $x \in Y$. Let the unit ball $K = \{y \in Y: ||y||_Y \leqslant 1\}$ be compact with respect to the norm $|| \; ||_X$. Let T be a continuous linear operator mapping X into Y and let $\tilde{T} = T|_Y$ be the restriction of T to the space Y. Then the operator $\tilde{T}$ regarded as an operator mapping Y into itself, is compact.*

PROOF. Let $\{y_n\}$ be an arbitrary sequence of elements of the set $\tilde{T}K = TK$. There exist $x_n \in K$ such that $Tx_n = y_n$ $(n = 1, 2, \ldots)$. Since the set K is compact in the norm $\|\ \|_X$, there is a subsequence $\{x_{n_k}\}$ of the sequence $\{x_n\}$ convergent to an element $x_0 \in K$ in the norm $\|\ \|_X$; i.e. $\lim_{k\to\infty}\|x_{n_k}-x_0\|_X = 0$. Since T is a continuous operator from X into Y, this implies that

$$\lim_{k\to\infty}\|y_{n_k}-y_0\|_Y = \lim_{k\to\infty}\|Tx_{n_k}-Tx_0\|_Y = 0, \quad \text{where } y_0 = Tx_0.$$

Hence the set TK is compact in the norm $\|\ \|_Y$. This implies that the operator $\tilde{T}$ is compact in the norm $\|\ \|_Y$, which was to be proved.

THEOREM 4.4. *If $B \in B(X)$ and $\|B\| < 1$, then the operator $I-B$ is invertible.*

PROOF. We shall show that

$$(I-B)^{-1}\sum_{i=0}^{\infty} = B^i. \tag{4.6}$$

Indeed, the series on the right-hand side of (4.6) is convergent. Moreover,

$$(I-B)\sum_{i=1}^{n} B^i = I-B^{n+1} \to I \quad \text{as } n \to \infty.$$

Hence Formula (4.6) is an immediate consequence of the continuity of the operator B. The series in Formula (4.6) is called the *Neumann series*.

This at once implies the following

THEOREM 4.5. *Let X be a Banach space and let $A = B+T$, where $B \in B(X)$, $\|B\| < 1$ and the operator T belongs to the ideal $T(X)$ of compact operators* (resp. *finite dimensional operators*). *Then the operator $I+A$ possesses a continuous simple regularizer $(I+B)^{-1}$ to that ideal.*

Indeed,

$$(I+A)(I+B)^{-1} = (I+B+T)(I+B)^{-1} = I+T(I+B)^{-1},$$

where the operator $T(I+B)^{-1}$ is clearly compact (resp. finite dimensional). The same result is also obtained by means of a left regularization.

EXAMPLE 4.5. Consider the equation

$$x(t)+ax(-t)+x(t_0)x_0(t) = x_1(t), \tag{4.7}$$

where a is a parameter and the given functions x_0 and x_1 are continuous and bounded throughout the real line. The point t_0 is fixed arbitrarily. We write

$$(Bx)(t) = ax(-t), \quad (Kx)(t) = x(t_0)x_0(t).$$

The operator K is thus one-dimensional. The space X of all functions continuous and bounded throughout real line is a Banach space under the norm $||x|| = \sup_{t \in R} |x(t)|$. If $|a| < 1$, we have $||B|| < 1$ and

$$[(I+B)^{-1}x](t) = \left[\sum_{k=0}^{\infty} (-1)^k B^k x\right](t) = \sum_{k=1}^{\infty} (-1)^k a^k x[(-1)^k t]$$

$$= \sum_{k=0}^{\infty} a^{2k} x(t) - \sum_{k=0}^{\infty} a^{2+1} x(-t) = \frac{1}{1-a^2} x(t) - \frac{a}{1-a^2} x(-t) = \frac{x(t)-ax(-t)}{1-a^2}.$$

Hence

$$[(I+B)^{-1}Kx](t) = [(I+B)^{-1}x(t_0)x_0](t) = \frac{x(t_0)}{1-a^2}[x_0(t)-ax_0(-t)].$$

Hence, if $|a| < 1$ equation (4.7) is equivalent by Theorem 4.5 to the equation

$$x(t) + \frac{x(t_0)}{1-a^2}[x_0(t)-ax_0(-t)] = \frac{x_1(t)-ax_1(-t)}{1-a^2}.$$

Since the last equation is of the form $(I+K_1)x = x_2$, where the operator K_1 is one-dimensional and x_2 is a given function, we finally obtain the following conclusions:

(1) if $|a| < 1$ and $[x_0(t_0)-ax_0(-t_0)]/(1-a^2) \neq 1$, equation (4.7) has the unique solution:

$$x(t) = \frac{x_1(t)-ax_1(-t)}{1-a^2} - \frac{x_1(t_0)-ax_1(-t_0)}{1-a^2} \cdot \frac{x_0(t)-ax_0(t)}{1-a^2-[x_0(t_0)-ax_0(-t_0)]},$$

(2) if $|a| < 1$ and $[x_0(t_0)-ax_0(-t_0)]/(1-a^2) = 1$, then equation (4.7) has a solution if and only if $x_2(t_0) = \frac{x_1(t_0)-ax_1(t_0)}{1-a^2} = 0$ and this solution is of the form

$$x(t) = \frac{x_1(t)-ax_1(-t)-C[x_0(t)-ax_0(-t)]}{1-a^2},$$

where C is an arbitrary constant.

If $|a| > 1$, we obtain an analogous solution on substituting $\tilde{t} = -t$, $\tilde{t}_0 = t_0$, $\tilde{a} = 1/a$, in equation (4.7).

If we are not able to solve the equation $(I+A)x = y$, i.e. $x = y - Ax$ directly, and if we want to find an approximate solution which is such that the error does not exceed a given value, we must know whether or not the solution of this equation is continuous with respect to the operator A. In other words, we want to know when "small" changes in A result in "small" changes in the solution. This question will be settled by means of the following theorem, in many cases more convenient than the preceding one.

THEOREM 4.6. *If X is a Banach space and if the operator $A \in B(X)$ is the limit (in the norm) of a sequence of uniformly bounded operators $\{A_n\} \subset B(X)$:*

$$||A_n|| \leqslant q < 1,$$

then the equation

$$x = y - Ax \qquad (y \in \boldsymbol{E}_A) \tag{4.8}$$

has a unique solution which is the limit of the sequence $\{x_n\}$ *of solutions of the approximating equations*:

$$x = \lim_{n\to\infty} x_n, \quad \textit{where } x_n = y - A_n x_n \ (n = 1, 2, \ldots). \tag{4.9}$$

PROOF. Since $q < 1$, each of equations (4.9) has a unique solution. Hence equation (4.8) also has a unique solution. Indeed, let ε be an arbitrary positive number. Then $||A|| \leqslant ||A-A_n|| + ||A_n|| \leqslant q+\varepsilon$ for sufficiently large n. Since ε is arbitrary, we conclude that $||A|| \leqslant q < 1$. Subtracting equation (4.8) from equation (4.9) we have

$$\begin{aligned} ||x-x_n|| = ||Ax - A_n x_n|| &\leqslant ||(A-A_n)x_n|| + ||A(x-x_n)|| \\ &\leqslant ||A-A_n|| \cdot ||x_n|| + ||A|| \cdot ||x-x_n||, \end{aligned}$$

i.e.

$$||x-x_n|| \leqslant \frac{||x_n||}{1-q} ||A-A_n||. \tag{4.10}$$

Hence for any given $\varepsilon > 0$ there is an integer N such that $||A-A_n|| < \dfrac{1-q}{||x_n||}\,\varepsilon$ implies $||x-x_n|| < \varepsilon$ for $n > N$.

Applying inequality (4.10), we observe that the error of the nth approximation is not greater than

$$\delta_n = \frac{||x_n||}{1-q} ||A-A_n||.$$

Let X and Y be Banach spaces. We recall that the graph of an operator $A \in L(X \to Y)$ is the set $W_A = \{(x, y) \colon x \in \boldsymbol{D}_A, y = Ax\} \subset X \times Y$. We say that an operator $A \in L(X \to Y)$ is *closed* if its graph is closed (in the space $X \times Y$). In other words, an operator $A \in L(X \to Y)$ is closed if the conditions $\{x_n\} \subset \boldsymbol{D}_A$, $x_n \to x$ and $Ax_n \to y$ imply $x \in \boldsymbol{D}_A$ and $y = Ax$. Note that every operator $A \in B(X \to Y)$ is closed, since $\boldsymbol{D}_A = X$.

A closed operator $A \in L(X \to Y)$ is said to be *normally solvable* (*normally resolvable*) if the set $\boldsymbol{E}_A$ of its values is closed. A normally solvable operator with finite d-characteristic (resp. nullity, deficiency) is called a *Φ-operator* (resp. *Φ_+-operator*, *Φ_--operator* (Gohberg and Krein [1]).

We denote by Y^+ the set of all continuous linear functionals defined on the space Y (and taking values in the field of scalars under consideration). This is clearly a total linear space. The corresponding conjugate to an operator $A \in B(X \to Y)$ will be denoted by A^+. In Section 3 we defined Φ_H-operators as operators whose d_H-characteristics are equal to their d-characteristics. According to this definition a normally solvable Φ_Y^+-operator is a Φ-operator, and conversely, every Φ-operator is a normally solvable Φ_Y^+-operator. Similar statements are true for Φ_+- and Φ_--operators.

For non-linear transformations we have

THEOREM 4.7 (BANACH FIXED-POINT THEOREM[1]). *Let A be a transformation of a Banach space into itself. If A satisfies the Lipschitz condition*

(4.11) $||Ax_1 - Ax_2|| \leqslant q||x_1 - x_2||$ *for all* $x_1, x_2 \in X$, *where* $0 < q < 1$,

then the equation

$$x = Ax + y \quad (y \in X) \tag{4.12}$$

has a unique solution, which is the limit in the norm of the sequence of successive approximations

$$x = \lim_{n\to\infty} x_n, \quad \text{where } x_{n+1} = Ax_n + y \text{ for } n = 1, 2, \ldots \tag{4.13}$$
$$\text{and } x_0 \in X \text{ is chosen arbitrarily.}$$

PROOF. Let $x_0 \in X$ be chosen arbitrarily and let $x_{n+1} = Ax_n + y$ for $n = 1, 2, \ldots$ Then $||x_2 - x_1|| = ||Ax_1 - Ax_0|| \leqslant q||x_1 - x_0||$ and successively

$$||x_{k+1} - x_k|| \leqslant q^k ||x_1 - x_0|| \quad (k = 1, 2, \ldots).$$

Therefore, by definition, for arbitrary $m > n$,

$$||x_m - x_n|| = ||\sum_{k=n}^{m-1} (x_{k+1} - x_k)|| \leqslant \sum_{k=n}^{m-1} ||x_{k+1} - x_k|| \leqslant \sum_{k=n}^{m-1} q^k ||x_1 - x_0||$$
$$= q^n \frac{1-q^m}{1-q} ||x_1 - x_0||.$$

Since $q < 1$, we conclude that as $n \to \infty$, $||x_m - x_n|| \to 0$. Hence the sequence $\{x_n\}$ is fundamental. Since the space X, being a Banach space, is complete, there is an $x \in X$ such that $x = \lim_{n\to\infty} x_n$. To prove that x satisfies equation (4.12) (i.e. that x is a *fixed-point* of the transformation A) note, that

$$||x - Ax - y|| \leqslant ||x - x_n|| + ||x_n - Ax_n - y|| + ||Ax - Ax_n|| \leqslant ||x - x_n|| + ||x_n - x_{n+1}|| +$$
$$+ q||x - x_n|| \leqslant (1+q)||x - x_n|| + q^n ||x_1 - x_0|| \to 0, \quad \text{when } n \to \infty.$$

Suppose now that the transformation A has two different fixed-points $\tilde{x}$ and x, i.e. that $\tilde{x} = A\tilde{x} + y$, $x = Ax + y$ and $\tilde{x} \neq x$. Then

$$||x - \tilde{x}|| = ||Ax - A\tilde{x}|| \leqslant q||x - \tilde{x}|| < ||x - \tilde{x}|| \quad \text{for } q < 1,$$

which is a contradiction. Thus $\tilde{x} = x$.

[1] See for instance Pogorzelski [1], p. 198.

Let us remark that the solution of equation (4.12) is continuous with respect to the transformation A and the given element y. Indeed, suppose we are given a sequence $\{A_n\}$ of transformations of a Banach space into itself such that

$$\lim_{n\to\infty} ||A_n x - Ax|| = 0 \quad \text{for all } x \in X \quad \text{and} \quad ||A_n x_1 - A_n x_2|| \leqslant q_n ||x_1 - x_2||,$$
$$\text{where } 0 < q_n \leqslant q < 1$$

($x_1, x_2 \in X$ are arbitrary). Then the solution x of equation (4.12) is the limit in the norm of the sequence $\{\tilde{x}_n\}$ of solutions of the equations

$$\text{(4.14)} \qquad \tilde{x} = A\tilde{x} + y \qquad (n = 1, 2, \ldots), \qquad \text{where } y \in X, \text{ and } y \to y \text{ as } n \to \infty.$$

Indeed, each of equations (4.14) has a unique solution x_n and

$$||x - \tilde{x}_n|| = ||(Ax + y) - (A_n \tilde{x}_n + y_n)|| \leqslant ||y - y_n|| + ||Ax - A_n x|| + ||A_n x - A_n \tilde{x}_n||.$$
$$\leqslant ||y - y_n|| + ||Ax - A_n x|| + q_n ||x - \tilde{x}_n|| \leqslant ||y - y_n|| + ||Ax - A_n x|| + q ||x - \tilde{x}_n||.$$

Hence, for $q < 1$,

$$||x - \tilde{x}_n|| \leqslant \frac{1}{1-q} ||Ax - A_n x|| + ||y - y_n|| \to 0 \quad \text{as } n \to \infty.$$

5. THE VOLTERRA INTEGRAL EQUATION

The following integral equation

$$\text{(5.1)} \qquad x(t) = \int_0^t N(t, s, x(s))\, ds + y(t), \qquad y \in C[0, T]$$

called the *Volterra integral equation of the second kind*, will play a very important role in subsequent parts of this book. We will solve this equation using the elegant method of Bielecki [1], [2]. This method makes it possible to apply the Banach fixed-point theorem (Theorem 4.7) without restrictions on the modulus of the function $N(t, s, x)$ of the type "if $N(t, s, x)$ is small enough...".

We will assume that $N(t, s, x)$ is a bounded real function defined for $0 \leqslant t, s \leqslant T$ and $x \in \boldsymbol{R}$, satisfying the Lipschitz condition with respect to x:

$$\text{(5.2)} \qquad |N(t, s, x) - N(t, s, y)| \leqslant L(t)|x - y| \quad \text{for all } x, y \in \boldsymbol{R},$$

where $L(t)$ is a non-negative function integrable over the interval $0 \leqslant t \leqslant T$. We write for $x \in C[0, T]$ (see Example 4.2)

$$\text{(5.3)} \qquad ||x||_p = \max_{0 \leqslant t \leqslant T} [e^{-p\int_0^t L(s)ds} |x(t)|] \quad (p = 0, 1, 2, \ldots).$$

For $p = 0$ this is the standard norm on $C[0, T]$. It is easy to verify that the $||\cdot||$ for $p \geqslant 1$ are also norms on $C[0, T]$. These norms are equivalent[1] since they are obtained from the norm $||\cdot||_0$ by means of multiplication by a non-vanishing function. We define the operator F acting in the space $C[0, T]$ by means of the equality

$$[F(x)](t) = \int_0^t N(t, s, x(s))ds, \qquad x \in C[0, T]. \tag{5.4}$$

We will show that for $p > 1$

$$||F(x)-F(y)||_p \leqslant \frac{1}{p}||x-y||_p \quad \text{for all } x, y \in C[0, T]. \tag{5.5}$$

Indeed, note that for all $x, y \in C[0, T]$ and $p \geqslant 0$ the function $L_1(t) = \int_0^t L(s)ds$ is non-negative, and consequently $\exp[p\int_0^t L(s)ds] = \exp[pL_1(t)] \geqslant 1$ and

$$|x(t)-y(t)| \leqslant e^{pL_1(t)}||x-y||_p \quad (p = 0, 1, 2, \ldots).$$

Since $L_1'(t) = L(t)$ and $L_1(0) = 0$, we find

$$\begin{aligned}
\exp[-pL_1(t)]|F(x)-F(y)| &= \exp[-pL_1(t)]\left|\int_0^t [N(t, s, x(s))-N(t, s, x(s))]ds\right| \\
&\leqslant \exp[-pL_1(t)]\int_0^t L(s)|x(s)-y(s)|\,ds \\
&\leqslant \exp[-pL_1(t)]\int_0^t L(s)\exp[pL_1(s)]||x-y||_p ds \\
&= \exp[-pL_1(t)]\int_0^t L_1(s)\exp[pL_1(s)]ds\cdot||x-y||_p \\
&= \frac{1}{p}\exp[-pL_1(t)]\exp[pL_1(s)]\Big|_0^t\cdot||x-y||_p \\
&= \frac{1}{p}\exp[-pL_1(t)]\{\exp[pL_1(t)]-1\}||x-y||_p \\
&= \frac{1}{p}||x-y||_p(1-\exp[-pL_1(t)]) \leqslant \frac{1}{p}||x-y||_p,
\end{aligned}$$

[1] Two norms $||\cdot||$ and $||\cdot||'$ are *equivalent* if every sequence converges in the norm $||\cdot||$ if and only if it converges in the norm $||\cdot||'$.

since $1-e^{-a} \leqslant 1$ for $a \geqslant 0$. This and the Banach fixed-point theorem (Theorem 4.7) together imply for $p > 1$ the following Bielecki theorem:

THEOREM 5.1. *If the real bounded function* $N(t, s, x)$ *satisfies condition* (5.2), *then equation* (5.1) *has a unique solution, which is the limit of a uniformly convergent sequence of successive approximations*:

$$x(t) = \lim_{n\to\infty} x_n(t), \qquad \textit{where } x_0(t) = y(t)$$

and

$$x_n(t) = y(t) + \int_0^t N\big(t, s, x_{n-1}(s)\big)ds \qquad (n = 1, 2, \ldots).$$

REMARK 5.1. This theorem could also be formulated for $T = +\infty$, in which case, instead of the space $C[0, T]$ we consider the space

$$X_p = \left\{x\colon \exp\left[-p\int_0^t L(s)\,ds\right]|x(t)| < \text{const}\right\}$$

for some $p > 1$.

REMARK 5.2. Inequality (5.5) and the proof of Theorem 4.7 show that by taking p greater we obtain a faster approximation for $q = 1/p$.

We now consider some special cases.

EXAMPLE 5.1. Let $N(t, s, x) = K(t, s)x$. Equation (5.1) becomes a linear equation

$$x(t) = \int_0^t K(t, s)x(s)ds + y(t), \qquad y \in C[0, T]. \tag{5.6}$$

Since $|N(t, s, x)-N(t, s, y)| = |K(t, s)|\cdot|x-y|$, assumption (5.2) is replaced by the assumption that $K(t, s)$ is a continuous function for $0 \leqslant t, s \leqslant T$ such that

(5.7) $\quad |K(t, s)| \leqslant L(t)$, where $L(t)$ is a non-negative function, integrable for $t \in [0, T]$;

for example we can take $L(t) = \sup_{0\leqslant s\leqslant T} |K(t, s)|$.

We conclude that equation (5.6) with the function $K(t, s)$ satisfying (5.7) has the unique solution

$$x(t) = \lim_{n\to\infty} x_n(t), \quad \text{where } x_0(t) = y(t),\ x_n(t) = y(t) + \int_0^t K(t, s)x_{n-1}(s)ds \ (n = 1, 2, \ldots).$$

We can write the sequence of successive approximations in a slightly different form. We first write

$$K_0(t, s) = 1, \quad K_1(t, s) = K(t, s), \quad K_{n+1}(t, s) = \int_0^t K_n(t, u)K(u, s)du,$$
$$(K_n x)(t) = \int_0^t K_n(t, s)x(s)ds \quad \text{for } x \in C[0, T] \text{ and } n = 1, 2, \ldots \tag{5.8}$$

We note that

$$(K^2x)(t) = \int_0^t K(t,u)\Big[\int_0^u K(u,s)x(s)ds\Big]du = \int_0^t\Big[\int_s^t K(t,u)K(u,s)du\Big]x(s)ds$$

$$= \int_0^t K_2(t,s)x(s)ds = (K_2x)(t)$$

and by simple induction

(5.9) $$K^n = K_n \quad (n = 0, 1, 2, \ldots), \text{ where } K_0 = I,\ K_1 = K.$$

We can therefore write the sequence of successive approximations in the following way:

$$x_1 = y+Ky = (I+K_1)y,$$
$$x_2 = y+Kx_1 = y+K(y+Ky) = y+Ky+K^2y = (I+K_1+K_2)y,$$
. . . .
$$x_n = y+Kx_{n-1} = y+Ky+ \ldots +K^ny = (I+K_1+K_2+ \ldots +K_n)y \quad (n = 1, 2, \ldots).$$

Since the sequence $\{x_n\}$ is convergent in the norm, we conclude that

(5.10) $$x(t) = y(t)+\int_0^t \mathscr{K}(t,s)y(s)ds,$$

where

$$\mathscr{K}(t,s) = \sum_{n=1}^{\infty} K_n(t,s), \quad K_n(t,s) = \int_s^t K_{n-1}(t,u)K(u,s)du,$$
$$K_1(t,s) = K(t,s) \quad (n = 2, 3, \ldots)$$

and the last series is absolutely and uniformly convergent. The function $\mathscr{K}(t,s)$ is called the *resolvent kernel* of equation (5.6).

EXAMPLE 5.2. Let the function $K(t,s)$ in Example 5.1 be of the form $K(t,s) = K(t-s)$. It is now enough to assume that $|K(t-s| \leqslant L(t)$, where $L(t)$ is an integrable function. However, this assumption is automatically satisfied if $K(t)$ is a continuous function.

EXAMPLE 5.3. Let $K(t,s) = p(s) \in C[0,T]$. Condition (5.7) is automatically satisfied, since $|p(s)| \leqslant \|p\|_0$. The unique solution of the equation

$$x(t)-\int_0^t p(s)x(s)ds = y(t)$$

is of the form

$$x(t) = y(t)+\int_0^t P(t,s)y(s)ds, \quad \text{where } P(t,s) = \sum_{n=1}^{\infty} p_n(t,s),$$

$$p_n(t,s) = p(s)\int_s^t p_{n-1}(t,u)du \quad (n = 2, 3, \ldots),\ p_1(s) = p(s).$$

EXAMPLE 5.4. Let $p(s) \equiv \lambda$, where λ is a constant. Then from Example 5.3 we immediately infer that the equation

(5.11) $$x(t)-\lambda\int_0^t x(s)ds = y(t)$$

has a unique solution, which is of the form

$$x(t) = y(t) - \lambda \int_0^t e^{\lambda(t-s)} y(s) ds.$$

Indeed,

$$p_1(s) = p(s) \equiv \lambda; \quad p_2(t, s) = \lambda^2 \int_s^t du = \lambda^2(t-s);$$

$$p_3(t, s) = \lambda^3 \int_s^t (t-u) du = \lambda^3 \frac{(t-s)^2}{2}; \quad \ldots$$

$$p_{n+1}(t, s) = \lambda^{n+1} \frac{(t-s)^n}{n!}; \quad \ldots$$

Hence

$$P(t, s) = \sum_{n=1}^{\infty} \lambda^n \frac{(t-s)^{n-1}}{(n-1)!} = \lambda \sum_{n=0}^{\infty} \lambda^n \frac{(t-s)^n}{n!} = \lambda e^{\lambda(t-s)}.$$

EXAMPLE 5.5. Consider now equation (5.11) in the case where $y(t) = k!$ ($k = 0, 1, \ldots$ is fixed arbitrarily). The corresponding solution is of the form

$$x_k(t) = \frac{t^k}{k!} - \lambda \int_0^t e^{\lambda(t-s)} \frac{s^k}{k!} ds = 2\frac{t^k}{k!} - \int_0^t e^{\lambda(t-s)} \frac{s^{k-1}}{(k-1)!} ds$$

$$= \frac{2t^k}{k!} + \frac{1}{\lambda} x_{k-1}(t) - \frac{t^{k-1}}{\lambda(k-1)!} \quad (k = 1, 2, \ldots).$$

Since

$$x_0(t) = 1 - \lambda \int_0^t e^{\lambda(t-s)} ds = 2 - e^{\lambda t},$$

we find successively

$$x_1(t) = 2t + \frac{1}{\lambda}(2 - e^{\lambda t}) - \frac{1}{\lambda} = \frac{1}{\lambda}[2\lambda t + 1 - e^{\lambda t}],$$

$$x_2(t) = \frac{2t^2}{2!} + \frac{1}{\lambda^2}[2\lambda t + 1 - e^{\lambda t}] - \frac{t}{\lambda} = \frac{1}{\lambda^2}[\lambda^2 t^2 + \lambda t + 1 - e^{\lambda t}],$$

and so on.

EXAMPLE 5.6. Write $(Rx)(t) = \int_0^t x(s) ds$. We shall discover the solution of the equation

$$(5.12) \qquad \left(\sum_{k=0}^{N} p_{N-k} R^k\right) x = y, \quad \text{where } y \in C[0, T],\ p_k \text{ are constants, } p_N = 1.$$

Note

$$(R^2 x)(t) = \int_0^t \left[\int_0^s x(u) du\right] ds = \int_0^t (t-s) x(s) ds,$$

and by simple induction

$$(R^{n+1}x)(t) = \int_0^t \frac{(t-s)^n}{n!} x(s)ds \qquad (n = 1, 2, \ldots).$$

We therefore rewrite equation (5.12) as follows:

$$x(t) + \int_0^t R(t, s)x(s)ds = y(t), \quad \text{where } R(t, s) = \sum_{k=1}^{N} p_{N-k} \frac{(t-s)^k}{k!}. \tag{5.13}$$

This equation has a unique solution, since $|t-s| \leqslant 2T$ if $0 \leqslant t, s \leqslant T$ (compare Example 5.2).

EXAMPLE 5.7. As in Example 5.6, we consider the equation

$$\sum_{k=0}^{N} R^k(p_{N-k}x) = y, \quad \text{where } y \in C[0, T], p_k = p_k(s) \in C[0, T]. \tag{5.14}$$

This equation can be written as follows:

$$x(t) + \int_0^t \tilde{R}(t, s)x(s)ds, \quad \text{where } \tilde{R}(t, s) = \sum_{k=1}^{N} \frac{(t-s)^k}{k!} p_{N-k}(s).$$

Hence equation (5.12) has a unique solution, since $|\tilde{R}(t, s)| \leqslant 2T \sum_{k=1}^{N} \frac{|p_{N-k}(s)|}{k!}$.

EXAMPLE 5.8. Let the function $K(t, s, x_0, \ldots, x_n)$ defined for $0 \leqslant t \leqslant T$, $-h \leqslant s \leqslant T$, $x_j \in \boldsymbol{R}$ satisfy the following condition:

$$|K(t, s, x_0, \ldots, x_n) - K(t, s, y_0, \ldots, y_n)| \leqslant L(t) \sum_{j=0}^{n} |x_j - y_j|, \tag{5.15}$$

where $L(t)$ is a non-negative function, integrable for $0 \leqslant t \leqslant T$. Then the equation

$$x(t) = \int_0^t K(t, s, x(s), x(s-h_1), \ldots x(s-h_n))ds + y(t), \qquad y \in C[0, T], \tag{5.16}$$

$h_j > 0$ and $h = \max_{1 \leqslant j \leqslant n} h_j$, with the initial condition

$$x(t) = \xi(t) \quad \text{for } t \in [-h, 0], \ \xi(0) = y(0), \tag{5.17}$$

has a unique solution, which is the uniform limit of successive approximations

$$x(t) = \lim_{n \to \infty} x_n(t), \qquad x_0(t) = \begin{cases} y(t) & \text{for } \quad 0 \leqslant t \leqslant T, \\ \xi(t) & \text{for } -h \leqslant t \leqslant 0, \end{cases}$$

$$x_n(t) = \begin{cases} \int_0^t K(t, s, x_{n-1}(s), \ldots, x_{n-1}(s-h_n))ds + y(t) & \text{for } 0 \leqslant t \leqslant T \ (n = 1, 2, \ldots), \\ \xi(t) & \text{for} -h \leqslant t \leqslant 0. \end{cases}$$

The proof is exactly the same as before, only we apply at each step the initial condition (5.17) since the function $x(t)$ must be defined on a larger interval.

CHAPTER II

Algebraic and almost Algebraic Operators

This chapter plays a special role: in it we give some properties of algebraic and almost algebraic operators and we solve the corresponding equations. These results will form a basis for our subsequent applications. Section 1 contains some fundamental lemmas.

1. HERMITE INTERPOLATION FORMULA. PARTITION OF UNITY. PROPERTIES OF NTH ROOTS OF UNITY

To begin with we will prove the following lemma on *interpolation with multiple knots.*

LEMMA 1.1 (Hermite [1]). *There exists a unique polynomial $W(t)$ of degree $N-1$ which, together with its derivatives, assumes given values y_{ki} at n different points t_i $(i = 1, 2, \ldots, n;\ k = 0, 1, \ldots, r_i-1;\ r_1+\ldots+r_n = N)$: $W^{(k)}(t_i) = y_{ki}$. The polynomial $W(t)$ is given by the following formula*:

$$(1.1)\qquad W(t) = \sum_{i=1}^{n} \frac{P(t)}{(t-t_i)^{r_i}} \sum_{k=0}^{r_i-1} y_{ki} \left\{\frac{(t-t_i)^{r_i}}{P(t)}\right\}_{(r_i-1-k;\, t_i)} \frac{(t-t_i)^k}{k!},$$

where

$$P(t) = \prod_{m=1}^{n} (t-t_m)^{r_m}, \qquad \{f(t)\}_{(k,s)} = \sum_{m=0}^{k} \frac{(t-s)^m}{m!} \cdot \left[\frac{d^m f(t)}{dt^m}\right]_{t=s}.$$

PROOF. To determine the polynomial $W(t)$ we decompose the rational function $W(t)/P(t)$ into vulgar fractions

$$\frac{W(t)}{P(t)} = \sum_{i=1}^{m} \sum_{j=1}^{r_i-1} \frac{\alpha_{i,j}}{(t-t_i)^{r_i-j}},$$

where the numbers $\alpha_{i,j}$ are unknown. If we write $F_i(t) = P(t)/(t-t_i)^{r_i}$ $(i = 1, 2, \ldots, n)$, then

$$\frac{W(t)}{F_i(t)} = \frac{W(t)}{P(t)}(t-t_i)^{r_i} = \sum_{j=0}^{r_i-1} \alpha_{i,j}(t-t_i)^j + (t-t_i)^{r_i} R_i(t),$$

where the function $R_i(t)$ is regular at the point t_i $(i = 1, 2, \ldots, n)$. We find

$$\alpha_{i,0} = \left[\frac{W(t)}{P(t)}\right]_{t=t_i},$$

$$\alpha_{i,1} = \frac{d}{dt}\left[\frac{W(t)}{F_i(t)}\right]_{t=t_i},$$

. . . .

. . . .

$$\alpha_{i,j} = \frac{1}{j!}\left[\frac{W(t)}{F_i(t)}\right]^{(j)}_{t=t_i}$$

$$= \frac{1}{j!}\sum_{k=0}^{j}\binom{j}{k} W^{(k)}(t_i)\left[\frac{1}{F_i(t)}\right]^{(j-k)}_{t=t_i} = \frac{1}{j!}\sum_{k=0}^{j}\binom{j}{k} y_{ki}\left[\frac{1}{F_i(t)}\right]^{(j-k)}_{t=t_i}.$$

Hence

$$W(t) = P(t)\sum_{i=1}^{n}\sum_{j=0}^{r_i-1}\frac{\alpha_{i,j}}{(t-t_i)^{r_i-j}} = \sum_{i=1}^{n}\sum_{j=0}^{r_i-1}\frac{P(t)}{(t-t_i)^{r_i}}(t-t_i)^j\alpha_{i,j}$$

$$= \sum_{i=1}^{n}\sum_{j=0}^{r_i-1} F_i(t)\frac{(t-t_i)^j}{j!}\sum_{k=0}^{j}\frac{j!}{k!(j-k)!} y_{ki}\left[\frac{1}{F_i(t)}\right]^{(j-k)}_{t=t_i}$$

$$= \sum_{i=1}^{n} F_i(t)\sum_{k=0}^{r_i-1} y_{ki}\left(\sum_{j=k}^{r_i-1}\frac{(t-t_i)^{j-k}}{(j-k)!}\left[\frac{1}{F_i(t)}\right]^{(j-k)}_{t=t_i}\frac{(t-t_i)^k}{k!}\right)$$

$$= \sum_{i=1}^{n} F_i(t)\sum_{k=0}^{r_i-1} y_{ki}\left(\sum_{m=0}^{r_i-1-k}\frac{(t-t_i)^m}{m!}\left[\frac{1}{F_i(t)}\right]^{(m)}_{t=t_i}\frac{(t-t_i)^k}{k!}\right)$$

$$= \sum_{i=1}^{n}\frac{P(t)}{(t-t_i)^{r_i}}\sum_{k=0}^{r_i-1} y_{ki}\left\{\frac{(t-t_i)^{r_i}}{P(t)}\right\}_{(r_i-1-k;\,t_i)}$$

and we obtain the required form of the polynomial $W(t)$. It is easy to check that $W(t)$ is a unique polynomial having the required properties.

Formula (1.1) is called the *Hermite interpolation formula.* Note that

$$\frac{P(t)}{(t-t_i)^{r_i}} = \prod_{m=1,\, m\neq i}^{n} (t-t_m)^{r_m}.$$

Using the preceding lemma we obtain the following

LEMMA 1.2 (Partition of unity). *If*

$$(1.2) \qquad \mathfrak{p}_i(t) = q_i(t) \prod_{m=1,\, m\neq i}^{n} (t-t_m)^{r_m}; \quad q_i(t) = \left\{\frac{(t-t_i)^{r_i}}{P(t)}\right\}_{(r_i-1;\, t_i)} \quad (i = 1, 2, \ldots, n),$$

then

$$(1.3) \qquad 1 \equiv \sum_{i=1}^{n} \mathfrak{p}_i(t),$$

and this representation is unique for fixed t_i and r_i.

PROOF. We take $W(t) \equiv 1$. Hence $W^{(0)}(t_i) = 1$, $W^{(k)}(t_i) = 0$ for $k \geqslant 1$ and $i = 1, 2, \ldots, n$. Hence, by the Hermite interpolation formula,

$$1 \equiv \sum_{i=1}^{n} \prod_{m=1,\, m\neq i}^{n} (t-t_m)^{r_m} \left\{\frac{(t-t_i)^{r_i}}{P(t)}\right\}_{(r_i-1;\, t_i)}.$$

If the t_i are single knots, the Hermite interpolation formula yields the *Lagrange interpolation formula.* Indeed, in this case $r_i = 1$ for $i = 1, \ldots, n$, hence

$$q_i(t) = \left\{\frac{(t-t_i)^{r_i}}{P(t)}\right\}_{(r_i-1;\, t_i)} = \left\{\frac{t-t_i}{P(t)}\right\}_{(0;\, t_i)} = \frac{t-t_i}{P(t)}\bigg|_{t=t_i}$$

$$= \left[\prod_{m=1,\, m\neq i}^{n} (t-t_m)\right]^{-1}\bigg|_{t=t_i} = \prod_{m=1,\, m\neq i}^{n} (t_i-t_m)^{-1}.$$

We thus obtain the following formula for the polynomial $W(t)$ putting $y_i = y_{0i}$:

$$(1.4) \qquad W(t) = \sum_{i=1}^{n} y_i \mathfrak{p}_i(t), \quad \text{where } \mathfrak{p}_i(t) = \prod_{m=1,\, m\neq i}^{n} \frac{t-t_m}{t_i-t_m} \quad (i = 1, \ldots, n).$$

This is the Lagrange interpolation formula. Observe that this formula yields the partition of unity in the case of simple knots:

$$(1.5) \qquad 1 \equiv \sum_{i=1}^{n} \prod_{m=1,\, m\neq i} \frac{t-t_m}{t_i-t_m}.$$

Next we will prove a lemma concerning Nth roots of unity.

LEMMA 1.3. *Let* $= e^{2\pi i/N}$, *where* $N \geqslant 2$. *Then*

(1) $\varepsilon^{kN} = \varepsilon^{Nk} = 1$ *for* $k = 0, 1, 2, \ldots$

(2) *All the Nth roots of unity are*: $1 = \varepsilon^0, \varepsilon, \varepsilon^2, \ldots, \varepsilon^{N-1}$.

(3) $\varepsilon^{-k} = \varepsilon^{N-k}$ *for* $k = 0, 1, \ldots, N$.

(4) *The following equalities hold*:

$$(1.6) \qquad \frac{1}{N}\sum_{j=0}^{N-1} \varepsilon^{jk} = \begin{cases} 1 & \textit{for } k = mN, \textit{ where } m = 0, \pm 1, \pm 2, \ldots, \\ 0 & \textit{otherwise}. \end{cases}$$

PROOF. (1) Since $e^{2\pi ik} = \cos 2\pi k + i \sin 2\pi k = 1$ for $k = 0, 1, 2, \ldots$, we find

$$\varepsilon^{kN} = (e^{2\pi i/N})^{Nk} = e^{2\pi ik} = 1 \qquad \text{for } k = 0, 1, 2, \ldots$$

(2) From (1) we have $(\varepsilon^k)^N = 1$ for $k = 0, 1, \ldots, N-1$. Moreover, $\varepsilon^k \neq \varepsilon^m$ for $k \neq m$ $(k,\ m = 0,\ 1, \ldots, N-1)$. Indeed $\varepsilon^m - \varepsilon^k = e^{2\pi im/N} - e^{2\pi ik/N} = e^{2\pi im/N}(1 - e^{2\pi i(k-m)/N}) = 0$ if and only if $k = m$.

(3) $\varepsilon^{-k} = \varepsilon^N \varepsilon^{-k} = \varepsilon^{N-k}$ for $k = 0, 1, \ldots, N$.

(4) Since $N \geqslant 2$, $\varepsilon \neq 1$ and we have for $k = \pm 1, \pm 2, \ldots$

$$\frac{1}{N}\sum_{j=0}^{N-1} \varepsilon^{jk} = \frac{1}{N}\sum_{j=0}^{N-1} (\varepsilon^k)^j = \frac{1}{N} \cdot \frac{1-(\varepsilon^k)^N}{1-\varepsilon} = \frac{1}{N} \cdot \frac{1-\varepsilon^{kN}}{1-\varepsilon} = 0,$$

because $\varepsilon^{kN} = 1$. If $k = 0$,

$$\frac{1}{N}\sum_{j=0}^{N-1} \varepsilon^{kj} = \frac{1}{N} \cdot N = 1.$$

This completes the proof for $k = 0, 1, \ldots, N-1$. For $k \geqslant N$ we have $\varepsilon^k = \varepsilon^{k'}$, where $k' \equiv k \pmod N$ (i.e. $k = k' + mN$, where $0 \leqslant k' \leqslant N-1$ and m is an integer). Hence the same pattern is repeated cyclically $(\operatorname{mod} N)$.

2. ALGEBRAIC AND ALMOST ALGEBRAIC ELEMENTS

Let $\mathscr{X}$ be a linear ring (over complex scalar field) with unit I. If there is a polynomial $P(t) = p_0 + p_1 t + \ldots + p_N t^N$ in a variable t with complex coefficients, satisfying the condition

$$(2.1) \qquad P(S) = T, \quad \text{where } S \in \mathscr{X},\ T \in J \subset \mathscr{X}$$

and J is a two-sided ideal in $\mathscr{X}$, then we say that S is an *almost algebraic element* (*with respect to the ideal J*). Whenever there is no danger of confusion we shall call S briefly an almost algebraic element.

Without loss of generality we shall assume once for all that $p_N = 1$. If S satisfies the polynomial identity $P(S) = T$ with a polynomial of degree N and does not satisfy any identity of degree less than N, we say that S is an *almost algebraic element of order N*. In this case we call $P(t)$ the *characteristic polynomial* of S and the roots of this polynomial the *characteristic roots* of S. If $T = 0$ in the identity (2.1), we assume that $J = \{0\}$ and in this case we call the element S *algebraic*.

Evidently, if an element S is almost algebraic with respect to an ideal $J \subset \mathscr{X}$, then the corresponding coset $[S]$ in the quotient ring $[\mathscr{X}] = \mathscr{X}/J$ is algebraic, and has the same characteristic polynomial and the same characteristic roots. This follows from the fact that the sum and the product of cosets correspond to the sum and the product of elements generating those cosets.

We shall see in the sequel that the properties of the characteristic polynomials in a sense determine the properties of the element S. Our subsequent studies will be based on these properties.

The following two cases are of great importance in various applications: $P(t) = t^2 - 1$ and $P(t) = t^N - 1$. In the first case we say that the element S is an *involution*, because it satisfies the identity $S^2 = I$. In the second case, S is said to be an *involution of order N*, because it satisfies the identity $S^N = I$. In the latter case we assume that $N > 2$. In both cases the characteristic roots are single and they are Nth roots of unity ($N \geqslant 2$).

Suppose we are given an algebraic element S of order N with the characteristic polynomial ($p_N = 1$)

$$P(t) = \prod_{m=1}^{n} (t - t_m)^{r_m}, \quad \text{where } r_1 + \ldots + r_n = N \tag{2.2}$$

and all the complex numbers t_m are different. We write

$$P_j = \mathfrak{p}_j(S) \quad (j = 1, 2, \ldots, n), \tag{2.3}$$

where $\mathfrak{p}_j(t)$ are the polynomials defined by Formulae (1.2). The elements P_j have the following properties, which are important for what follows.

PROPERTY 2.1. *The sum of the elements P_j is equal to the unit of the ring $\mathscr{X}$, i.e. these elements give a partition of unity*:

$$\sum_{j=1}^{n} P_j = I. \tag{2.4}$$

This follows immediately from the definition of the polynomials $\mathfrak{p}_j(t)$ and from Lemma 1.2 on the partition of unity, because an identity with respect to t gives rise to an identity with respect to S if we substitute S for t.

PROPERTY 2.2. *The elements P are idempotent and disjoint, i.e.*

$$P_j P_k = \begin{cases} P_j & \text{if } k = j \\ 0 & \text{if } k \neq 0 \end{cases} \quad (k, j = 1, 2, \ldots, n). \tag{2.5}$$

Indeed, let $k \neq j$. Taking into account the fact that all polynomials of the element S whose coefficients are numbers are commutative, we obtain

$$P_j P_k = q_j(S) q_k(S) \Big[\prod_{m=1, m\neq j}^{n} (S - t_m I)^{r_m} \Big] \Big[\prod_{\mu=1,\, \mu\neq k}^{n} (S - t_\mu I)^{r_\mu} \Big]$$

$$= q_j(S) q_k(S) \prod_{\substack{m\neq 1 \\ m\neq j,\, m\neq k}}^{n} (S - t_m I) P(S) = 0.$$

Hence, by Formula (2.4) we find

$$P_j = P_j \sum_{k=1}^{n} P_k = \sum_{k=1}^{n} P_j P_k = P_j^2 \quad (j = 1, 2, \ldots, n). \tag{2.6}$$

Formula (2.6) implies also that every element P_j is algebraic with characteristic polynomial $t^2 - t$ and, consequently, is of order 2 and has characteristic roots 0 and 1.

PROPERTY 2.3. *For a fixed m* $(m = 1, 2, \ldots, n)$

$$(S - t_m I)^{r_m} P_m = 0. \tag{2.7}$$

Indeed,

$$(S - t_m I)^{r_m} P_m = q(S)(S - t_m I)^{r_m} \prod_{k=1, k\neq m}^{n} (S - t_k I)^{r_k} = q_m(S) P(S) = 0.$$

Suppose now that the characteristic roots of the element S are simple, i.e. that $r_1 = \ldots = r_n = 1$ and $n = N$. From Formula (2.7) we have $(S - t_m I) P_m = 0$ and from (2.4) we obtain

$$S = \sum_{j=1}^{n} t_j P_j, \tag{2.8}$$

where

$$P_j = \mathfrak{p}_j(S) = \prod_{m=1,\, m\neq j} \frac{S - t_m I}{t_j - t_m} \quad (j = 1, 2, \ldots, n). \tag{2.9}$$

From simple induction we conclude that

$$S^k = \sum_{j=1}^{n} t_j^k P_j \quad \text{for } k = 1, 2, \ldots \tag{2.10}$$

PROPERTY 2.4. *If an algebraic element has simple roots only, then*

$$q(S) = \sum_{j=1}^{n} q(t_j) P_j \tag{2.11}$$

for an arbitrary polynomial $Q(t)$ with constant complex coefficients. Hence the element $q(S)$ is also algebraic with characteristic roots $q(t_j)$, provided that $q(t_j) \neq q(t_k)$ for $j \neq k$.

Indeed, if $q(t) = \sum_{k=0}^{n-1} q_k S^k$ (it is enough to consider polynomials of degree not greater than $n-1$: if it is not the case, we can reduce such polynomials by means of the identity $P(S) = 0$), then Formulae (2.4) and (2.10) imply

$$q(S) = \sum_{k=0}^{n-1} q^k S^k = \sum_{k=0}^{n-1} q_k S^k \Big(\sum_{j=1}^{n} P_j\Big) = \sum_{j=1}^{n} \Big(\sum_{k=0}^{n-1} q_k S^k P_j\Big)$$

$$= \sum_{j=1}^{n} \Big(\sum_{k=0}^{n-1} q_k t_j^k P_j\Big) = \sum_{j=1}^{n} \Big(\sum_{k=0}^{k-1} q_k t_j^k\Big) P_j = \sum_{j=1}^{n} q(t_j) P_j .$$

PROPERTY 2.5. *If an algebraic element S has simple roots only, then*

$$q(S) r(S) = \sum_{j=1}^{n} \big(q(t_j) r(t_j)\big) P_j; \quad q(S) + r(S) = \sum_{j=1}^{n} [q(t_j) + r(t_j)] P_j ,$$

where

$$q(t) = \sum_{k=0}^{n-1} q_k^k t^k \quad \text{and} \quad r(t) = \sum_{k=0}^{n-1} r_k t^k$$

are arbitrary polynomials (of a complex variable t) with constant complex coefficients. Hence $q(S)r(S)$ and $q(S)+r(S)$ are algebraic elements with characteristic roots $q(t_j)r(t_j)$ and $q(t_j)+r(t_j)$, respectively, provided that $q(t_j)r(t_j) \neq q(t_k)r(t_k)$ and $q(t_j)+r(t_j) \neq q(t_k)+r(t_k)$, respectively, for $j \neq k$.

Indeed, Formula (2.5) and the preceding property imply that

$$q(S) + r(S) = \sum_{j=1}^{n} q(t_j) P_j + \sum_{j=1}^{n} r(t_j) P_j = \sum_{j=1}^{n} [q(t_j) + r(t_j)] P_j$$

and

$$q(S)r(S) = \Big[\sum_{j=1}^{n} q(t_j)P_j\Big]\Big[\sum_{k=1}^{n} r(t_k)P_j\Big] = \sum_{j,k=1}^{n} q(t_j)r(t_k)P_jP_k$$

$$= \sum_{k,j=1}^{n} q(t_j)r(t_k)\,\delta_{jk}P_j = \sum_{j=1}^{n} q(t_j)r(t_j)P_j,$$

where δ_{jk} is the *Kronecker symbol*, i.e. $\delta_{jj} = 1$ and $\delta_{jk} = 0$ if $k \neq j$.

COROLLARY 2.1. *Under the assumption of Property 2.5, $q(S)r(S) = aI$, where a is some (fixed) complex number, if and only if $q(t_j)r(t_j) = a$ for $j = 1, 2, \ldots, n$.*

COROLLARY 2.2. *Under the assumption of Property 2.4, $[q(S)]^{-1}$ exist if and only if $q(t_j) \neq 0$ for $j = 1, 2, \ldots, n$ and*

$$[q(S)]^{-1} = \sum_{j=1}^{N} [q(t_j)]^{-1}P_j.$$

Hence $[q(S)]^{-1}$ is an algebraic element with characteristic roots $[q(t_j)]^{-1}$.

COROLLARY 2.3. *Under the assumption of Property 2.4,*

$$q(S)P_j = q(t_j)P_j \quad \text{for } j = 1, 2, \ldots, n.$$

Indeed

$$q(S)P_j = \Big[\sum_{k=1}^{n} q(t_k)P_k\Big]P_j = \sum_{k=1}^{n} q(t_k)P_kP_j = \sum_{k=1}^{n} q(t_k)\,\delta_{jk}P_k = q(t_j)P_j.$$

REMARKS. If S is an almost algebraic element, $P(S) = T$, and if there is a polynomial $q(t)$ such that $q(S)T = 0$ (or $Tq(S) = 0$), then S is an algebraic element. Indeed, $q(S)P(S) = q(S)T = 0$ (resp. $P(S)q(S) = Tq(S) = 0$).

If S is an almost algebraic element, $P(S) = T$, and if there is a polynomial $q(t)$ such that $q(T) = 0$, then S is an algebraic element, since $q\big(P(S)\big) = q(T) = 0$.

If S is an almost algebraic element: $P(S) = T$, then $ST - TS = 0$. Indeed, $ST = SP(S) = P(S)S = T$.

Without loss of generality we can consider in the preceding properties and corollaries only polynomials of degree not greater than $n-1$. Indeed, if we have a polynomial of degree $M \geqslant n$, it can be reduced by means of the identity

$$S^N = -p_0I + p_1S + \ldots + p_{N-1}S^{N-1},$$

where $P(t) = \sum_{k=0}^{n} p_kt^k$ is the characteristic polynomial of S ($p_N = 1$).

Some properties of polynomials in algebraic elements having simple roots have been included here only because these properties will be used further. For properties of algebraic operators with multiple roots see the book [6] of the author and Rolewicz.

We now consider the polynomials

$$A(S) = \sum_{m=0}^{N-1} A_m S^m, \quad {}^{\boldsymbol{\cdot}}A(S) = \sum_{m=0}^{N-1} S^m A_m, \tag{2.12}$$

where S is an algebraic element of the linear ring $\mathscr{X}$ and $A_m \in \mathscr{X}$. If the coefficients A_m commute with S: $A_m S - S A_m = 0$ for $m = 0, 1, \ldots, N-1$, then the polynomials $A(S)$ and ${}^{\boldsymbol{\cdot}}A(S)$ are identical. In any cases we have the following equality:

$${}^{\boldsymbol{\cdot}}A(S) = A(S) + \sum_{m=0}^{N-1} (S^m A_m - A_m S^m). \tag{2.13}$$

The bilinear form

$$[A, B] = AB - BA$$

is called the *commutator* of the elements A and B. The following formulae are easily verified:

$$\begin{aligned} &[B, A] = -[A, B], \\ &[A+B, C] = [A, C] + [B, C], \\ &[AB, C] = A[B, C] + [A, C]B, \\ &[A, [B, C]] + [B, [C, A]] + [C, [A, B]] = 0. \end{aligned} \tag{2.14}$$

THEOREM 2.1. *If S is an algebraic element of an algebra $\mathscr{X}$ and if*

$$b(S) = \sum_{m=0}^{N-1} b_m S^m, \quad A(S) = \sum_{m=0}^{N-1} A_m S^m,$$

where the coeefficients b_m are complex numbers, $A_m \in \mathscr{X}$ and $[A_m, S] \in J$, J being a two-sided ideal in the algebra $\mathscr{X}$, then

$$[b(S), A(S)] \in J.$$

PROOF. Let $0 \leqslant m \leqslant N-1$. The properties of commutators imply

$$[S, A_m S^m] = -[A_m S^m, S] = -A_m[S^m, S] - [A_m, S]S^m = -[A_m, S]S^m \in J.$$

Hence

$$[S, A(S^m)] = \sum_{m=0}^{N-1} [S, A_m S^m] \in J.$$

Suppose that, for an arbitrary k, $[S_k, A(S)] \in J$. Then

$$[S^{k+1}, A(S)] = S[S^k, A(S)] + [S, A(S)]S^k \in J.$$

Then $[S^k, A(S)] \in J$ for $k = 1, 2, \ldots$ Consequently,

$$[b(S), A(S)] = \sum_{m=0}^{N-1} [b_m S^m, A(S)] = \sum_{m=0}^{N-1} b_m [S^m, A(S)] \in J.$$

REMARK. This theorem is not true in general if the coefficients b_m are not numbers. In such cases it is necessary to suppose additionally that $[b_m, A_k] = 0$ for $k, m = 0, 1, \ldots, N-1$.

PROPERTY 2.6. *If S is an algebraic element of order N in an algebra $\mathscr{X}$ and if $[A_m, S] \in J$ for $m = 0, 1, \ldots, N-1$, J being a two-sided ideal in $\mathscr{X}$, then*

$$^{\boldsymbol{\cdot}}A(S) = A(S) + T, \quad \textit{where } T \in J$$

(*where* $^{\boldsymbol{\cdot}}A(S)$ *is given by Formula* (2.12).

Indeed, Formula (2.13) implies that it is enough to show that

$$\sum_{m=0}^{N-1} (S^m A_m - A_m S^m) \in J.$$

Fix $0 \leqslant m \leqslant N-1$. By assumption $SA_m - A_m S \in J$, and, from the proof of 2.1 we now conclude that $S^k A_m - A_m S^k \in J$ for $k = 0, 1, 2, \ldots$ Since n is fixed arbitrarily, we obtain the required formula.

3. REGULARIZATION IN ALGEBRAIC AND ALMOST ALGEBRAIC ELEMENTS

In this section we consider only algebraic and almost algebraic elements with simple characteristic roots, as this is sufficient for our purposes [1].

Suppose we are given an algebra $\mathscr{X}$ and a two-sided ideal $J \subset \mathscr{X}$. As in Section 3 of Chapter I, we can define regularizers of an element in $\mathscr{X}$. Namely, we say that an element $a \in \mathscr{X}$ has a *simple* (*left*, *right*) *regularizer* if the corresponding coset $[a]$ in the quotient ring $\mathscr{X}/J$ is invertible (resp. left invertible, right invertible), i.e. if there is a $b \in \mathscr{X}$ such that $ab = I + d_1$, $ba = I + d_2$, where $d_1, d_2 \in J$ (resp. $ba = I + d_2$, $ab = I + d_1$).

Regularizers of polynomials in algebraic and almost algebraic elements can easily be determined as polynomials whose coefficients depend on the coefficients of given the polynomials.

[1] For regularization of polynomials in algebraic and almost algebraic elements with multiple characteristic roots see the present author [5] and the present author and Rolewicz [6].

THEOREM 3.1. *If S is an algebraic element in an algebra $\mathscr{X}$ with the characteristic polynomial* $P(t) = \sum_{m=1}^{N} (t-t_m)$ *and if*

$$A(S) = \sum_{m=0}^{N-1} A_m S^m, \qquad B(S) = \sum_{m=0}^{N-1} B_m S^m,$$

where $A_m, B_m \in \mathscr{X}$, $[A_m, S] \in J \subset \mathscr{X}$ *for* $m = 0, 1, \ldots, N-1$, J *being a two-sided ideal, then for every* $T_1, T_2 \in J$

$$[B(S)+T_2]\,[A(S)+T_1] = C(S)+T_3,$$

where

$$C(S) = \sum_{k=0}^{N-1} C_k S^k \quad \text{and} \quad C_k = \frac{1}{V}(-1)^{k+m+1} d_{m,k} B(t_m) A(t_m),$$

$$[C_k, S] \in J \quad \text{for } k = 0, 1, \ldots, N-1,$$

$$T_3 = \sum_{k,\, m=1}^{N} B(t_m)\,[P_m, A(t)]\,P_k + T_2 A(S) + B(S) T_1 \in J,$$

V *denotes the Vandermonde determinant of the numbers* $t_1, t_2, \ldots, t_N$ *and* $d_{m,k}$ *is the minor of this determinant obtained by cancelling its mth row and* $(k+1)$*th column.*

PROOF. From Property 2.2 and from our assumptions we obtain

$$\begin{aligned}
BA &= [B(S)+T_2]\,[A(S)+T_1] \\
&= \Big(\sum_{m=1}^{N} B(t_m) P_m\Big) \cdot \Big(\sum_{k=1}^{N} A(t_k) P_k\Big) + T_2 A(S) + B(S) T_1 \\
&= \sum_{k,\, m=1}^{N} B(t_m) P_m A(t_k) P_k + T_2 A(S) + B(S) T_1 \\
&= \sum_{k,\, m=1}^{N} B(t_m) A(t_k) P_m P_k + \sum_{k,\, m=1}^{N} B(t_m)\,[P_m, A(t_k)]\,P_k + T_2 A(S) + B(S) T_1 \\
&= \sum_{m=1}^{N} B(t_m) A(t_m) P_m + T_3 .
\end{aligned}$$

We now have to show that $T_3 \in J$. From the second Formula (2.14) we find

$$[P_m, A(t_k)] \in J \quad \text{for } k, m = 0, 1, \ldots, N-1.$$

Hence $B(t_m)[P_m, A(t_k)]P_k \in J$ for $k, m = 0, 1, \ldots, N-1$ and also $T_2 A(S) \in J$, $B(S)T_1 \in J$. Hence $T_3 \in J$.

We put

$$B(t)A(t) = C(t) = \sum_{k=0}^{N-1} t^k C_k,$$

so that $BA = \sum_{m=1}^{N} C(t_m)P_m + T_3$. We now solve the system

$$B(t_m)A(t_m) = C(t_m) = \sum_{k=0}^{N-1} C_k t^k \quad (m = 1, 2, \ldots, N-1) \tag{3.1}$$

of N linear equations with respect to the N unknown coefficients C_k. The determinant of this system is the Vandermonde determinant of the numbers $t_1, \ldots, t_N$. Since t_k are simple roots, $t_k \neq t_j$ for $j \neq k$ and $V \neq 0$. Hence

$$C_k = \frac{1}{V} \sum_{m=1}^{N} (-1)^{k+m+1} B(t_m)A(t_m)d_{m,k},$$

where the $d_{m,k}$ are as defined above. We show that $[C_k, S] \in J$. Indeed,

$$[C_k, S] = \frac{1}{V} \sum_{m=1}^{N} (-1)^{k+m+1} d_{m,k}[B(t_m)A(t_m), S]$$

$$= \frac{1}{V} \sum_{m=1}^{N} (-1)^{k+m+1} d_{m,k}\{B(t_m)[A(t_m), S] + [B(t_m), S]A(t_m)\}.$$

Moreover, $[A(t_m), S] = \sum_{k=0}^{N-1} t_m^k [A_k, S] \in J$ and also $[B(t_m), S] \in J$. This implies that $[C_k, S] \in J$.

THEOREM 3.2. *If S and $A(S)$ satisfy the condition of Theorem 3.1 and $A(t_m)$ is invertible for $m = 0, 1, \ldots, N-1$, then $A(S)$ possesses a simple regularizer to the ideal J of the form*

$$R_{A(S)} = \sum_{k=0}^{N-1} \tilde{A}_k S_k = \sum_{m=1}^{N} [A(t_m)]^{-1} P_m, \tag{3.2}$$

where

$$\tilde{A}_k = \frac{1}{V} \sum_{m=1}^{N} (-1)^{k+m+1} d_{m,k}[A(t_m)]^{-1}$$

(V and $d_{m,k}$ are defined in Theorem 3.1). *Moreover,*

$$R_{A(S)}A(S) = I+T', \quad A(S)R_{A(S)} = I+T'',$$

where

$$T' = \sum_{m,k=1}^{N} [A(t_k)]^{-1}[P_k, A(t_m)]P_m \in J,$$

$$T'' = \sum_{m,k=1}^{N} [A(t_k)][P_k,[A(t_m)]^{-1}]P_m \in J.$$

We write $R_{A(S)}A(S) = C(S)+T_3$ and we put $C(S) = I$. We now obtain the following equalities from Theorem 3.1 and Corollary 2.1:

$$\tilde{A}(t_m)A(t_m) = C(t_m) = I \quad \text{for } m = 1, 2, \ldots, N, \tag{3.3}$$

where we write $\tilde{A}(t_m) = \sum_{k=0}^{N-1} t_m^k \tilde{A}_k$. This implies that $\tilde{A}(t_m) = [A(t_m)]^{-1}$ for $m = 1, 2, \ldots, N$ and that $\tilde{A}(S) = \sum_{m=1}^{N} [A(t_m)]^{-1}P_m$. On the other hand, we obtain from (3.3) the following system of linear equations with respect to the unknowns $\tilde{A}_k$:

$$\sum_{k=0}^{N-1} t_m^k \tilde{A}_k = [A(t_m)]^{-1} \quad (m = 1, 2, \ldots, N).$$

The determinant V of the last system is the Vandermonde determinant of the numbers $t_1, \ldots, t_N$ and by assumption $V \neq 0$ (cf. the proof of the preceding theorem). Hence

$$\tilde{A}_k = \frac{1}{V}\sum_{m=1}^{N} (-1)^{k+m+1} d_{m,k}[A(t_m)]^{-1} \quad (k = 0, 1, \ldots, N-1),$$

where the $d_{m,k}$ are defined in Theorem 3.1. From this theorem we get that

$$T_3 = T' = \sum_{m,k=1}^{N} [A(t_k)]^{-1}[P_k, A(t_m)]P_m \in J.$$

Changing the roles of $A(S)$ and $R_{A(S)}$ we obtain the corresponding result for the superposition $A(S)R_{A(S)}$.

COROLLARY 3.1. *If S and $A(S)$ satisfy the condition of Theorem 3.2, then for every $T \in J$ there exists a simple regularizer of the element $A(S)+T$ and*

$$R_{A(S)+T} = R_{A(S)}.$$

This is an immediate consequence of Theorem 3.2, along the same lines as in Property 3.4 of Chapter I.

COROLLARY 3.2. *If S and $A(S)$ satisfy the conditions of Theorem* 3.2, *then there exists a simple regularizer of the element* $\dot{A}(S) = \sum_{k=0}^{N-1} S^k A_k$ *and* $R_{\dot{A}(S)} = R_{A(S)}$.

Indeed, from Property 2.6 it follows that $\dot{A}(S) = A(S)+T$, where $T \in J$. Hence from the preceding corollary $R_{\dot{A}(S)} = R_{A(S)+T} = R_{A(S)}$.

The last corollary allows us to ignore polynomials of the type $\dot{A}(S)$ in our subsequent studies.

COROLLARY 3.3. *Let S and $A(S)$ satisfy the conditions of Theorem* 3.1. *If* $[A_m, S] = 0$ $(m = 0, 1, \ldots, N-1)$ *and $A(t_m)$ is invertible for* $m = 1, \ldots, N$, *then the element $A(S)$ is invertible and* $[A(S)]^{-1} = R_{A(S)} = \sum_{m=1}^{N} P_m [A(t_m)]^{-1}$.

Indeed, if we put $[A_m, S] = 0$ in Theorem 3.2, we obtain $T' = T'' = 0$.

THEOREM 3.3. *Let S be an almost algebraic element in an algebra $\mathscr{X}$, i.e.* $P(S) = T$, *where* $P(t) = \prod_{m=1}^{N} (t-t_m)$, *and T belongs to a two-sided ideal $J \subset \mathscr{X}$. If a polynomial* $A(S) = \sum_{k=0}^{N} A_k S^k$ *satisfies the following conditions*:

(i) $[A_k, S] \in J$ *for* $k = 0, 1, \ldots, N-1$,

(ii) *the elements $A(t_m)$ have simple regularizers $B(t_m)$ to the ideal J* $(m = 1, \ldots$ $\ldots, N)$, *then there exists a simple regularizer $R_{A(S)}$ of the element $A(S)$ to be ideal J which is given by Formula* (3.2).

PROOF. Let $[A]$ be the coset in the quotient ring $\mathscr{X}/J$ defined by the element A. Evidently, if S is an almost algebraic element in the algebra $\mathscr{X}$, then $[S]$ is an algebraic element in $\mathscr{X}/J$. It is easily verified that the polynomial $[A]([S]) = \sum_{m=0}^{N-1} [A_m][S]^m$ satisfies the conditions of Corollary 3.3. Hence the element $\{[A]([S])\}^{-1} = R_{[A]([S])}$ exists. But $R_{A(S)} \in R_{[A]([S])}$. Thus

$$R_{A(S)} A(S) = I+T', \quad A(S) R_{A(S)} = I+T'', \quad \text{where } T', T'' \in J.$$

This proves that the element $R_{A(S)}$ is a simple regularizer of the element $A(S)$.

COROLLARY 3.4. *If the conditions of Theorem* 3.3 *are satisfied and if*

(ii′) *elements* $[A(t_m)]^{-1}$ *exist for* $m = 1, 2, \ldots, N$,

then the element $A(S)$ has a simple regularizer $R_{A(S)}$ given by Formula (3.2).

4. ALGEBRAIC AND ALMOST ALGEBRAIC OPERATORS

Let X be a linear space over the complex numbers and let $L_0(X)$ denote the algebra of all linear operators acting in X and defined on the whole space X (see Section 2 of Chapter I). An operator $S \in L_0(X)$ is said to be *algebraic* (resp. *almost algebraic*) if it is an algebraic (resp. almost algebraic) element of some algebra $\mathscr{X}(X) \subset L_0(X)$. Let us observe that in this case the operators P_j determined by Formulae (2.3) are disjoint projectors giving rise to a partition of unity.

We now give some simple examples of algebraic operators. Commencing in Chapter IV we will consider algebraic and almost algebraic operators in differential, integral, functional-differential and functional-integral equations.

Note that an operator conjugate to an algebraic operator is also algebraic, as follows from the equality

$$0 = [P(S)]' = \Big[\sum_{m=0}^{N} p_m S^m\Big]' = \sum_{m=0}^{N} \bar{p}_m (S)'.$$

The first person to consider algebraic operators which are *selfadjoint*, i.e. such that $A' = A$, was Dirac in [1]. For instance, the projections of the spin of an electron on the axes of coordinates are involutions.

EXAMPLE 4.1. Every linear transformation S of a finite dimensional vector space is algebraic. The characteristic polynomial of S coincides with the minimal polynomial of the matrix corresponding to the transformation S.

EXAMPLE 4.2. Every finite dimensional operator S defined on an infinite dimensional space X is algebraic since the space SX is finite dimensional.

PROPERTY 4.1. *If an operator $S \in \mathscr{X}(X)$ is algebraic of order N, then the space X is a direct sum of projections*:

$$X = \bigoplus_{j=1}^{n} X_j,$$

where $X_j = P_j X = \{P_j X : x \in X\}$ $\big(N = r_1 + \ldots + r_n, \prod_{m=1}^{n} (S - t_m I)^{r_m} = 0\big)$.

To establish this property note that, by Property 2.1, every element $x \in X$ can be written as a sum

$$x = \sum_{j=1}^{n} x_{(j)}, \quad \text{where } x_{(j)} = P_j x \quad (j = 1, 2, \ldots, n). \tag{4.1}$$

Property 2.2 implies that this representation is unique.

Properties 2.3 and 4.1 imply that

$$(S - t_j I)^{r_j} x = 0 \quad \text{for an arbitrary } x \in X \text{ and for } j = 1, 2, \ldots, n. \tag{4.2}$$

The last formula shows that the operator S is an algebraic operator on each of the spaces X_j with only one root t_j which is of order r_j.

THEOREM 4.1. *If S belong to an algebra* $\mathscr{X}(X) \subset L_0(X)$, *then the following conditions are equivalent*:

(a) *S is an algebraic operator with the characteristic polynomial* $\prod_{m=1}^{n} (t-t_m)^{r_m}$, *and is of order* $N = r_1 + \ldots + r_n$.

(b) *There exist n linear operators* $P_j \in \mathscr{X}(X)$ *such that*

$$P_j P_k = \delta_{jk} P_j, \qquad \sum_{j=1}^{n} P_j = I \quad \text{and} \quad (S-t_j I)^{r_j} P_j = 0 \quad (j, k = 1, 2, \ldots, n)$$

(δ_{jk} *is the Kronecker symbol*).

(c) *The space X is the direct sum of n subspaces* X_j *such that*

$$(S-t_j I)^{r_j} x = 0 \quad \text{for } x \in X_j \ (j = 1, \ldots, n).$$

If all the roots t_j *are simple, i.e. if* $r_j = 1$ *for* $j = 1, \ldots, n = N$, *then each of the spaces* X_j *is an eigenspace of the operator S corresponding to the eigenvalue* t_j.

PROOF. Implication (a) → (b) → (c) follows immediately from Properties 2.1–2.3 and 4.1. We therefore prove only that (a) follows from (c). Indeed, let the space X be the direct sum of n subspaces X_n such that $(S-t_m I)^{r_m} x = 0$ for $x \in X_m$ and $m = 1, 2, \ldots, n$. We put $P(S) = \prod_{m=1}^{n} (S-t_m I)^{r_m}$. By assumption, $x = \sum_{j=1}^{n} x_j$, where $x_j \in X_j$, for every $x \in X$. Hence

$$P(S)x = \sum_{j=1}^{n} P(S)x_j = \sum_{j=1}^{n} \Big[\prod_{m=1,\, m\neq j}^{n} (S-t_m I)^{r_m} \Big] (S-t_j I)^{r_j} x_j = 0 \quad \text{for } x \in X.$$

Consequently, the operator S is algebraic of order $N = r_1 + \ldots + r_n$ with the characteristic polynomial $P(t)$. Thus conditions (a), (b), (c) are equivalent. If the characteristic roots of an algebraic operator are simple, then from Formula (4.2)

$$Sx = t_j x \quad \text{for any } x \in X_j \ (j = 1, 2, \ldots, N). \tag{4.3}$$

Hence each of the spaces X_j is an eigenspace for the operator S, corresponding to the eigenvalue t_j.

EXAMPLE 4.3. Let X be an arbitrary space of functions defined on the real line. If $(Sx)(t) = x(-t)$ for $x \in X$, then $S^2 = I$. X_1 is the space of all even functions, i.e. functions satisfying the condition $x(-t) = x(t)$ and X_2 is the space of all odd functions, i.e. functions $x(t)$ such that $x(-t) = -x(t)$. Hence it follows that every function can be written uniquely as the sum of an even and an odd function (Fig. 4).

EXAMPLE 4.4. Let X be the space of all square matrices of order n with real elements. The operation of transposition of a matrix is an involution. X_1 is the space of all symmetric matrices, i.e. matrices whose elements satisfy the condition $a_{ik} = a_{ki}$. X_2 is the space of all antisymmetric matrices, i.e. matrices satisfying $a_{ki} = -a_{ik}$. This implies that every matrix can be written uniquely as the sum of a symmetric matrix and an antisymmetric matrix.

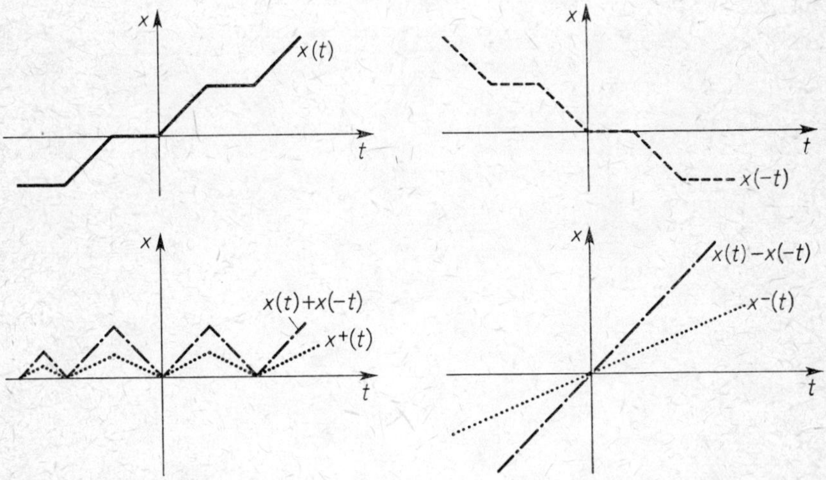

Fig. 4. Decomposition of a function into even and odd parts

It follows from the considerations of Section 3, that under certain assumptions the simple regularizers of polynomials in algebraic and almost algebraic operators are easy to determine. We obtain the following

THEOREM 4.2. *Let S be an almost algebraic operator in an algebra $\mathscr{X}(X) \subset L_0(X)$ and let $P(S) = T_0$, where $P(t) = \prod_{m=1}^{n} (t - t_m)$ is the characteristic polynomial and T_0 belongs to a quasi-Fredholm ideal $J \subset \mathscr{X}(X)$. If the polynomial $A(S) = \sum_{m=1}^{N-1} A_m S^m$ satisfies the following conditions*:

(1) *$A_m \in \mathscr{X}(X)$ and $[A_m, S] \in J$ for $m = 0, 1, \ldots, N-1$,*

(2) *there exist simple regularizers $B(t_j)$ of the operators $A(t_j)$ to the ideal J $(j = 1, \ldots, N)$;*

then for every $T \in J$ *the operator* $A(S)+T$ *has finite d-characteristic. Moreover, if* J *is a Fredholm ideal and the algebra* $\mathscr{X}(X)$ *is regularizable to a Fredholm ideal* $J_1 \subset \mathscr{X}(X)$, *then*

$$\varkappa_{A(S)+T} = \varkappa_{A(S)} \quad \textit{for every } T \in J.$$

PROOF. Theorem 3.3 and the above conditions together imply that the operator $A(S)$ has a simple regularizer to the ideal J. Corollary 3.1 implies that the operator $A(S)+T$ also has a simple regularizer to the ideal J equal to the preceding one. Hence the operator $A(S)+T$ has finite d-characteristic (cf. Section 3 of Chapter I) for every $T \in J$. Moreover, if both J and J_1 are Fredholm ideals and the algebra $\mathscr{X}(X)$ is regularizable to the ideal J_1, then Theorem 3.1 of Chapter I implies that every $T \in J$ is a perturbation of the class of all operators with finite d-characteristic in $\mathscr{X}(X)$ which leaves the index unchanged. Hence $\varkappa_{A(S)+T} = \varkappa_{A(S)}$ for all $T \in J$.

THEOREM 4.3. *Let* S *be an algebraic operator in an algebra* $\mathscr{X}(X)$, *with the characteristic polynomial* $P(t) = \prod_{m=1}^{N} (t-t_m)$, *and suppose that the polynomial* $A(S)$ *satisfies the conditions*:

(1) $A_m \in \mathscr{X}(X)$, $[A_m, S] \in J \subset \mathscr{X}(X)$ $(m = 0, 1, \ldots, N-1)$, *where* J *is a quasi-Fredholm ideal*;

(2) *the operators* $[A(t_j)]^{-1}$ $(j = 1, \ldots, N)$ *exist.*

Then the operator $A(S)+T$ *has finite d-characteristic for every* $T \in J$.

Moreover, if J *is a Fredholm ideal, and if the algebra* $\mathscr{X}(X)$ *is regularizable to a Fredholm ideal* J_1, *then* $\varkappa_{A(S)+T} = \varkappa_{A(S)}$ *for every* $T \in J$.

PROOF. It follows from Theorem 3.2 that the operator $A(S)$ has a simple regularizer $R_{A(S)}$ and further more the proof follows the same lines as that of Theorem 4.2.

From Theorem 3.3 of Chapter I we immediately obtain

COROLLARY 4.1. *If* S *and* $A(S)$ *satisfy the conditions of Theorem* 4.2 *and if* J *is a* Ξ*-quasi-Fredholm ideal, then* $A(S)+T$ *is a* Φ_{Ξ}*-operator for all* $T \in J$.

COROLLARY 4.2. *If* S *and* $A(S)$ *satisfy the conditions of Theorem* 4.3 *and if* J *is a* Ξ*-quasi-Fredholm ideal, then* $A(S)+T$ *is a* Φ_{Ξ}*-operator for all* $T \in J$.

Some results for the case where S is an involution and not all the $A(t_i)$ are invertible will be given in Section 5 of Chapter IX.

5. EQUATIONS WITH COEFFICIENTS COMMUTING WITH AN ALGEBRAIC OPERATOR

To begin with, we consider polynomials in algebraic operators with constant coefficients. If an operator $S \in L_0(X)$ is algebraic, and if

$$a(S) = a_0 I + a_1 S + \ldots + a_{N-1} S^{N-1},$$

where $a_0, a_1, \ldots, a_{N-1}$ are complex numbers, then of course $a_i S - S a_i = 0$ for $i = 0, 1, \ldots, N-1$. Hence, if $a(t_i) \neq 0$ $(i = 1, \ldots, N)$, Corollary 3.3 implies that the operator $[a(S)]^{-1}$ exists and $[a(S)]^{-1} = R_{a(S)}$, where $R_{a(S)}$ is given by Formula (3.2). However, in the case of constant coefficients we can apply Theorem 4.1 and obtain in addition results in the case where some of the $a(t_i)$ are equal to zero, as is shown by the following

THEOREM 5.1. *Let the operator $S \in L_0(X)$ be algebraic with characteristic polynomial $P(t) = \prod_{k=1}^{N} (t - t_k)$ and let $a(t) = a_0 + \ldots + a_{N-1} S^{N-1}$, where $a_0, a_1, \ldots, a_{N-1}$ are complex constants. Then the equation*

$$a(S)x = y, \quad y \in X \tag{5.1}$$

is equivalent to N independent equations

$$a(t_j) x_{(j)} = y_{(j)} \quad (j = 1, 2, \ldots, N), \tag{5.2}$$

where we write $x_{(j)} = P_j x$, and $x = \sum_{j=1}^{N} x_{(j)}$ for every $x \in X$.

PROOF. Theorem 4.1 together with our assumptions imply that the space X is the direct sum of the spaces $X_j = P_j X$ such that $S x_{(j)} = t_j x_{(j)}$ for all $x_{(j)} \in X_j$ and $j = 1, 2, \ldots, N$. Since by Property 2.4

$$a(S)x = \Big[\sum_{j=1}^{N} a(t_j) P_j\Big] x = \sum_{j=1}^{N} a(t_j) P_j x = \sum_{j=1}^{N} a(t_j) x_{(j)},$$

where the $a(t_j)$ are numbers, we can rewrite equation (5.1) as follows, noting first that $y = \sum_{j=1}^{N} y_{(j)}$, where $y_{(j)} = P_j y$:

$$\sum_{j=1}^{N} a(t_j) x_{(j)} = \sum_{j=1}^{N} y_{(j)}.$$

On the other hand, it follows from Theorem 4.1 that X is the direct sum of the spaces $X_j = P_j X$. Hence the last equation is equivalent to the system of independent equations (5.2).

THEOREM 5.2. *Under the same conditions as those of Theorem* 5.1, *a necessary and sufficient condition for equation* (5.1) *to have a solution is*

$$P_m y = 0 \quad \textit{for all } m \textit{ such that } a(t_m) = 0.$$

If this condition is satisfied, the solution is of the form

$$x = \sum_{m:a(t_m)\neq 0} \frac{1}{a(t_m)} P_m y + \sum_{m:a(t_m)=0} z_{(m)},$$

where the $z_{(m)} \in X_m$ *are arbitrary. (The first sum runs over all* m *such that* $1 \leqslant m \leqslant N$ *and* $a(t_m) \neq 0$, *the second sum over all* m *such that* $1 \leqslant m \leqslant N$ *and* $a(t_m) = 0$.)

PROOF. It follows from Theorem 5.1 that equation (5.1) is equivalent to the N independent equations (5.2) : $a(t_j)x_{(j)} = y_{(j)}$ $(j = 1, 2, \ldots, N)$. If $a(t_m) = 0$ for some m, then the equation $a(t_m)x_{(m)} = y_{(m)}$ has a solution if and only if $P_m y = y_{(m)} = 0$. Under this condition every element $z_{(m)} \in X_m$ is a solution of the equation in question. If $a(t_m) \neq 0$, then $x_{(m)} = [1/a(t_m)]y_{(m)}$. However, from Theorem 5.1 follows that X is the direct sum of spaces $X_m = P_m x$. Hence each solution of equation (5.1) is of the form $x = \sum_{m=1}^{N} x_m$, where x_m is a solution of the mth equation 5.2. This completes the proof.

We now show that Theorems 5.1 and 5.2 can be extended to arbitrary polynomials whose coefficients commute with S. To do this, it suffices to apply the following

LEMMA 5.1. *If a linear operator* A *acting in a linear space* X *commutes with an algebraic operator* $S \in L_0(X)$ *of order* N, *then*

$$A(D_A \cap X_m) \subset X_m \quad \textit{for } m = 1, 2, \ldots, N.$$

Indeed, if A commutes with S, then A commutes also with P_j for $j = 1, 2, \ldots, N$ (since the P_j are polynomials in S with constant coefficients). Let $x \in D_A$ be arbitrary. Then $x_{(j)} = P_j x \in X_j$ for $j = 1, 2, \ldots, N$ and

$$Ax_{(j)} = AP_j x = P_j Ax = (Ax)_{(j)} \in X_j \quad \text{for } j = 1, 2, \ldots, N.$$

This implies the following

THEOREM 5.3. *Let* $S \in L_0(X)$ *be an algebraic operator with characteristic polynomial* $P(t) = \prod_{m=1}^{N} (t - t_m)$ *and let* $A(S) = \sum_{k=0}^{N-1} A_k S^k$, *where* $A_0, A_1, \ldots, A_{N-1}$ *are linear operators acting in* X *and satisfying the following conditions*:

(1) A_m *commutes with* S $(m = 0, 1, \ldots, N-1)$;

(2) $A(t_m)$ *is either invertible or equal to zero* $(m = 1, 2, \ldots, N)$. *Then*

(i) *The equation*

$$A(S)x = y, \quad y \in X \tag{5.3}$$

is equivalent to the system of N *independent equations*

$$A(t_m)x_{(m)} = y_{(m)} \quad (m = 1, 2, \ldots, N). \tag{5.4}$$

(ii) *Equation* (5.3) *has a solution if and only if*

$$P_m y = 0 \quad \text{for all } m \text{ such that } A(t_m) = 0.$$

If this condition is satisfied, then the solution of equation (5.3) *is of the form*

$$x = \sum_{m:A(t_m)\neq 0} [A(t_m)]^{-1} P_m y + \sum_{m:A(t_m)=0} z_{(m)},$$

where $z_{(m)} \in X_m$ *is arbitrary. (The first sum runs over all* m *such that* $1 \leqslant m \leqslant N$ *and* $A(t_m) \neq 0$ *and the second sum runs over all* m *such that* $1 \leqslant m \leqslant N$ *and* $A(t_m) = 0$).

PROOF. We apply Lemma 5.1 to the operator $A(S)$, which clearly commutes with S by our condition (1). From Property 2.4, for every $x_{(m)} \in \boldsymbol{D}_{A(S)} \cap X_m$ and for $m = 1, 2, \ldots, N$

$$A(S)x_{(j)} = \sum_{m=1}^{N} A(t_m)P_m x_{(j)} = \sum_{m=1}^{N} A(t_m)P_m P_j x = A(t_j)P_j x = A(t_j)x_{(j)} \in X_j.$$

Furthermore the proof is along the same lines as the proofs of Theorems 5.1 and 5.2.

This Theorem immediately implies

COROLLARY 5.1. *Let* X *be a linear space. Let*

$$A(S) = \sum_{k=0}^{n}\sum_{j=0}^{m} A_{kj} D^k S^j,$$

where

1° S, D, A_{kj} $(k = 0, 1, \ldots, n;\ j = 0, 1, \ldots, m \leqslant N-1)$ *are linear operators acting in* X,

2° S *is an algebraic operator with characteristic polynomial* $\prod_{k=1}^{N} (t-t_k)$,

3° $SD-DS = 0$ *and* $SA_{kj}-A_{kj}S = 0$ $(k = 0, 1, \ldots, n;\ j = 0, 1, \ldots, m)$.

Then the equation

$$A(S)x = y, \quad y \in X, \tag{5.5}$$

is equivalent to the N independent equations

$$A(t_k)x_{(k)} = y_{(k)} \quad (k = 1, 2, \ldots, N), \text{ where } A(t) = \sum_{k=0}^{n}\sum_{j=0}^{m} A_{kj} D^k t^j. \tag{5.6}$$

If each of these equations has a solution belonging to X, then the solution of equation (5.5) *is of the form* $x = \sum_{j=1}^{N} x_{(j)}$, *where* $x_{(j)} \in X_j$ *is a solution of the jth equation* (5.6).

It is enough for the proof to write $A_j = \sum_{k=0}^{n} A_{kj} D^k$.

REMARK 5.1. The equation $A(S)x = 0$ always has a solution belonging to X, since the jth equation (5.6) at least has the solution $x_{(j)} = 0$.

6. INVOLUTIONS OF ORDER N

In this section we study some special properties of operators which are involutions of order N. We recall that an algebraic operator is an involution of order N $(N \geqslant 2)$ if its characteristic polynomial is of the form $P(t) = t^N - 1$, i.e. if

$$S^N = I \quad \text{on } X. \tag{6.1}$$

If $N = 2$, S is simply called an involution (cf. Section 2).

Let S be an involution of order N on a linear space X. The characteristic roots of the operator S are the Nth roots of unity:

$$\varepsilon = e^{2\pi i/N}, \varepsilon^2, \ldots, \varepsilon^{N-1}, \varepsilon^N = 1$$

(see Section 1). Applying the properties of the Nth roots of unity, we obtain some useful new results. We remark to begin with that the projectors $P_1, \ldots, P_N$ associated with an involution S of order N can be written in a more convenient form. Namely we have

PROPERTY 6.1. *If S is an involution of order N acting in a linear space X, then the projectors* $P_1, \ldots, P_N$ *giving a partition of unity have the following form*:

$$P_j = \frac{1}{N}\sum_{k=0}^{N-1} \varepsilon^{-kj} S^k, \quad \text{where } \varepsilon = e^{2\pi i/N} \ (j = 1, 2, \ldots, N). \tag{6.2}$$

PROOF. Formulae (2.9) imply

$$(6.3)\qquad P_j = \mathfrak{p}_j(S), \quad \text{where } \mathfrak{p}_j(t) = \prod_{k=1,\, k\neq j}^{n} \frac{t-\varepsilon^k}{\varepsilon^j - \varepsilon^k} \qquad (j = 1, 2, \ldots, N).$$

Now write

$$a_j(t) = \prod_{k=1,\, k\neq j}^{N} (t-\varepsilon^k); \quad b_j(t) = \frac{1}{N}\sum_{k=0}^{N-1} \varepsilon^{-kj} t^k \qquad (j = 1, 2, \ldots, N).$$

Since $\varepsilon, \varepsilon^2, \ldots, \varepsilon^N = 1$ are the different Nth roots of unity, we find for $t \neq \varepsilon^j$ $(j = 1, 2, \ldots, N)$

$$b_j(t) = \frac{1}{N}\sum_{k=0}^{N-1} \varepsilon^{-kj} t^k = \frac{1}{N}\cdot\frac{(\varepsilon^{-j}t)^N - 1}{\varepsilon^{-j}t - 1} = \frac{1}{N}\,\frac{\varepsilon^{-Nj}t^N - 1}{\varepsilon^{-j}t-1}$$

$$= \frac{\varepsilon^j}{N}\,\frac{t^N-1}{t-\varepsilon^j} = \frac{\varepsilon^j}{N}\,\frac{1}{t-\varepsilon^j}\prod_{k=1}^{N}(t-\varepsilon^k) = \frac{\varepsilon^j}{N}\prod_{k=1,\, k\neq j}^{N}(t-\varepsilon^k) = \frac{\varepsilon^j}{N}a_j(t).$$

Moreover,

$$b_j(\varepsilon^j) = \frac{1}{N}\sum_{k=0}^{N-1}\varepsilon^{-kj}\varepsilon^{kj} = N\cdot\frac{1}{N} = 1; \quad a_j(\varepsilon^j) = \prod_{k=1,\, k\neq j}^{N}(\varepsilon^j - \varepsilon^k)$$

$$(j = 1, 2, \ldots, N).$$

On the other hand, the functions $a_j(t)$, $b_j(t)$, being polynomials, are continuous at the points ε^j. Thus

$$1 = b_j(\varepsilon^j) = \lim_{t\to\varepsilon^j} b_j(t) = \frac{\varepsilon^j}{N}\lim_{t\to\varepsilon^j} a_j(t) = \frac{\varepsilon^j}{N}a_j(\varepsilon^j) \qquad (j = 1, 2, \ldots, N).$$

Therefore

$$\prod_{k=1,\, k\neq j}^{N}(\varepsilon^j - \varepsilon^k) = a_j(\varepsilon^j) = N\varepsilon^{-j} \qquad (j = 1, 2, \ldots, N).$$

Hence Formulae (6.3) imply

$$\mathfrak{p}_j(t) = \frac{a_j(t)}{a_j(\varepsilon^j)} = \frac{N\varepsilon^{-j}b_j(t)}{N\varepsilon^{-j}} = b_j(t) = \frac{1}{N}\sum_{k=0}^{N-1}\varepsilon^{-kj}t^k \qquad (j = 1, 2, \ldots, N)$$

and

$$P_j = \mathfrak{p}_j(S) = \frac{1}{N}\sum_{k=0}^{N-1} \varepsilon^{-kj} S^k \qquad (j = 1, 2, \ldots, N)$$

which was to be proved.

When $N = 2$, i.e. when S is an involution, we write the corresponding projectors as follows

(6.4) $$P^+ = \tfrac{1}{2}(I+S), \quad P^- = \tfrac{1}{2}(I-S)$$

and we also write

(6.5) $$X^+ = P^+X, \quad X^- = P^-X, \quad x^+ = P^+x, \quad x^- = P^-x \quad \text{for } x \in X.$$

Observe that every involution of order N is invertible, and that

(6.6) $$S^{-1} = S^{N-1} = \sum_{j=1}^{N} \varepsilon^{j(N-1)} P_j = \sum_{j=1}^{N} \varepsilon^{-j} P_j,$$

where $S^N = I$.

PROPERTY 6.2. *If S is an involution of order N acting in a linear space X and if $q(t)$ is an arbitrary polynomial with constant complex coefficients, then*

$$q(\varepsilon^m S) = \sum_{j=1}^{N} q(\varepsilon^{m+j}) P_j \qquad (m = 0, \pm 1, \pm 2, \ldots).$$

Indeed,

$$q(\varepsilon^m S) = \sum_{k=0}^{N-1} q_k(\varepsilon^m S)^k = \sum_{k=0}^{N-1} q_k \varepsilon^{mk} S^k \Big(\sum_{k=0}^{N} P_j\Big) = \sum_{j=1}^{N} \Big(\sum_{k=0}^{N-1} q_k \varepsilon^{mk} S^k P_j\Big)$$

$$= \sum_{j=1}^{N} \Big(\sum_{k=0}^{N-1} q_k \varepsilon^{mk} \varepsilon^{jk} P_j\Big) = \sum_{j=1}^{N} \Big(\sum_{k=0}^{N-1} q_k \varepsilon^{(j+m)k}\Big) P_j = \sum_{j=1}^{N} q(\varepsilon^{j+m}) P_j.$$

Suppose now that X is a unitary space (see Example 4.1 of Chapter I) under the inner product (x, y). We now consider the properties of involutions of order N with respect to the inner product (x, y). A linear operator S acting in a unitary space X is said to be *unitary* if $(Sx, Sy) = (x, y)$ for all $x, y \in X$.

PROPERTY 6.3. *Let S be a unitary operator acting in a unitary space X. If k and m are the least positive integers such that $(S^*)^k = S^m$ (where S^* denotes the operator conjugate to S with respect to the inner product (x, y), i.e. such that $(x, S^*y) = (Sx, y)$ for all $x, y \in X$), then S is an involution of order $k+m$.*

Indeed, it follows from our assumptions that for all $x, y \in X$

$$(S^{k+m}x, y) = (S^m x, (S^k)^* y) = (S^m x, (S^*)^k y) = (Sx, Sy) = (x, y).$$

Since x and y are arbitrary, it follows that $S^{k+m} = I$.

THEOREM 6.1. *Let S be an involution of order N acting in a unitary space X with inner product (x, y). Then S is unitary if and only if the spaces X_j, where $X_j = P_j X$, are pairwise orthogonal ($j = 1, 2, \ldots, N$) with respect to the inner product (x, y).*

PROOF. Let S be unitary. Since $\bar{\varepsilon} = \varepsilon^{-1}$ ($\bar{a}$ denotes the complex number conjugate to a) and since $Sx_{(j)} = \varepsilon^j x_{(j)}$, we have for every $x \in X$ and for $j, m = 1, 2, \ldots, N$

$$(x_{(j)}, x_{(m)}) = (\varepsilon^{-j}\varepsilon^j x_{(j)}, \varepsilon^{-m}\varepsilon^m x_{(m)}) = \overline{\varepsilon^{-m}}\varepsilon^{-j}(Sx_{(j)}, Sx_{(m)}) = \varepsilon^{m-j}(x_{(j)}, x_{(m)}).$$

But $\varepsilon^{m-j} \neq 1$ for $m \neq j$, and since x is arbitrary this implies that

$$(x_{(j)}, x_{(m)}) = 0 \quad \text{if } j \neq m \quad \text{for all } x \in X. \tag{6.7}$$

Let $x \in X_j$ and $y \in X_m$ be arbitrary. Putting $u = x+y$, we obtain $u_{(j)} = x$ and $u_{(m)} = y$. From (6.7) we then obtain

$$(x, y) = (u_{(j)}, u_{(m)}) = 0 \quad \text{if } j \neq m.$$

Hence the spaces X_j and X_m are orthogonal if $j \neq m$.

Conversely, suppose that X_j is orthogonal to X_m for $m \neq j$ ($j, m = 1, 2, \ldots, N$). For arbitrary $x \in X$ we have

$$\begin{aligned}(Sx, Sy) &= \Big(\sum_{j=1}^{N} Sx_{(j)}, \sum_{m=1}^{N} Sy_{(m)}\Big) = \Big(\sum_{j=1}^{N} \varepsilon^j x_{(j)}, \sum_{m=1}^{N} \varepsilon^m y_{(m)}\Big) \\ &= \sum_{j,m=1}^{N} \varepsilon^{j-m}(x_{(j)}, y_{(m)}) = \sum_{j=1}^{N} \varepsilon^j \varepsilon^{-j}(x_{(j)}, y_{(j)}) \\ &= \sum_{j=1}^{N} (x_{(j)}, y_{(j)}).\end{aligned}$$

Since $\sum_{k=1, k\neq j}^{N} (x_{(j)}, y_{(k)}) = 0$, we finally obtain

$$\begin{aligned}(Sx, Sy) &= \sum_{j=1}^{N} (x_{(j)}, y_{(j)}) = \sum_{j=1}^{N} (x_{(j)}, y_{(j)}) + \sum_{k=1, k\neq j}^{N} (x_{(j)}, y_{(k)}) \\ &= \sum_{j=1}^{N} \Big(x_{(j)}, \sum_{k=1}^{N} y_{(k)}\Big) = \sum_{j=1}^{N} (x_{(j)}, y) = \Big(\sum_{j=1}^{N} x_{(j)}, y\Big) = (x, y).\end{aligned}$$

Hence S is unitary, which was to be proved.

We define now a new inner product

$$(x, y)_S = \sum_{m=1}^{N} (S^m x, S^m y) \quad \text{for all } x, y \in X. \tag{6.8}$$

It is easy to check that an involution of order N is a unitary operator with respect to the inner product $(x, y)_S$. This yields the following

COROLLARY 6.1. *Under the conditions of Theorem* 6.1 *the spaces X_j are pairwise orthogonal with respect to the inner product* (6.7).

For some applications we need the following

PROPOSITION 6.1. *Let S be an involution acting in a linear space X. Let the operators A and B acting in X commutate with S but be such that $AB-BA \neq 0$. If the operator A is invertible, then*

$$(A-\tilde{B}S)(A+BS) = A^2-\tilde{B}B, \quad \textit{where } \tilde{B} = ABA^{-1}.$$

Moreover, if $A^2-\tilde{B}B = A^2-ABA^{-1}B$ is invertible, then the operator $A^2-A\tilde{B}A^{-1}\tilde{B}$ is also invertible.

Indeed, since $S^2 = I$ and $AS-SA = BS-SB = 0$, we have

$$\begin{aligned}(A-\tilde{B}S)(A+BS) &= (A-ABA^{-1}S)(A+BS)\\ &= A^2+ABS-ABA^{-1}SA-ABA^{-1}SBS\\ &= A^2+ABS-ABA^{-1}AS-ABA^{-1}BS^2\\ &= A^2+ABS-ABS-ABA^{-1}B = A^2-\tilde{B}B.\end{aligned}$$

If the operator $A^2-\tilde{B}B$ is invertible, then the operator

$$\begin{aligned}A^2-A\tilde{B}A^{-1}\tilde{B} &= A^2-A^2BA^{-1}A^{-1}ABA^{-1} = A^2-A^2BA^{-1}BA^{-1}\\ &= A^2-A\tilde{B}BA^{-1} = A(A^2-\tilde{B}B)A^{-1}\end{aligned}$$

is invertible being the superposition of three invertible operators.

From Corollary 5.1 we immediately obtain

THEOREM 6.2. *Let X be a linear space. Let*

$$A(S) = \sum_{k=0}^{n}\sum_{j=0}^{m} A_{kj} D^k S^{\eta_j},$$

where

1° *S, D and A_{kj} $(k = 0, 1, \ldots, n;\ j = 0, 1, \ldots, m)$ are linear operators acting in X,*

2° $SA_{kj}-A_{kj}S = 0$ ($k = 0, 1, \ldots, n$; $j = 0, 1, \ldots, m$) and $SD-DS = 0$,

3° $\eta_0 = 0$, $\eta_1, \ldots, \eta_m$ are integers, not all zero.

Let N be a common multiple (not necessarily the l.c.m.) of the integers $\eta_1, \ldots, \eta_m$ and suppose that there is a subspace $\tilde{X} \subset X$ such that S is an involution of order N on $\tilde{X}$. Then the equation

$$A(S)x = y, \quad y \in \tilde{X} \tag{6.9}$$

is equivalent to the N independent equations

$$A(\varepsilon^j)x_{(j)} = y_{(j)} \quad (j = 1, 2, \ldots, N), \tag{6.10}$$

where $y_{(j)} = P_j y \in \tilde{X}_j = P_j\tilde{X}$. If each of these equations has a solution in $\tilde{X}$, then the solution of equation (6.9) is of the form $x = \sum_{j=1}^{N} x_j$, where $x_j \in \tilde{X}_j$ is a solution of the jth equation (6.10).

For the proof it suffices to remark that for all positive integers k, $S^{-k} = S^{N-k}$. This follows from Formulae (6.6) and Corollary 2.2, since

$$S^{-k} = \sum_{j=1}^{N} \varepsilon^{-jk} P_j = \sum_{j=1}^{N} \varepsilon^{jN}\varepsilon^{-jk} P_j = \sum_{j=1}^{N} \varepsilon^{(N-k)j} P_j = S^{N-k}.$$

REMARK 6.1. The homogeneous equation $A(S)x = 0$ always has a solution belonging to X. In fact, the jth equation (6.10) at least has the solution zero (cf. Remark 5.1).

7. MULTI-INVOLUTIONS

Suppose that we are given q involutions $S_1, \ldots, S_q$ of orders $N_1, \ldots, N_q$ respectively acting in a linear space X. We write

$$\varepsilon_p = e^{2\pi i/N_p}, \quad P_{p,r} = \frac{1}{N_p} \sum_{k=0}^{N_p-1} \varepsilon^{-kr} S_p^k \quad (p = 1, 2, \ldots, q;\ r = 1, 2, \ldots, N_p).$$

From our foregoing studies (Property 6.1) it follows that

$$\sum_{r=1}^{N_p} P_{p,r} = I, \quad P_{p,r_1} P_{p,r_2} = \delta_{r_1 r_2} P_{p,r_1}, \quad S_p P_{p,r} = \varepsilon_p^r P_{p,r} \tag{7.1}$$

($\delta_{r_1 r_2} = 1$ for $r_1 = r_2$ and $= 0$ otherwise, $r, r_1, r_2 = 1, 2, \ldots, N_p$; $p = 1, 2, \ldots$ $\ldots, q$).

In order to simplify the theorems to be given later on, we now introduce multi-involutions.

We consider q-dimensional *multi-indices* $k = (k_1, \ldots, k_q)$ and $m = (m_1, \ldots$ $\ldots, m_q)$, where the k_p and the m_p are integers $(p = 1, 2, \ldots, q)$. We write in the usual way,

$$|k| = |k_1| + \ldots + |k_q|; \quad k+m = (k_1+m_1, \ldots, k_q+m_q),$$

$$\lambda k = (\lambda k_1, \ldots, \lambda k_q) \quad \text{for any integer } \lambda,$$

$$km = (k_1 m_1, \ldots, k_q m_q), \quad k^m = (k_1^{m_1}, \ldots, k_q^{m_q}).$$

We also write $k = (n)_q$ if $k_p = n$ for $p = 1, 2, \ldots, q$, where n is some integer. If all the $m_p \neq 0$, we write $k/m = (k_1/m_1, \ldots, k_q/m_q)$. We further write:

$$k \leqslant m \text{ if and only if } k_p \leqslant m_q \quad \text{for } p = 1, 2, \ldots, q,$$

$$k = m \text{ if and only if } k_p = m_q \quad \text{for } p = 1, 2, \ldots, q.$$

Let $N = (N_1, \ldots, N_q)$ and $\varepsilon = (\varepsilon_1, \ldots, \varepsilon_q)$, where $\varepsilon_p = e^{2\pi i/N_p}$ $(p = 1, 2, \ldots, q)$. We write

$$\varepsilon^k = (\varepsilon_1^{k_1}, \ldots, \varepsilon_q^{k_q}) \quad \text{for any multi-index } k.$$

By definition $\varepsilon^{-jk} = \varepsilon^{(N-j)k}$, where N is defined above and k, j are arbitrary multi-indices.

A superposition $S = S_1 \ldots S_q$ acting in a linear space X is called a *multi-involution of order* $N = (N_1, \ldots, N_q)$ if S_p is an involution of order N_p and S_p commute with S_r for $r = 1, 2, \ldots, N_p$; $p = 1, 2, \ldots, q$. If we write

$$S^k = S_1^{k_1} \ldots S_q^{k_q}, \quad \text{where } k = (k_1, \ldots, k_q),$$

then

$$S^N = S_1^{N_1} \ldots S_q^{N_q} = I.$$

If we write

$$P_j = P_{1,j_1} \ldots P_{q,j_q}, \quad j = (j_1, \ldots, j_q), \quad (1)_q \leqslant j \leqslant N,$$

$$x_{(j)} = P_j x, \quad X_j = P_j X,$$

we find that the following formulae (which are analogous to those given in Property 6.1) hold:

(7.2) $$\sum_{(1)_q \leqslant j \leqslant N} P_j = I, \quad \text{where} \sum_{(1)_q \leqslant j \leqslant N} = \sum_{j_1=1}^{N_1} \ldots \sum_{j_q=1}^{N_q},$$

(7.3) $$P_j P_k = \delta_{jk} P_j \quad (\delta_{jk} = 1 \text{ for } j = k \text{ and } 0 \text{ otherwise}),$$

(7.4) $$S^k P_j = \varepsilon^{kj} P_j, \quad k = (k_1, \ldots, k_q),\ j = (j_1, \ldots, j_q),\ (1)_q \leqslant j, k \leqslant N.$$

Indeed, from the first formula of (6.1)

$$\sum_{(1)_q \leqslant j \leqslant N} P_j = \sum_{(1)_q \leqslant j \leqslant N} P_{1,j_1} \dots P_{q,j_q} = \sum_{j_1=1}^{N_1} \dots \sum_{j_q=1}^{N_q} P_{1,j_1} \dots P_{q,j_q}$$

$$= \sum_{j_1=1}^{N_1} \dots \sum_{j_{q-1}-1}^{N_{q-1}} P_{1,j_1} \dots P_{q-1,j_{q-1}} = \dots = \sum_{j_1=1}^{N_1} P_{1,j_1} = I.$$

From the second formula of (7.1) we obtain

$$P_j P_k = \Big(\prod_{1 \leqslant p \leqslant q} P_{p,j_p}\Big)\Big(\prod_{1 \leqslant p \leqslant q} P_{p,k_p}\Big) = \prod_{1 \leqslant p \leqslant q} P_{p,j_p} P_{p,k_p} = \prod_{1 \leqslant p \leqslant q} \delta_{j_p k_p} P_{p,j_p}$$

$$= \begin{cases} 1 & \text{if } k_p = j_p \ (p = 1, 2, \dots, q), \text{ i.e. if } j = k, \\ 0 & \text{otherwise.} \end{cases}$$

Finally

$$S^k P_j = S_1^{k_1} \dots S_q^{k_q} P_{1,j_1} \dots P_{q,j_q} = (S_1^{k_1} P_{1,j_1}) \dots (S_q^{k_q} P_{q,j_q})$$

$$= (\varepsilon_1^{k_1 j_1} P_{1,j_1}) \dots (\varepsilon_q^{k_q j_q} P_{q,j_q}) = \varepsilon_1^{k_1 j_1} \dots \varepsilon_q^{k_q j_q} P_{1,j_1} \dots P_{q,j_q} = \varepsilon^{kj} P_j.$$

Note that the space X is a direct sum

(7.5) $$X = \bigoplus_{(1)_q \leqslant j \leqslant N} X_j, \quad \text{where } X_j = P_j X.$$

If a linear operator acting in the space X commutes with the involutions $S_1, \dots, S_q$ of orders $N_1, \dots, N_q$ respectively, then A also commutes with the multi-involution $S = S_1 \dots S_q$ of order $N = (N_1, \dots, N_q)$ and

(7.6) $$A(\boldsymbol{D}_A \cap X_j) \subset X_j \quad \text{for } (1)_q \leqslant j \leqslant N.$$

Indeed, let $x \in \boldsymbol{D}_A$. Then $x_{(j)} = P_j x \in X_j$ for $(1)_q \leqslant j \leqslant N$ and $Ax_{(j)} = AP_j x = P_j(Ax) \in X_j$, since P_j commutes with S. This yields the required formula (7.6).

We denote by $A(t)$ an arbitrary polynomial in the variables $t = (t_1, \dots, t_q)$ which we shall also write in any one of the following ways:

$$A(t) = A(t_1, \dots, t_q) = \sum_{(0)_q \leqslant k \leqslant m} a_{k_1 \dots k_q} t_1^{k_1} \dots t_q^{k_q} = \sum_{(0)_q \leqslant k \leqslant m} a_k t^k.$$

Let $S = S_1 \dots S_q$ be a multi-involution of order $N = (N_1, \dots, N_q)$, acting in a linear space X. Let

$$A(S) = \sum_{(0)_q \leqslant k \leqslant N-(1)_q} a_k S^k.$$

Then

(7.7) $$A(S) = \sum_{(1)_q \leqslant j \leqslant N} A(\varepsilon^j) P_j.$$

Indeed,

$$A(S) = \sum_{(1)_q \leqslant j \leqslant N} A(S) P_j = \sum_{(1)_q \leqslant j \leqslant N} \sum_{(0)_q \leqslant k \leqslant N-(1)_q} a_k S^k P_j$$
$$= \sum_{(1)_q \leqslant j \leqslant N} \sum_{(0)_q \leqslant k \leqslant N-(1)_q} a_k \varepsilon^{kj} P_j = \sum_{(1)_q \leqslant j \leqslant N} A(\varepsilon^j) P_j.$$

This implies that any equation

(7.8) $$A(S_1, \ldots, S_q)x = y, \quad y \in X$$

is equivalent to a system of independent equations

$$A(\varepsilon^j) x_{(j)} = y_{(j)} \quad \text{for } (1)_q \leqslant j \leqslant N.$$

The number of equations is $N_0 = N_1 \ldots N_q$.

THEOREM 7.1. *Let X be a linear space. Let*

(7.9) $$A(S) = \sum_{(0)q \leqslant j \leqslant m} A_j S^{\eta_j},$$

where

1° $S = S_1 \ldots S_q$ *and* $S_1, \ldots, S_q, A_{(0)_q}, \ldots, A_m$ *are linear operators acting in* X;

2° $S_p S_r - S_r S_p = 0$, $S_p A_j - A_j S_p = 0$ *for* $p, r = 1, 2, \ldots, q$ *and* $(0)_q \leqslant j \leqslant m$,

3° $\eta_j = (\eta_{1,j}, \ldots, \eta_{q,j})$, $\eta_{p,j}$ *are integers,* $\sum_{p=1}^{q} \eta_{p,j} > 0$ $((0)_q \leqslant j \leqslant m)$, *and* $\eta_{(0)_q} = (0)_q$.

Let N_p be a common multiple of the integers $\eta_{p,j}$ $((0)_q \leqslant j = m)$ and let us suppose that there is a subspace $\tilde{X} \subset X$ such that S is a multi-involution of order $N = (N_1, \ldots, N_q)$ on $\tilde{X}$. Then

(7.10) $$A(S) = \sum_{(1)_q \leqslant j \leqslant N} A(\varepsilon^j) P_j \quad \text{on } \tilde{X},$$

where $\varepsilon = (\varepsilon_1, \ldots, \varepsilon_q)$, $\varepsilon_p = e^{2\pi i/N_p}$, $P_j = P_{1,j_1}, \ldots, P_{q,j_q}$;

$$P_{p,j_p} = \frac{1}{N_p} \sum_{k=0}^{N-1} \varepsilon_p^{-kj_p} S_p^k.$$

PROOF. Since S_p and A_j commute, we find

$$A(S) = A(S) \sum_{(1)_q \leqslant \nu \leqslant N} P_\nu = \sum_{(0)_q \leqslant j \leqslant m} A_j S^{\eta_j} \sum_{(1)_q \leqslant \nu \leqslant N} P_\nu$$

$$= \sum_{(0)_q \leqslant j \leqslant m} \sum_{(1)_q \leqslant \nu \leqslant N} A_j S^{\eta_j} P_\nu = \sum_{(0)_q \leqslant j \leqslant m} \sum_{(1)_q \leqslant \nu \leqslant N} A_j \varepsilon^{\nu\eta_j} P_\nu$$

$$= \sum_{(1)_q \leqslant j \leqslant N} \Big(\sum_{(0)_q \leqslant j \leqslant m} \varepsilon^{\nu\eta_j} A_j \Big) P_\nu = \sum_{(1)_q \leqslant \nu \leqslant N} A(\varepsilon^\nu) P_\nu,$$

which was to be proved.

THEOREM 7.2. *Under the conditions of Theorem* 7.1 *the equation*

$$A(S)x = y, \quad y \in \tilde{X}, \tag{7.11}$$

is equivalent to the $N_0 = N_1 \ldots N_q$ *independent equations*

$$A(\varepsilon^\nu) x_{(\nu)} = y_{(\nu)}, \quad (1)_q \leqslant \nu \leqslant N, \tag{7.12}$$

where $y_{(\nu)} = P_\nu y \in \tilde{X}_\nu = P_\nu \tilde{X}$, *and if each of equations* (7.12) *has a solution* $x_\nu \in \tilde{X}_\nu$, *then the solution of equation* (7.11) *is given by the formula*

$$x = \sum_{(1)_q \leqslant \nu \leqslant N} x_\nu . \tag{7.13}$$

PROOF. Since the operator S, which is the superposition of operators $S_1, \ldots, S_q$, commutes with A_j for $(0)_q \leqslant j \leqslant m$, each space $\tilde{X}_\nu = P_\nu \tilde{X}$ is invariant with respect to the operator $A(\varepsilon^\nu)$. Hence equation (7.12) can also be written for $y \in \tilde{X}$ in the following way:

$$0 = A(S)x - y = \sum_{(1)_q \leqslant \nu \leqslant N} A(\varepsilon^\nu) P_\nu x - \sum_{(1)_q \leqslant \nu \leqslant N} P_\nu y = \sum_{(1)_q \leqslant \nu \leqslant N} [A(\varepsilon^\nu) x_{(\nu)} - y_{(\nu)}].$$

Since the space $\tilde{X}$ is a direct sum of the spaces $\tilde{X}_\nu$, we infer that equation (7.11) is equivalent to the $N_0 = N_1 \ldots N_q$ independent equations

$$A(\varepsilon^\nu) x_{(\nu)} = y_{(\nu)}, \quad (1)_q \leqslant \nu \leqslant N,$$

and that if x is a solution of equation (7.11) belonging to $\tilde{X}$, then each of these equations has $x_{(\nu)}$ as a solution.

Conversely, suppose that each of the equations $A(\varepsilon^\nu)x = y_{(\nu)}$ has a solution. Denote by x_ν the solution of the νth equation. Since $A(\varepsilon^\nu)$ leaves the subspace $\tilde{X}_\nu$ invariant, we have $x_\nu \in \tilde{X}_\nu$ for $(1)_q \leqslant \nu \leqslant N$. Writing $x = \sum_{(1)_q \leqslant \nu \leqslant N} x_\nu$, we obtain $x_{(\nu)} = P_\nu x = P_\nu x_\nu = x_\nu$ and

$$A(S)x = A(S) \sum_{(1)_q \leqslant \nu \leqslant N} x_\nu = \sum_{(1)_q \leqslant \nu \leqslant N} \Big(\sum_{(1)_q \leqslant \mu \leqslant N} A(\varepsilon^\mu) P_\mu \Big) x_\nu$$

$$= \sum_{(1)_q \leqslant \mu \leqslant N} A(\varepsilon^\mu) \Big(\sum_{(1)_q \leqslant \nu \leqslant N} \delta_{\mu\nu} P_\nu x_\mu \Big) = \sum_{(1)_q \leqslant \nu \leqslant N} P_\nu \big(A(\varepsilon^\nu) x_\nu \big)$$

$$= \sum_{(1)_q \leqslant \nu \leqslant N} P_\nu y_{(\xi)} = \sum_{(1)_q \leqslant \nu \leqslant N} y_{(\nu)} = y,$$

which proves that x is a solution of equation (7.11).

This Theorem immediately implies

COROLLARY 7.1. *Let X be a linear space. Let*

$$A(S) = \sum_{(0)_q \leqslant k \leqslant n} \sum_{(0)_q \leqslant j \leqslant m} A_{kj} D^k S^{\eta_j},$$

where

1° $S = S_1 \dots S_q$ and $S_1, \dots, S_q$, A_{kj}, D are linear operators acting in X,

2° $S_p S_r - S_r S_p = 0$, $S_p A_{kj} - A_{kj} S_p = 0$, $S_p D - D S_p = 0$, $S A_{kj} - A_{kj} = 0$ for $p, r = 1, 2, \dots, q$, $(0)_q \leqslant k \leqslant n$, $(0)_q \leqslant j \leqslant m$,

3° $\eta_j = (\eta_{1,j}, \dots, \eta_{q,j})$, $\eta_{(0)_q} = (0)_q$, $\eta_{p,j}$ are integers and $\sum_{p=1}^{q} \eta_{p,j} > 0$, for $(0)_q < j \leqslant m$.

Let N_q be a common multiple of the numbers $\eta_{p,j}$, $(0)_q \leqslant j = m$ and suppose that there is a subspace $\tilde{X} \subset X$ such that S is a multi-involution of order $N = (N_1, \dots, N_q)$ on $\tilde{X}$. Then the equation

$$A(S)x = y, \quad y \in \tilde{X} \tag{7.14}$$

is equivalent to the $N_0 = N_1 \dots N_q$ independent equations

$$A(\varepsilon^\nu) x_{(\nu)} = y_{(\nu)}, \quad (1)_q \leqslant \nu \leqslant N, \tag{7.15}$$

where $y_{(\nu)} = P_\nu y$. Moreover, if each of these equations has a solution, then a solution of equation (7.14) belonging to $\tilde{X}$ is given by the formula

$$x = \sum_{(1)_q \leqslant \nu \leqslant N} x_\nu, \tag{7.16}$$

where $x_\nu \in \tilde{X}_\nu = P_\nu \tilde{X}$ is a solution of the νth equation (7.15).

For the proof it suffices to write

$$A_j = \sum_{(0)_q \leqslant k \leqslant N} A_{kj} D^k, \qquad (0)_q \leqslant j \leqslant m$$

and to apply Theorem 7.1.

REMARK 7.1. A homogeneous equation always has a solution belonging to the space $\tilde{X}$, because the νth equation has at least the solution $x_\nu = 0$ (cf. Remarks 5.1 and 6.1).

CHAPTER III

Equations with Operators Permuting an Involution of Order N

In this chapter we consider equations with an operator D which do not commute with an involution S of order N but which satisfy the relation

$$SD+\varepsilon DS = 0, \quad \text{where } \varepsilon = e^{2\pi i/N}.$$

D is said to be a *permuting operator*. If $N = 2$ we have $SD+DS = 0$ and we say that D is *anticommutative* or that D *anticommutes* with S.

1. EQUATIONS WITH OPERATORS ANTICOMMUTING WITH INVOLUTION

Let X be a linear space (over the complex field) and let S be an involution: $S^2 = I$ on X. Let $P^+ = \frac{1}{2}(I+S)$, $P^- = \frac{1}{2}(I-S)$. The following properties of an involution, given in Chapter II, Section 6, will be used in the sequel:

1° The operators P^+ and P^- are disjoint projectors giving a partition of unity:

(1.1) $$P^+P^- = P^-P^+ = 0, \quad (P^+)^2 = P^+, \quad (P^-)^2 = P^-, \quad P^++P^- = I.$$

Moreover, $P^+-P^- = S$, $SP^+ = P^+$, $SP^- = P^-$.

2° the eigenvalues of the operator S are $+1$, -1 and the respective eigenspaces are $X^+ = P^+X$ and $X^- = P^-X$, i.e. if we write $x^+ = P^+x$, $x^- = P^-x$ for any $x \in X$, we have

(1.2) $$Sx^+ = x^+, \quad Sx^- = x^-.$$

3° The space X is the direct sum of the spaces X^+ and X^-,

(1.3) $$X = X^++X^-,$$

which implies that any element $x \in X$ can be written uniquely as a sum

(1.4) $$x = x^++x^-, \quad \text{where } x^+ \in X^+,\ x^- \in X^-, \text{ and } Sx = x^+-x^-.$$

Let S be an involution in the space X and let D be a linear operator acting in X and *anticommutative* with S, i.e. such that

(1.5) $$SD+DS = 0.$$

Note that the operator D^2 commutes with S:

(1.6) $$SD^2-D^2S = 0.$$

Indeed, $SD^2 = (SD)D = -(DS)D = -D(SD) = -D(-DS) = D^2S$.

The following property will play a very important role in our subsequent considerations:

PROPERTY 1.1. *If* $S^2 = I$ *and* $SD+DS = 0$ *on a linear space* X, *then*

(1.7) $$P^+D = DP^-, \quad P^-D = DP^+.$$

Indeed, since $SD = -DS$, we have

$$P^+D = \tfrac{1}{2}(I+S)D = \tfrac{1}{2}(D+SD) = \tfrac{1}{2}(D-DS) = \tfrac{1}{2}D(I-S) = DP^-,$$
$$P^-D = \tfrac{1}{2}(I-S)D = \tfrac{1}{2}(D-SD) = \tfrac{1}{2}(D+SD) = \tfrac{1}{2}D(I+S) = DP^+.$$

Property 1.1 implies that

(1.8) $$Dx^+ = (Dx)^- \in X^-, \quad Dx^- = (Dx)^+ \in X^+,$$

because

$$Dx^+ = DP^+x = P^-Dx = (Dx)^-, \quad Dx^- = DP^-x = P^+Dx = (Dx)^+.$$

Now the operator D interchanges the roles of the spaces X^+, X^-, and the operator D^2, since it commutes with S, maps each of these spaces into itself. Thus an operator D which anticommutes with an involution is an example of permuting operators, whose properties in the general case will be studied in Sections 3 and 4. The special properties of anticommuting operators distinguish then from the class of permuting operators.

Suppose we are given, in the space X, an operator

$$A = (a_0I+b_0S)+(a_1I+b_1S)D,$$

where S is an involution in X, D is a linear operator acting in X and anticommutative with S and a_0, b_0, a_1, b_1 are scalars. In order to solve the equation $Ax = y$, $y \in X$, we consider a few special cases, which will be treated in different ways.

A. Suppose that

(1.9) $$a_0^2-b_0^2 \neq 0, \quad a_1^2-b_1^2 \neq 0 \quad \text{and } D \text{ is not an involution.}$$

PROPOSITION 1.1. *Let*

$$B = (a_0 I - b_0 S) - (a_1 I + b_1 S) D, \qquad R_A = -(a_1^2 - b_1^2)^{-1} B. \tag{1.10}$$

Then

$$AR_A = R_A A = D^2 - \lambda I, \tag{1.11}$$

where

$$\lambda = \frac{a_0^2 - b_0^2}{a_1^2 - b_1^2} \neq 0. \tag{1.12}$$

PROOF. Note that $SDS = -S^2 D = -D$ and $SDSD = -DS^2 D = -D^2$ since $DS = -SD$ and $S^2 = I$. Hence

$$\begin{aligned} BA &= (a_0 I - b_0 S)(a_0 I + b_0 S) - (a_1 I + b_1 S) D (a_0 I + b_0 S) + \\ &\quad + (a_0 I - b_0 S)(a_1 I + b_1 S) D - (a_1 I + b_1 S) D (a_1 I + b_1 S) D \\ &= (a_0^2 - b_0^2) I + (a_0 a_1 - b_0 b_1) D + (a_0 b_1 - a_1 b_0) SD - (a_0 a_1 - b_0 b_1) D - \\ &\quad - (a_0 b_1 - a_1 b_0) SD - (a_1^2 - b_1^2) D^2 = (a_0^2 - b_0^2) I - (a_1^2 - b_1^2) D^2 \\ &= -(a_1^2 - b_1^2)(D^2 - \lambda I). \end{aligned}$$

Similarly, we can show that $AB = -(a_1^2 - b_1^2)(D^2 - \lambda I)$, if we now put $R_A = -(a_1^2 - b_1^2)^{-1} B$, we obtain the required Formulae (1.11).

PROPOSITION 1.2. $Z_A \subset Z_{D^2 - \lambda I}$ *(similarly,* $Z_{R_A} \subset Z_{D^2 - \lambda I}$*).*

Indeed, let $x \in Z_A$. Then $Ax = 0$ and $(D^2 - \lambda I)x = R_A(Ax) = 0$, which implies that $x \in Z_{D^2 - \lambda I}$. Hence, $Z_A \subset Z_{D^2 - \lambda I}$. We have a similar proof for Z_{R_A}.

THEOREM 1.1. $Z_{D^2 - \lambda I} = \{z \in X: z = z_1 + S z_2,\ z_1, z_2 \in Z_{D - \sqrt{\lambda} I}\}$.

PROOF. Suppose that $z = z_1 + S z_2$, where $z_1, z_2 \in Z_{D - \sqrt{\lambda} I}$. Then

$$\begin{aligned} (D^2 - \lambda I) z &= (D^2 - \lambda I) z_1 + (D^2 - \lambda I) S z_2 = (D^2 - \lambda I) z_1 + S (D^2 - \lambda I) z_2 \\ &= (D + \sqrt{\lambda} I)(D - \sqrt{\lambda} I) z_1 + S (D + \sqrt{\lambda} I)(D - \sqrt{\lambda} I) z_2 = 0 \end{aligned}$$

as $SD^2 = D^2 S$. Hence $z \in Z_{D^2 - \lambda I}$.

Conversely, suppose that $z \in Z_{D^2 - \lambda I}$. The operator $\frac{1}{\sqrt{\lambda}} D$ is an involution on the space $Z_{D^2 - \lambda I}$. Indeed, since $\lambda \neq 0$, we have for any $z \in Z_{D^2 - \lambda I}$

$$\left(\frac{1}{\sqrt{\lambda}} D\right)^2 z - z = 0, \quad \text{i.e.} \left(\frac{1}{\sqrt{\lambda}} D\right)^2 = I \text{ on } Z_{D^2 - \lambda I}.$$

This leads to the decomposition of $Z_{D^2 - \lambda I}$ into the direct sum

$$Z_{D^2 - \lambda I} = Z_{D + \sqrt{\lambda} I} \oplus Z_{D - \sqrt{\lambda} I}.$$

Hence $z = z_1 + z_2'$, where $z_1 \in Z_{D-\sqrt{\lambda}I}$, $z_2' \in Z_{D+\sqrt{\lambda}I}$ are linearly independent. We have to show that $z_2' = Sz_2$, where $z_2 \in Z_{D+\sqrt{\lambda}I}$. Since $z_2' \in Z_{D+\sqrt{\lambda}I}$, we have $Dz_2' = -\sqrt{\lambda}z_2'$. Hence $\sqrt{\lambda}Sz_2' = S(\sqrt{\lambda}z_2') = -SDz_2' = DSz_2'$, which implies $(D-\sqrt{\lambda}I)Sz_2' = 0$ and $z_2 = Sz_2' \in Z_{D-\sqrt{\lambda}I}$. But $z_2' = S^2z_2' = S(Sz_2') = Sz_2$, which gives Z in the required form.

THEOREM 1.2. $Z_A = \{z \in X\colon z = [(a_0 - a_1\sqrt{\lambda})I - (b_0 + b_1\sqrt{\lambda})S]z_1,\ z_1 \in Z_{D-\sqrt{\lambda}I}\}$.

PROOF. Proposition 1.2 implies that $Z_A \subset Z_{D^2-\lambda I}$. Theorem 1.1 implies that any element of $Z_{D^2-\lambda I}$ is of the form $z = \tilde{z}_1 + S\tilde{z}_2$, where $\tilde{z}_1, \tilde{z}_2 \in Z_{D-\sqrt{\lambda}I}$. We choose $\tilde{z}_1$ and $\tilde{z}_2$ in such a way that $Az = 0$. Let $\tilde{z}_1, \tilde{z}_2 \in Z_{D-\sqrt{\lambda}I}$. Then $S\tilde{z}_1, S\tilde{z}_2 \in Z_{D+\sqrt{\lambda}I}$ (compare the proof of Theorem 1.1). Hence

$$\begin{aligned}
A\tilde{z}_1 &= (a_0I+b_0S)\tilde{z}_1 + (a_1I+b_1S)D\tilde{z}_1 = (a_0I+b_0S)z_1 + (a_1I+b_1S)\sqrt{\lambda}\tilde{z}_1 \\
&= [(a_0+a_1\sqrt{\lambda})I + (b_0+b_1\sqrt{\lambda})S]\tilde{z}_1, \\
AS\tilde{z}_2 &= (a_0I+b_0S)S\tilde{z}_2 + (a_1I+b_1S)DS\tilde{z}_2 = (a_0I+b_0S)Sz_2 + \\
&\quad + (a_1I+b_1S)(-\sqrt{\lambda})S\tilde{z}_2 = [(a_0-a_1\sqrt{\lambda})I + (b_0-b_1\sqrt{\lambda})S]S\tilde{z}_2, \\
Az &= A(\tilde{z}_1+S\tilde{z}_2) = [(a_0+a_1\sqrt{\lambda})I + (b_0+b_1\sqrt{\lambda})S]\tilde{z}_1 + \\
&\quad + [(a_0-a_1\sqrt{\lambda})I + (b_0-b_1\sqrt{\lambda})S]S\tilde{z}_2 \\
&= (a_0+a_1\sqrt{\lambda})\tilde{z}_1 + (b_0+b_1\sqrt{\lambda})S\tilde{z}_1 + (a_0-a_1\sqrt{\lambda})S\tilde{z}_2 + (b_0-b_1\sqrt{\lambda})\tilde{z}_2.
\end{aligned}$$

Now the space $Z_{D^2-\lambda I}$ is the direct sum of the spaces $Z_{D-\sqrt{\lambda}I}$ and $Z_{D+\sqrt{\lambda}I}$ (compare the proof of Theorem 1.1) and $\tilde{z}_1, \tilde{z}_2 \in Z_{D-\sqrt{\lambda}I}$, $S\tilde{z}_1, S\tilde{z}_2 \in Z_{D+\sqrt{\lambda}I}$. The equality $Az = 0$ thus holds if and only if

$$(a_0+a_1\sqrt{\lambda})\tilde{z}_1 + (b_0-b_1\sqrt{\lambda})\tilde{z}_2 = 0, \qquad (b_0+b_1\sqrt{\lambda})S\tilde{z}_1 + (a_0-a_1\sqrt{\lambda})S\tilde{z}_2 = 0.$$

Transforming the second equation by means of S and using the property $S^2 = I$, we obtain the following system of equations:

$$\begin{aligned}
&(a_0+a_1\sqrt{\lambda})\tilde{z}_1 + (b_0-b_1\sqrt{\lambda})\tilde{z}_2 = 0, \\
&(b_0+b_1\sqrt{\lambda})z_1 + (a_0-a_1\sqrt{\lambda})\tilde{z}_2 = 0.
\end{aligned} \tag{1.13}$$

From these equations it follows that $\tilde{z}_1$ and $\tilde{z}_2$ are linearly dependent. Indeed, suppose that $\tilde{z}_1$ and $\tilde{z}_2$ are linearly independent. Then all the coefficients of the system (1.13) are equal to zero:

$$a_0+a_1\sqrt{\lambda} = 0, \quad a_0-a_1\sqrt{\lambda} = 0, \quad b_0+b_1\sqrt{\lambda} = 0, \quad b_0-b_1\sqrt{\lambda} = 0.$$

Since $\lambda \neq 0$, we find $a_0 = a_1 = b_0 = b_1 = 0$, which is a contradiction. Hence there are scalars c_1 and c_2 such that $c_1\tilde{z}_1 + c_2\tilde{z}_2 = 0$ and $|c_1|+|c_2| > 0$. Using the last equality, we write (1.13) in the following form

$$[-(a_0+a_1\sqrt{\lambda})c_2+(b_0-b_1\sqrt{\lambda})c_1]\tilde{z}_2 = 0,$$

$$[-(b_0+b_1\sqrt{\lambda})c_2+(a_0-a_1\sqrt{\lambda})c_1]\tilde{z}_2 = 0.$$

These equalities hold if and only if

$$(1.14)\qquad \begin{aligned} &(a_0-a_1\sqrt{\lambda})c_1-(b_0+b_1\sqrt{\lambda})c_2 = 0,\\ &(b_0-b_1\sqrt{\lambda})c_1-(a_0+a_1\sqrt{\lambda})c_2 = 0. \end{aligned}$$

The determinant of system (1.14) is

$$\Delta = -(a_0^2-a_1^2\lambda)+(b_0^2-b_1^2\lambda) = -[a_0^2-b_0^2-\lambda(a_1^2-b_1^2)]$$

$$= -\left[a_0^2-b_0^2-\frac{a_0^2-b_0^2}{a_1^2-b_1^2}(a_1^2-b_1^2)\right] = 0.$$

Hence system (1.14) has a non-trivial solution:

$$c_1 = c(b_0+b_1\sqrt{\lambda}), \qquad c_2 = c(a_0-a_1\sqrt{\lambda}), \qquad \text{where } c \text{ is an arbitrary scalar.}$$

We have further

$$c(b_0+b_1\sqrt{\lambda})\tilde{z}_1+c(a_0-a_1\sqrt{\lambda})\tilde{z}_2 = 0.$$

Taking the formula for the solutions of system (1.14) from the second equation of (1.14), we check that the first equation of (1.13) is also satisfied.

This implies that

$$z = c[(a_0-a_1\sqrt{\lambda})\tilde{z}_1-(b_0+b_1\sqrt{\lambda})S\tilde{z}_1].$$

Since $\tilde{z}_1 \in Z_{D-\sqrt{\lambda}I}$ is arbitrary, we obtain the required formula by putting $z_1 = c\tilde{z}_1$.

In a similar way we can determine the set

$$Z_{R_A} = \{z \in X:\ z = [(a_0+a_1\sqrt{\lambda})I+(b_0+b_1\sqrt{\lambda})S]z_1,\ z_1 \in Z_{D-\sqrt{\lambda}I}\}.$$

PROPOSITION 1.3. *If $\tilde{x}$ a solution of the equation $(D^2-\lambda I)\tilde{x} = y$, then $\tilde{x} = R_A\tilde{x}$ is a solution of the equation $Ax = y$.*

Indeed, let $\tilde{x}$ satisfy the equation $(D^2-\lambda I)\tilde{x} = y$. Then

$$Ax = AR_A\tilde{x} = (D^2-\lambda I)\tilde{x} = y.$$

Similarly, $u = A\tilde{x}$ is a solution of the equation $R_A u = y$.

Finally, we obtain the following theorem concerning the general form of the solution of the equations $Ax = y$ and $R_A u = y$:

THEOREM 1.3. *Suppose that* $A = (a_0 I + b_0 S) + (a_1 I + b_1 S) D$, *where* $S^2 = I$ *and* $SD + DS = 0$ *on* X, *and, moreover,* D *is not an involution and* $a_0^2 - b_0^2 \neq 0 \neq a_1^2 - b_1^2$. *Let* $\tilde{x}$ *be a solution of the equation* $(D^2 - \lambda I)\tilde{x} = y$, $y \in X$. *Then every solution of the equation* $Ax = y$ *is of the form*

$$x = R_A \tilde{x} + [(a_0 - a_1 \sqrt{\lambda}) I - (b_0 + b_1 \sqrt{\lambda}) S] z_1,$$

where z_1 *is any solution of the equation* $(D - \sqrt{\lambda} I) z_1 = 0$; λ *and* R_A *are given by Formulae* (1.11) *and* (1.12). *Every solution of the equation* $R_A u = y$ *is of the form*

$$u = Ax + [(a_0 + a_1 \sqrt{\lambda}) I + - (b_0 + b_1 \sqrt{\lambda}) S] z_1.$$

This follows immediately from Theorem 1.1, Proposition 1.3 and the linearity of A.

B. Suppose D is not an involution, as in A. We have so far considered the case, where $a_0^2 - b_0^2 \neq 0$, $a_1^2 - b_1^2 \neq 0$. To study other cases we first remark that according to (1.1)

$$\begin{aligned} A &= (a_0 I + b_0 S) + (a_1 I + b_1 S) D \\ &= (a_0 I + b_0 S)(P^+ + P^-) + (a_1 I + b_1 S)(P^+ + P^-) D \\ &= (a_0 + b_0) P^+ + (a_0 - b_0) P^- + (a_1 + b_1) P^+ D + (a_1 - b_1) P^- D. \end{aligned} \tag{1.15}$$

We now consider the following cases:

I. If $a_0 + b_0 = a_0 - b_0 = a_1 + b_1 = a_1 - b_1 = 0$, then $A = 0$.

II. If $a_1 + b_1 = a_1 - b_1 = 0$, we have a case which is completely solved by Theorem II.5.2[1].

III. If $a_0 + b_0 = a_0 - b_0 = 0$, but $a_1^2 - b_1^2 \neq 0$, then $A = (a_1 I + b_1 S) D$ and we solve the equation $Ax = y$ with respect to the unknown Dx in the same way as in II. We obtain an equation without involution: $Dx = y_1$, where $y_1 = (a_1^2 - b_1^2)^{-1} (a_1 I - b_1 S) y$.

IV. Let $a_1 + b_1 = 0$, but $a_1 - b_1 \neq 0$. Then

$$A = (a_0 + b_0) P^+ + (a_0 - b_0) P^- + (a_1 - b_1) P^- D.$$

Since the space X is the direct sum of the spaces X^+ and X^-, the equation $Ax = y$, $y \in X$, is equivalent to the two equations

$$(a_0 + b_0) x^+ = y^+, \quad (a_0 - b_0) x^- + (a_1 - b_1)(Dx)^- = y^-.$$

---

[1] Theorem II.5.2 means Theorem 5.2 of Chapter II.

Moreover, from Formula (1.8) we have $(Dx)^- = Dx^+$. Hence we have two equations of the form

$$(1.16) \qquad (a_0+b_0)x^+ = y^+,$$

$$(1.17) \qquad (a_0-b_0)x^- + (a_1-b_1)Dx^+ = y^-,$$

(a) Let $a_0^2-b_0^2 \neq 0$. Then $x^+ = (a_0+b_0)^{-1}y^+$ and $(a_0-b_0)x^- = y^- - (a_1-b_1)(a_0^2-b_0^2)^{-1}y$.

Hence $x^- = (a_0-b_0)^{-1}y^- - (a_1-b_1)(a_0^2-b_0^2)^{-1} Dy^+$ and

$$\begin{aligned} x = x^+ + x^- &= (a_0+b_0)^{-1}y^+ + (a_0-b_0)^{-1}y^- - (a_1-b_1)(a_0^2-b_0^2)^{-1}Dy^+ \\ &= (a_0^2-b_0^2)^{-1}[(a_0-b_0)P^+ + (a_0+b_0)P^- - (a_1-b_1)DP^+]y \\ &= (a_0^2-b_0^2)^{-1}[(a_0I-b_0S) - (a_1-b_1)P^-D]y. \end{aligned}$$

(b) Let $a_0+b_0 = 0$ but $a_0-b_0 \neq 0$. Then a solution exists if and only if $y^+ = 0$ (compare with Theorem II.5.2), i.e. if $(I+S)y = 0$. Under this condition the solution is of the form

$$x = x_1^+ + x^- = x_1^+ + (a_0-b_0)^{-1}y - (a_1-b_1)Dx_1^+,$$

where $x_1^+ \in X^+$ is arbitrary.

(c) Let $a_0-b_0 = 0$ but $a_0+b_0 \neq 0$. Then $x^+ = (a_0+b_0)^{-1}y^+$ and equation (1.17) is of the form

$$(1.18) \qquad (a_0-b_0)x^- = y^- - (a_0+b_0)^{-1}(a_1-b_1)Dy^+.$$

A solution of equation (1.18) exists (compare Theorem II.5.2) if and only if $y^- - (a_0+b_0)^{-1}(a_1-b_1)Dy^+ = 0$. This condition can be written as follows:

$$(I-S)[I-(a_0+b_0)^{-1}(a_1-b_1)D]y = 0.$$

Under this condition the solution is

$$x = x^+ + x^- = (a_0+b_0)^{-1}y^+ + x_1^-, \quad \text{where } x_1^- \in X^- \text{ is arbitrary.}$$

(d) $a_0+b_0 = a_0-b_0 = 0$. From equation (1.16) we have $y^+ = 0$. From equation (1.17) we find that $Dx^+ = (a_1-b_1)^{-1}y^-$ and x^- is arbitrary.

V. Let $a_1-b_1 = 0$ but $a_1-b_1 \neq 0$. Changing the roles of a_1-b_1 and a_1+b_1 and, respectively, of the spaces X^- and X^+, we obtain a result analogous to that of IV.

C. We now examine the case where the operator D anticommuting with an involution S is also an involution on the same space X. This means that we assume the following:

$$(1.19) \qquad SD+DS = 0, \quad S^2 = I, \quad D^2 = I \quad \text{on } X.$$

I. Let $a_0^2-b_0^2 \neq 0 \neq a_1^2-b_1^2$ and let $\lambda = (a_1^2-b_0^2)/(a_1^2-b_1^2) = 1$. In this case we have $(D^2-I)x = 0$ for all $x \in X$. The assumption $\lambda = 1$ implies that

$$0 \neq a_0^2-b_0^2 = a_1^2-b_1^2. \tag{1.20}$$

(a) Let $b_1 = b_0 \neq 0$. From equality (1.21) we find $a_1^2 = a_0^2$. If $a_1 = a_0$, then

$$A = (a_0 I+b_0 S)+(a_0 I+b_0 S)D = (a_0 I+b_0 S)(I+D).$$

Since $a_0^2-b_0^2 \neq 0$, the equation $Ax = y$ is equivalent to the equation

$$(I+D)x = (a_0^2-b_0^2)^{-1}(a_0 I-b_0 S)y$$

(Theorem II.5.2), and applying Theorem II.5.2 once more, which is possible as D is an involution, we find that

$$x = \tfrac{1}{2}(a_0^2-b_0^2)^{-1}(a_0 I-b_0 S)y+(I-D)z, \quad \text{where } z \in X \text{ is arbitrary,}$$

under the necessary and sufficient condition that

$$(a_0^2-b_0^2)^{-1}(I-D)(a_0 I-b_0 S)y = 0.$$

Since in our case $(I-D)(a_0 I-b_0 S) = (a_0 I+b_0 S)(I+D) = A$, we conclude that a necessary and sufficient condition for the equation $Ax = y$ to have a solution is that $Ay = 0$ (in the case where $b_1 = b_0$, $a_1 = a_0$).

If $a_1 = -a_0$, then

$$A = (a_0 I+b_0 S)+(-a_0 I+b_0 S)D = (a_0 I+b_0 S)-D(a_0 I+b_0 S)$$
$$= (I-D)(a_0 I+b_0 S)$$

and from Theorem II.5.2.

$$(a_0 I+b_0 S)x = \tfrac{1}{2}y+(I+D)z, \quad \text{where } z \in X \text{ is arbitrary,}$$

under the necessary and sufficient condition, that $(I+D)y = 0$. Since $a_0^2-b_0^2 \neq 0$, we conclude applying Theorem II.5.2 again that the equation $Ax = y$ has a solution

$$x = (a_0^2-b_0^2)^{-1}(a_0 I-b_0 S)[\tfrac{1}{2}y+(I+D)z], \quad \text{where } z \in X \text{ is arbitrary,}$$

if and only if $(I+D)y = 0$.

(b) $b_1 = -b_0$. From equality (1.21) we find $a_1^2 = a_0^2$. If $a_1 = a_0$, then

$$A = (a_0 I+b_0 S)+(a_0 I-b_0 S)D = (a_0 I+b_0 S)+D(a_0 I+b_0 S)$$
$$= (I+D)(a_0 I+b_0 S).$$

As in the second part of (a) we conclude that the equation $Ax = y$ has a solution

$$x = (a_0^2-b_0^2)^{-1}(a_0 I-b_0 S)[\tfrac{1}{2}y+(I-D)z], \quad \text{where } z \in X \text{ is arbitrary,}$$

under the necessary and sufficient condition that $(I-D)y = 0$.

If $a_1 = -a_0$, then

$$A = (a_0 I + b_0 S) + (-a_0 I - b_0 S)D = (a_0 I + b_0 S)(I-D)$$

and, as in the first part of (a), we conclude that the equation $Ax = y$ has a solution

$$x = \tfrac{1}{2}(a-b)(a_0 I - b_0 S) + y(I+D)z, \quad \text{where } z \in X \text{ is arbitrary,}$$

under the necessary and sufficient condition that $Ay = 0$.

(c) $b_1^2 = b_0^2 = 0$. Then from equality (1.21) we have $a_1^2 = a_0^2$. Without loss of generality we can assume that $a_0 \neq 0$, since we would otherwise have $a_0 = a_1 = b_0 = b_1 = 0$, which means that $A = 0$. If $a_1 = a_0$, then $A = a_0(I+D)$, and so a solution of the equation $Ax = y$ is (by Theorem II.5.2)

$$x = \frac{1}{2a_0} y + (I-D)z, \quad \text{where } z \in X \text{ is arbitrary,}$$

under the necessary and sufficient condition that $(I-D)y = 0$.

If $a_1 = -a_0$, then, similarly, we obtain $A = a_0(I-D)$. Hence the equation $Ax = y$ has a solution

$$x = \frac{1}{2a_0} y + (I+D)z, \quad \text{where } z \in X \text{ is arbitrary,}$$

under the necessary and sufficient condition that $(I+D)y = 0$.

II. Let $a_0^2 - b_0^2 \neq 0 \neq a_1^2 - b_1^2$ and let $\lambda = (a_0^2 - b_0^2)/(a_1^2 - b_1^2) \neq 1$. In this case we can apply Theorem 1.3, which does not work when both $\lambda = 1$ and D is an involution simultaneously. Since now $D^2 = I$, we infer from Proposition 1.3 that $x = R_A \tilde{x}$ is a solution of the equation $Ax = y$, where R_A is determined by Formula (1.11) and x is a solution of the equation $(D^2 - \lambda I)\tilde{x} = y$, i.e. of the equation $(1-\lambda)x = y$. Since $\lambda \neq 1$, we have $x = \frac{1}{1-\lambda} y$ and the equation $(D - \lambda I)z_1 = 0$ only has the one solution $z_1 = 0$. Therefore the equation $Ax = y$ has the unique solution

$$x = \frac{1}{1-\lambda} R_A y.$$

III. The case $a_1 + b_1 = a_1 - b_1 = 0$ eliminates the operator D from our equation; it was considered in B.I and B.II.

IV. If $a_0 + b_0 = a_0 - b_0 = 0$, but $a_1^2 - b_1^2 \neq 0$, then $A = (a_1 I + b_1 S)D$ and we have an equivalent equation

$$Dx = (a_1^2 - b_1^2)^{-1}(a_1 I - b_1 S)y$$

which has the unique solution

$$x = (a_1^2-b_1^2)^{-1}D(a_1I-b_1S)y = (a_1^2-b_1^2)^{-1}(a_1I+b_1S)Dy = (a_1^2-b_1^2)^{-1}Ay.$$

V. Let $a_1+b_1 = 0$ but $a_1-b_1 \neq 0$. The cases where $a_0^2-b_0^2 \neq 0$, $a_0+b_0 = 0$ but $a_0-b_0 \neq 0$, $a_0+b_0 \neq 0$ but $a_0-b_0 = 0$ can be solved as in B.IV.(a), (b), (c) respectively.

In the case $a_0+b_0 = a_0-b_0 = 0$ we find from B.IV.(d) that $x = D(I+S)z$, where $z \in X$ is arbitrary, under the necessary and sufficient condition that $y = 0$.

VI. Let $a_1-b_1 = 0$ but $a_1+b_1 \neq 0$. Interchanging the roles of a_1-b_1 and a_1+b_1 and of the spaces X^- and X^+, respectively, we obtain results analogous to those of V.

2. EQUATIONS WITH POLYNOMIALS IN AN OPERATOR ANTICOMMUTING WITH AN INVOLUTION

Let X be a linear space (over the complex scalar field). We consider the equation

$$(2.1) \qquad (a_0I+b_0S)+(a_1I+b_1S)D+\ldots+(a_nI+b_nS)D^nx = y, \qquad y \in X,$$

where S is an involution on X and D is a linear operator transforming X into itself and anticomuting with S; $a_0, \ldots, a_n, b_0, \ldots, b_n$ are complex numbers. All the notation of the preceding section is preserved. Our assumptions and Formula (1.6) immediately imply that the operator D has the following properties:

1° S commutes with the operator D^{2k} for $k = 0, 1, 2, \ldots$:

$$(2.2) \qquad SD^{2k}-D^{2k}S = 0 \quad (k = 0, 1, 2, \ldots).$$

2° S anticommutes with the operator D^{2k+1} for $k = 0, 1, 2, \ldots$:

$$(2.3) \qquad SD^{2k+1}+D^{2k+1}S = 0 \quad (k = 0, 1, 2, \ldots).$$

We write

$$A = \sum_{k=0}^{n} (a_kI+b_kS)D.$$

THEOREM 2.1[1]. *Let the coefficients of equation* (1.1) *satisfy the following conditions*:

1° $a_n^2-b_n^2 \neq 0$;

2° $a_{j-k}a_k-b_{j-k}b_k \neq 0$ $(k = 0, 1, \ldots, n;\ j = k+1, \ldots, k+n)$;

[1] All the theorems of this section were given by Mażbic-Kulma [1]. Here they are slightly reformulated.

3° *the polynomial* $\Lambda(t) = \sum_{j=0}^{n} \lambda_{2j} t^j$ *has simple roots* u_k *only* $(k = 1, \ldots, n)$, *where*

$$\lambda_j = \begin{cases} \sum_{k=0}^{n} c_{jk} & \text{for } 0 \leqslant j \leqslant n, \\ \sum_{k=j-n}^{n} c_{jk} & \text{for } n < j = 2n, \end{cases} \qquad c_{jk} = (-1)^{n+j-k}(a_{j-k}a_k - b_{j-k}b_k)\,(a_n^2 - b_n^2)^{-1}$$

$(k = 0, 1, \ldots, n;\ j = k+1, \ldots, k+n)$. *Let*

$$B = \sum_{m=0}^{n} [(-1)^m a_m I - b_m S] D^m \quad \text{and} \quad R_A = (-1)^n\,(a_n^2 - b_n^2)^{-1} B.$$

Then

$$AR_A = R_A A = \Lambda(D^2).$$

PROOF. We have

$$\begin{aligned} BA &= \sum_{k=0}^{n}\sum_{m=0}^{n} [(-1)^m a_m I - b_m S] D^m (a_k I + b_k S) D^k \\ &= \sum_{k=0}^{n}\sum_{m=0}^{n} [(-1)^m a_m I - b_m S]\,[a_k D^m + (-1)^m b_k S D^m] D^k \\ &= \sum_{k=0}^{n}\sum_{m=0}^{n} [(-1)^m a_m I - b_m S]\,[a_k I + (-1)^m b_k S] D^{m+k} \\ &= \sum_{k=0}^{n}\sum_{m=0}^{n} [(-1)^m\,(a_m a_k - b_m b_k) I + (a_m b_k - a_k b_m) S] D^{m+k}. \end{aligned}$$

Observe that

$$(a_m b_k - a_k b_m) S D^{m+k} = -(a_k b_m - a_m b_k) S D^{m+k},$$

and hence

$$\sum_{k=0}^{n}\sum_{m=0}^{n} (a_m b_k - a_k b_m) S D^{m+k} = 0.$$

This implies that

$$BA = \sum_{k=0}^{n}\sum_{m=0}^{n} (-1)^m (a_m a_k - b_m b_k) D^{m+k}.$$

Similarly we can show that $AB = BA$. Putting $m = j-k$ in the last sum we obtain

$$BA = \sum_{k=0}^{n}\sum_{j=k}^{-n+k} (-1)^{j-k}(a_{j-k}a_k - b_{j-k}b_k) = (-1)^n (a_n^2 - b_n^2) \sum_{j=0}^{2n} \lambda_j D^j.$$

It is easy to check by induction that the last polynomial contains only even powers of D. Hence

$$AB = BA = (-1)^2\,(a_n^2-b_n^2)\sum_{j=0}^{n}\lambda_{2j}D^{2j} = (-1)^n(a_n^2-b_n^2)\Lambda(D^2)$$

and finally

$$AR_A = R_A A = (-1)^n(a_n^2-b_n^2)^{-1}BA = \Lambda(D^2),$$

which was to be proved.

COROLLARY 2.1. *Under the conditions of Theorem* 2.1

$$Z_A \subset Z_{\Lambda(D^2)} \quad \textit{and} \quad Z_{R_A} \subset Z_{\Lambda(D^2)}.$$

Indeed, if $x \in Z_A$, then $Ax = 0$ and $\Lambda(D^2)x = R_A(Ax) = 0$, hence $x \in Z_{\Lambda(D^2)}$. We similarly obtain the second inclusion.

In the sequel we will make use of condition 3°, under which the polynomial $\Lambda(t)$ has only simple roots. In Section 1, the roots are simple for $n = 1$ because the corresponding polynomial is of the form $t^2-\lambda$, where $\lambda \neq 0$. For $n = 2$ this polynomial may have multiple roots, so that our condition is necessary here.

Since we assume that the polynomial $\Lambda(t)$ has only simple roots, we can write

$$\Lambda(D^2) = \prod_{q=1}^{n}(D^2-u_qI), \quad \text{where } u_q \text{ denotes the } q\text{th root of } \Lambda(t),$$

because the coefficient λ_{2n} of this polynomial is equal to one.

THEOREM 2.2. *Under the conditions of Theorem* 2.1

$$(2.4) \qquad Z_{\Lambda(D^2)} = \Big\{z \in X\colon\ z = \sum_{q=1}^{n}(z_q+Sz_q'),\ z_q, z_q' \in Z_{D-\sqrt{u_q}I}\Big\},$$

where u_q denote the roots of the polynomial $\Lambda(t)$.

PROOF. Suppose that $z = \sum_{q=1}^{n}(z_q+Sz_q'),\ z_q, z_q' \in Z_{D-\sqrt{u_q}I}$. Then

$$\Lambda(D^2)z = \Lambda(D^2)\sum_{q=1}^{n}(z_q+Sz_q') = \Lambda(D^2)\Big[\sum_{q=1}^{n}z_q+S\sum_{q=1}^{n}z_q'\Big]$$

$$= \Lambda(D^2)\sum_{q=1}^{n}z_q+S\Lambda(D^2)\sum_{q=1}^{n}z_q' = 0,$$

since the polynomial $\Lambda(D^2)$ contains the factor $D-\sqrt{u_q}I$. Hence $z \in Z_{\Lambda(D^2)}$.

Conversely, suppose that $z \in Z_{A(D^2)}$. We can decompose this space into the direct sum

$$Z_{A(D^2)} = \bigoplus_{q=1}^{n} [Z_{D-\sqrt{u_q}I} \oplus Z_{D+\sqrt{u_q}I}],$$

because D is an algebraic operator on the space $Z_{A(D^2)}$ with simple characteristic roots (see Theorem 4.1 of Chapter II). Hence $z = \sum_{q=1}^{n} (z_q+z_q'')$, where $z_q \in Z_{D-\sqrt{u_q}I}$ and $z_q'' \in Z_{D+\sqrt{u_q}I}$ for $q = 1, 2, \ldots, n$. We have to prove that $z_q'' = Sz_q'$, where $z_q' \in Z_{D-\sqrt{u_q}I}$ for $q = 1, 2, \ldots, n$. But $z_q'' \in Z_{D-\sqrt{u_q}I}$, hence $Dz_q'' = -\sqrt{u_q}z_q''$ and $\sqrt{u_q}\ Sz_q'' = S(\sqrt{u_q}\ z_q'') = -SDz_q'' = DSz_q''$. Therefore $(D-\sqrt{u_q}I)Sz_q'' = 0$ and $z_q' = Sz_q'' \in Z_{D-\sqrt{u_q}I}$ $(q = 1, 2, \ldots, n)$. But $z_q'' = S^2z_q'' = S(Sz_q'') = Sz_q'$ and this gives the required form of z.

THEOREM 2.3. *Under the conditions of Theorem* 2.1

$$Z_A = \Big\{z \in X:\ z = \sum_{q=1}^{n}\sum_{k=0}^{n} [(-1)^k a_k I - b_k S] u_q^{k/2} z_q,\ z_q \in Z_{D-\sqrt{u_q}I}\Big\}.$$

PROOF. From Corollary 2.1 we have $Z_A \subset Z_{A(D^2)}$. From Theorem 2.2 we infer that every $z \in Z_{A(D^2)}$ is of the form $z = \sum_{q=1}^{n} (y_q+Sy_q')$, where $y_q,\ y_q' \in Z_{D-\sqrt{u_q}I}$. Since $Dy_q = \sqrt{u_q}\ y_q$ and $DSy_q' = -\sqrt{u_q}Sy_q'$, we have

$$D^k y_q = u_q^{k/2} y_q, \qquad D^k Sy_q' = (-1)^k u_q^{k/2} Sy_q' \qquad (k = 0, 1, \ldots;\ q = 1, 2, \ldots, n).$$

Then

$$Ay_q = \sum_{k=0}^{n} (a_k I + b_k S) D^k y_q = \sum_{k=0}^{n} (a_k I + b_k S) u_q^{k/2} y_q,$$

$$ASy_q' = \sum_{k=0}^{n} (a_k I + b_k S) D^k Sy_q' = \sum_{k=0}^{n} (a_k I + b_k S)(-1)^k u_q^{k/2} Sy_q'.$$

Hence

$$Az = \sum_{q=1}^{n}\sum_{k=0}^{n} u_q^{k/2} [a_k y_q + b_k Sy_q + (-1)^k b_k y_q' + (-1)^k a_k Sy_q'].$$

We recall that

$$Z_{A(D^2)} = \bigoplus_{q=1}^{n} [Z_{D-\sqrt{u_q}I} \oplus Z_{D+\sqrt{u_q}I}] \quad \text{and} \quad y_q, y_q' \in Z_{D-\sqrt{u_q}I}, \quad Sy_q, Sy_q' \in Z_{D+\sqrt{u_q}I}.$$

The equality $Az = 0$ thus holds if and only if

$$(2.5)\quad \sum_{k=0}^{n} u_q^{k/2}[a_k y_q + (-1)^k b_k y_q'] = 0, \qquad \sum_{k=0}^{n} u_q^{k/2}[b_k Sy_q + (-1)^k a_k Sy_q'] = 0.$$

We can write this system in a different form if we put

$$(2.6)\quad \begin{aligned} C_{1q} &= \sum_{k=0}^{n} a_k u_q^{k/2}, & C_{2q} &= \sum_{k=0}^{n} (-1)^k a_k u_q^{k/2} \\ C_{3q} &= \sum_{k=0}^{n} (-1)^k b_k u_q^{k/2}, & C_{4q} &= \sum_{k=0}^{n} b_k u_q^{k/2} \end{aligned} \qquad (q = 1, 2, \ldots, n),$$

and if we transform the second equation of (2.5) by means of the operator S using the identity $S^2 = I$. We then obtain the following system of equations

$$(2.7)\quad \begin{aligned} C_{1q} y_q + C_{3q} y_q' &= 0 \\ C_{4q} y_q + C_{2q} y_q' &= 0 \end{aligned} \qquad (q = 1, 2, \ldots, n).$$

To show that the system (2.7) has a non-trivial solution we shall not repeat all the arguments, which have already appeared in the proof of Theorem 1.2. We will only show, that the determinant of the system (2.7) is

$$\Delta_q = C_{1q} C_{0q} - C_{2q} C_{3q} = 0 \quad \text{for } q = 1, 2, \ldots, n.$$

Indeed,

$$\Delta_q = C_{1q} C_{4q} - C_{2q} C_{3q} = \sum_{k=0}^{n} a_k u_q^{k/2} \sum_{j=0}^{n} (-1)^j a_j u_q^{j/2} - \sum_{k=0}^{n} b_k u_q^{k/2} \sum_{j=0}^{n} (-1)^j b_j u_q^{j/2}$$

$$= \sum_{k=0}^{n} \sum_{j=0}^{n} (-1)^j (a_k a_j - b_k b_j) u_q^{(k+j)/2} = \sum_{k=0}^{n} \sum_{m=k}^{n} (-1)^{m-k} (a_{m-k} a - b_{m-k} b_k) u_q^{m/2}$$

$$= (-1)^n (a_n^2 - b_n^2) \sum_{m=0}^{2n} \lambda_m u_q^{m/2}.$$

As in the proof of Theorem 2.1, we conclude that the last polynomial only contains components with even indices m. Hence

$$\Delta_q = (-1)^n (a_n^2 - b_n^2) \sum_{j=0}^{n} \lambda_{2j} u_q^j = (-1)^n (a_n^2 - b_n^2) \Lambda(u_q) = 0 \quad \text{for } q = 1, 2, \ldots, n,$$

because the u_q are roots of the polynomial $\Lambda(t)$. From the second equation of (2.7) we find

$$y_q = C_{2q} z_q, \quad y_q' = -C_{4q} z_q, \quad \text{where } z_q \in Z_{D-\sqrt{u_q} I} \text{ is arbitrary } (q = 1, 2, \ldots, n).$$

Therefore,

$$z = \sum_{q=1}^{n} (C_{2q}I - C_{4q}S)z_q = \sum_{k=0}^{n}\sum_{q=1}^{n} [(-1)^k a_k I - b_k S] u_q^{k/2} z_q$$

which was to be proved.

THEOREM 2.4. *Under the conditions of Theorem* 2.1 *every solution of the equation* $Ax = y$ *is of the form*

$$x = R_A\tilde{x} + \sum_{q=1}^{n}\sum_{k=0}^{n} [(-1)^k a_k I - b_k S] u_q^{k/2} z_q,$$

where $\tilde{x}$ *is a solution of the equation* $\Lambda(D^2)\tilde{x} = y$ *and* $z_q \in Z_{D-\sqrt{u_q}I}$ *for* $q = 1, 2, \ldots, n$.

For the proof it suffices to remark that every solution of the equation $\Lambda(D^2)\tilde{x} = y$ satisfies the equality $A(R_A\tilde{x}) = \Lambda(D^2)\tilde{x} = y$ and to apply Theorem 2.3.

Similar results can be obtained for the operator R_A.

As at the end of Section 1, we can consider some other cases by writing the operator A in the form:

$$A = \sum_{k=0}^{n} (a_k I + b_k S) D^k = \sum_{k=0}^{n} [(a_k + b_k) P^+ + (a_k - b_k) P^-] D^k \tag{2.8}$$

(see Formula (2.11) in Chapter II). We obtain some particular results, but a full discussion of solvability, as was given in Section 1 is not possible. For example

(1) If $a_k + b_k = 0$, $a_k - b_k = 0$ for $k = 0, 1, \ldots, n$, then $A = 0$.

(2) If $a_k + b_k = 0$ for $k = 0, 1, \ldots, n$, but not all $a_k - b_k = 0$, then by using Formula (2.8) we can write the equation $Ax = y$ as follows:

$$P^- \sum_{k=0}^{n} [(a_k - b_k) D^k x] = y.$$

From Theorem 5.3 of Chapter II we then conclude that a necessary and sufficient condition for the solvability of this equation is $P^+ y = 0$ and that under this condition we have to solve the equation $\sum_{k=0}^{n} (a_k - b_k) D^k x = y + z^+$, where $z^+ \in X^+$ is arbitrary. For the applications of the last two sections see:

Chapter VII, Section 1. Ordinary differential equations with reflection.

Chapter XI, Section 1. Integral and integro-differential equations with reflection.

Chapter XIII, Section 1. Singular integral equations with reflection.

Chapter XVII, Section 3. Characterization of commutators with singular integral operators.

3. HOMOGENEOUS EQUATIONS WITH A PERMUTING OPERATOR

Let S be an involution of order N acting in a linear space over complex scalars:

(3.1) $$S^N = I \quad \text{on } X \quad (N \geqslant 2).$$

Following the notations used in Section 6 of Chapter II we write

(3.2) $$P_j = \frac{1}{N}\sum_{k=0}^{N-1} \varepsilon^{-kj}S^k, \quad \text{where } \varepsilon = e^{2\pi i/N} \ (j = 1, 2, \ldots, N).$$

We recall that the operators P_j are disjoint projectors giving a partition of unity:

(3.3) $$P_jP_k = \delta_{jk}P_j, \quad \sum_{j=1}^{N} P_j = I, \quad \text{where } \delta_{jk} \text{ is the Kronecker symbol.}$$

Moreover,

(3.4) $$SP_j = \varepsilon^j P_j \quad (j = 1, 2, \ldots, N).$$

From this it follows that the space X can be decomposed into the direct sum

(3.5) $$X = X_1 \oplus \ldots \oplus X_N$$

of the spaces X such that

(3.6) $$X_j = P_j X \quad \text{and} \quad Sx = \varepsilon^j x \quad \text{for } x \in X_j \quad (j = 1, 2, \ldots, N)$$

and that every element $x \in X$ can be written uniquely in the form

(3.7) $$x = x_{(1)} + \ldots + x_{(N)}, \quad \text{where } x_j = P_j x \quad (j = 1, 2, \ldots, N).$$

A linear operator D acting in the linear space X *permutes an involution S of order N* acting in X if the superpositions SD and DS both exist and

(3.8) $$DS = \varepsilon SD, \quad \text{where } \varepsilon = e^{2\pi i/N}.$$

We will say briefly that D is a *permuting operator*. To begin with, we give some properties of permuting operators. For $N = 2$ permuting operators anti-commute with the involution S; they were considered in Sections 1 and 2.

PROPERTY 3.1. *If D permutes an involution S of order N, then for arbitrary positive integers k and m*

(3.9) $$D^m S^k = \varepsilon^{mk} S^k D^m.$$

PROOF by induction. By assumption, equality (3.9) is true for $k = m = 1$. Suppose now that (3.9) is true for $m = 1$. Then $DS^{k+1} = (DS)S^k = \varepsilon SDS^k = \varepsilon^{k+1}S^{k+1}D$. Let k be fixed arbitrarily. Then, supposing (3.9) to be true, we obtain

$$D^{m+1}S^k = D(D^mS^k) = D(\varepsilon^{km}S^kD^m) = \varepsilon^{km}(DS^k)D = \varepsilon^{km}\varepsilon^kS^kD^{m+1} = \varepsilon^{k(m+1)}S^kD^{m+1}.$$

PROPERTY 3.2. *If D permutes an involution S of order N, then*

$$P_j D = DP_{j+1} \quad \textit{for } j = 1, 2, \ldots, N-1,$$
$$P_N D = DP_1.$$

PROOF. By definition and Property 3.1, we have, for $j = 1, 2, \ldots, N$

$$P_j D = \left(\frac{1}{N}\sum_{k=0}^{N-1} \varepsilon^{-kj} S^k\right) D = \frac{1}{N}\sum_{k=0}^{N-1} \varepsilon^{-kj} S^k D = \frac{1}{N}\sum_{k=0}^{N-1} \varepsilon^{-kj}\varepsilon^{-k} D S^k$$

$$= D\left(\frac{1}{N}\sum_{k=0}^{N-1} \varepsilon^{-\qquad)} S^k\right),$$

and for $j = N$ we find $\varepsilon^{-k(N+1)} = \varepsilon^{-k}$. Hence $P_N D = DP_1$. For $j = 1, 2, \ldots$ $\ldots, N-1$ we thus have $P_j D = DP_{j+1}$.

Similarly we can prove the following

PROPERTY 3.3. *If D permutes an involution S of order N, then*

$$P_j D^k = \begin{cases} D^k P_{j+k} & \textit{for } 1 \leqslant j+k \leqslant N \\ D^k P_{j+k-N} & \textit{for } N < j+k \leqslant 2N \end{cases} \qquad (j, k = 1, 2, \ldots, N),$$

$$D^k P_j = \begin{cases} P_{j-k} D^k & \textit{for } k < j \\ P_{N+j-k} & \textit{for } k \geqslant j \end{cases} \qquad (j, k = 1, 2, \ldots, N).$$

PROPERTY 3.4. *If D permutes un involution S of order N, then the operator D^N commutes with S and P_j:*

$$D^N S = SD^N \quad \textit{and} \quad D^N P_j = P_j D^N \quad (j = 1, 2, \ldots, N).$$

PROPERTY 3.5. *Let $q(t)$ be an arbitrary polynomial (in a complex variable t) with constant complex coefficients. If the operator D permutes an involution S of order N, then $Dq(S) = q(\varepsilon S) D$.*

Indeed, by Property 3.3,

$$Dq(S) = D\sum_{j=1}^{N} q(\varepsilon^j) P_j = \sum_{j=1}^{N} q(\varepsilon^j) DP_j = q(\varepsilon^j) DP_1 + \sum_{j=2}^{N} q(\varepsilon^j) DP_j$$

$$= q(\varepsilon) P_N D + \sum_{j=2}^{N} q(\varepsilon^j) P_{j-1} D = [q(\varepsilon^{N+1}) P_N + \sum_{k=1}^{N-1} q(\varepsilon^{k+1}) P_k] D$$

$$= \sum_{k=1}^{N} q(\varepsilon^{k+1}) P_k D = q(\varepsilon S) D.$$

COROLLARY 3.1. *Under the condition of Property* 3.5

$$D^m q(S) = q(\varepsilon^m S) D \quad \textit{for } m = 1, 2, \ldots, N-1,$$
$$D^N q(S) = q(S) D^N.$$

Consider now the operator $A = a(S) - b(S) D$, where D is a permuting operator with an involution S of order N, and $a(t) = \sum_{k=0}^{N-1} a_k t^k$, $b(t) = \sum_{k=0}^{N-1} b_k t^k$ are arbitrary polynomials with constant complex coefficients[1].

In this section we assume that

$$(3.10) \qquad a(\varepsilon^j) \neq 0, \quad b(\varepsilon^j) \neq 0 \quad \text{for } j = 1, 2, \ldots, N.$$

Under this condition we shall determine the set Z_A.

LEMMA 3.1. *The equation $Ax = 0$ is equivalent to the following system of equations*:

$$(3.11) \qquad \begin{aligned} Dx_{(1)} &= c_N x_{(N)} \\ Dx_{(m+1)} &= c_m x_{(m)} \end{aligned} \qquad (m = 1, 2, \ldots, N-1),$$

where $c_m = a(\varepsilon^m)/b(\varepsilon^m) \neq 0$ *by assumption* $(m = 1, 2, \ldots, N)$.

PROOF. Properties 6.2 and 6.1 of Chapter II imply that

$$(3.12) \qquad P_m a(S) = \sum_{j=1}^{N} a(\varepsilon^j) P_m P_j = a(\varepsilon^m) P_m \quad (m = 1, 2, \ldots, N)$$

and, similarly, $P_m b(S) = b(\varepsilon^m) P_m$. Hence

$$P_m A = P_m [a(S) - b(S) D] = P_m a(S) - P_m b(S) D = a(\varepsilon^m) P - b(\varepsilon^m) P_m D$$

and, by Property 3.2

$$(3.12') \qquad P_m A = \begin{cases} a(\varepsilon^m) P_m - b(\varepsilon^m) D P_{m+1} & \text{for } m = 1, 2, \ldots, N-1, \\ a(\varepsilon^N) P_m - b(\varepsilon^N) D P_1 & \text{for } m = N. \end{cases}$$

Applying Formulae (3.5), (3.6) and (3.7), we infer that the equation $Ax = 0$ is equivalent to the system of equations $P_m A x = 0$ $(m = 1, 2, \ldots, N)$. According to (3.12′), this last system can be written as follows:

$$[a(\varepsilon^m) P_m - b(\varepsilon^m) D P_{m+1}] x = 0 \quad \text{for } m = 1, 2, \ldots, N-1,$$
$$[a(\varepsilon^N) P_N - b(\varepsilon^N) D P_1] x = 0 \quad \text{for } m = 0.$$

[1] We here write: $A = a(S) - b(S) D$ instead of $a(S) + b(S) D$, as in the preceding section merely to simplify the calculations.

Since $P_m x = x_{(m)}$ and $b(\varepsilon^m) \neq 0$ for $m = 1, 2, \dots, N$, we finally obtain system (3.11).

LEMMA 3.2. $Z_A \subset Z_{D^N-\lambda I}$, *where* $\lambda = c_1 c_2 \dots c_N = \prod_{j=1}^{N} \frac{a(\varepsilon^m)}{b(\varepsilon^m)} \neq 0$.

PROOF. Let $x \in Z_A$, i.e. $Ax = 0$. According to Lemma 3.1, the equation $Ax = 0$ is equivalent to system (3.11). Consider $x_{(1)} = P_1 x$. From system (3.11) we obtain successively

$$D x_{(1)} = c_N x_{(N)},$$
$$D^2 x_{(1)} = c_N D x_{(N)} = c_N c_{N-1} x_{(N-1)},$$
$$D^3 x_{(1)} = c_N c_{N-1} D x_{(N-1)} = c_N c_{N-1} c_{N-2} x_{(N-2)},$$
$$\cdots\cdots\cdots\cdots\cdots\cdots\cdots\cdots\cdots\cdots$$
$$\cdots\cdots\cdots\cdots\cdots\cdots\cdots\cdots\cdots\cdots$$
$$D^N x_{(1)} = D(D^{N-1} x_{(1)}) = c_N c_{N-1} \dots c_2 D x_{(2)} = c_N c_{N-1} \dots c_2 c_1 x_{(1)} = \lambda x_{(1)}.$$

Hence $(D^N - \lambda I) x_{(1)} = 0$ and $x_{(1)} \in Z_{D^N-\lambda I}$. Similarly we can show that $x_{(m)} = P_m x \in Z_{D^N-\lambda I}$. Hence $x = \sum_{m=1}^{N} x_{(m)} \in Z_{D^N-\lambda I}$, which was to be proved.

LEMMA 3.3. $Z_{D^N-\lambda I} \bigoplus_{0 \leqslant k \leqslant N-1} Z_{D-\lambda_k I}$, *where the* λ_k *are the Nth roots of* λ:

$$(3.13)\quad \lambda_k = \sqrt[N]{|\lambda|}\, e^{(2\pi k - \varphi)i/N}, \qquad \varphi = \operatorname{Arg} \lambda \ (0 \leqslant \varphi < 2\pi),\ k = 0, 1, \dots, N-1.$$

PROOF. Note that $D^N - \lambda I = (D - \lambda_0 I) \dots (D - \lambda_{N-1} I)$. The operator D is algebraic of order N on the space $Z_{D^N-\lambda I}$. Hence Theorem II.4.1 yields the required decomposition.

THEOREM 3.1.

$$(3.14)\quad Z_{D^N-\lambda I} = \left\{ z \in X\colon z = \sum_{k=0}^{N-2} a_k S^k z_k,\ z_k \in Z_{D-\lambda_0 I},\ \textit{where the } a_k \textit{ are scalars} \right\}.$$

PROOF. We first remark that

$$(3.15)\qquad \lambda_k = \lambda_0 \varepsilon^k \quad \text{for } k = 1, 2, \dots, N-1.$$

Indeed,

$$\lambda_k = \sqrt[N]{|\lambda|}\, e^{(2\pi k + \varphi)i/N} = \sqrt[N]{|\lambda|}\, e^{i\varphi/N} (e^{2\pi i/N})^k = \lambda_0 \varepsilon^k \quad (k = 0, 1, \dots, N-1).$$

Suppose now that $z \in Z_{D-\lambda_k I}$. We show that $z = S^k u$, where $u \in Z_{D-\lambda_0 I}$. Indeed, $Dz = \lambda_k z = \lambda_0 \varepsilon^k z$ and $S^{N-k} Dz = \lambda_0 \varepsilon^k S^{N-k} z$. However, Property 1.1 implies $S^{N-k} Dz = \varepsilon^{-(N-k)} D S^{N-k} z$. Hence

$$D S^{N-k} z = \lambda_0 \varepsilon^k \varepsilon^{N-k} z = \lambda_0 S^{N-k} z.$$

Therefore $u = S^{N-k}z \in Z_{D-\lambda_0 I}$. But $S^N z = S^k S^{N-k} z = S^k u$. Conversely, we show that for any $z \in Z_{D-\lambda_0 I}$ we have $S^k z \in Z_{D-\lambda_k I}$. Indeed,

$$DS^k z = \varepsilon^k S^k Dz = \lambda_0 \varepsilon^k S^k z = \lambda_k S^k z.$$

Hence $S^k z \in Z_{D-\lambda_k I}$.

To find the general form of the set Z_A we shall first determine this set in a special case.

PROPOSITION 3.4. *If* $\dim Z_{D-\lambda_0 I} = 1$, *then*

$$Z_A = \Big\{ z \in X\colon z = \sum_{k=0}^{N-1} d_k S^k z_0,\ z_0 \in Z_{D-\lambda_0 I},\ d_k = \sum_{m=1}^{N} \lambda_0^{-m} V_{k,m} c_1 c_2 \dots c_m \Big\},$$

where $c_m = a(\varepsilon^m)/b(\varepsilon^m)$ *and where by* $V_{k,m}$ *we mean the subdeterminant obtained by cancelling the* $(k+1)$*th column and the* m*th row of the Vandermonde determinant* V *of the numbers* $\varepsilon^2, \varepsilon^3, \dots, \varepsilon^N, \varepsilon$.

PROOF. Since $Z_A \subset Z_{D^N-\lambda I}$ (Lemma 3.2) and $\dim Z_{D-\lambda_0 I} = 1$, we have $z \in Z_A$ if and only if $z = \sum_{k=0}^{N-1} \tilde{a}_k S^k \tilde{z}_0$, where $\tilde{z}_0 \in Z_{D-\lambda_0 I}$ is arbitrary and the coefficients $\tilde{a}_k$ are chosen suitably. We write $\tilde{a}(S) = \sum_{k=0}^{N-1} a_k S^k$.

Then $z = a(S)\tilde{z}_0 = \sum_{j=1}^{N} \tilde{a}(\varepsilon^j) P_j \tilde{z}_0$ and $P_m z = \tilde{a}(\varepsilon^m) P_m \tilde{z}_0$. Hence, by Formula (3.12) in Lemma 3.1

$$[c_m \tilde{a}(\varepsilon^m) P_m - \tilde{a}(\varepsilon^{m+1}) D P_{m+1}]\tilde{z}_0 = 0, \qquad m = 1, 2, \dots, N-1,$$
$$[c_N \tilde{a}(\varepsilon^N) P_N - \tilde{a}(\varepsilon) D P_1]\tilde{z}_0 = 0,$$

where $c_m = a(\varepsilon^m)/b(\varepsilon^m)$. But

$$DP_{m+1}\tilde{z}_0 = P_m D z_0 = P_m \lambda_0 \tilde{z}_0 = \lambda_0 P_m \tilde{z}_0 \quad \text{for } m = 1, 2, \dots, N-1,$$
$$DP_1 \tilde{z}_0 = P_N D\tilde{z}_0 = P_m \lambda_0 \tilde{z}_0 = \lambda_0 P_N \tilde{z}_0.$$

Hence the last system can be written as follows:

$$\begin{aligned} &[c_m \tilde{a}(\varepsilon^m) - \lambda_0 \tilde{a}(\varepsilon^{m+1}) P_m]\tilde{z}_0 = 0 \quad (m = 1, 2, \dots, N-1), \\ &[c_N \tilde{a}(\varepsilon^N) - \lambda_0 \tilde{a}(\varepsilon) P_N]\tilde{z}_0 = 0. \end{aligned} \tag{3.16}$$

Note that $P_m \tilde{z}_0 \neq 0$ for $m = 1, 2, \dots, N$, if $\tilde{z}_0 \neq 0$. Indeed, suppose that for some m we have $P_m \tilde{z}_0 = 0$. This means that $\sum_{k=0}^{N-1} \varepsilon^{-km} S^k z_0 = 0$, and this implies the linear dependence of the elements $\tilde{z}_0, S\tilde{z}_0, \dots, S^{N-1}\tilde{z}_0$. Since $S^k \tilde{z}_0 \in Z_{D-\lambda_k I}$ and the space $Z_{D^N-\lambda I}$ is a direct sum of the spaces $Z_{D-\lambda_k I}$ we find $S^k \tilde{z}_0 = 0$ for

$k = 0, 1, \ldots, N-1$. In particular $\tilde{z}_0 = 0$, which is a contradiction. Hence, since $\tilde{z}_0$ is arbitrary, equalities (3.16) hold if and only if

$$c_m \tilde{a}(\varepsilon^m) - \lambda_0 \tilde{a}(\varepsilon^{m+1}) = 0 \quad \text{for } m = 1, 2, \ldots, N-1, \qquad (3.17)$$
$$c_N \tilde{a}(\varepsilon^N) - \lambda_0 \tilde{a}(\varepsilon) = 0.$$

We have finally obtained a system of N homogeneous equations in the N unknowns $\tilde{a}(\varepsilon), \ldots, a(\varepsilon^N)$. The determinant of this system is

$$\Delta = \begin{vmatrix} c_1 & -\lambda_0 & 0 & 0 & \ldots & 0 & 0 \\ 0 & c_2 & -\lambda_0 & 0 & \ldots & 0 & 0 \\ 0 & 0 & c_3 & -\lambda_0 & \ldots & 0 & 0 \\ \cdots & \cdots & \cdots & \cdots & \cdots & \cdots & \cdots \\ \cdots & \cdots & \cdots & \cdots & \cdots & \cdots & \cdots \\ 0 & 0 & 0 & 0 & \ldots & c_{N-1} & -\lambda_0 \\ -\lambda_0 & 0 & 0 & 0 & \ldots & 0 & c_N \end{vmatrix}.$$

The expansion of Δ with respect to the last row gives

$$\Delta = (-1)^{N+1}(-\lambda_0) \begin{vmatrix} -\lambda_0 & 0 & 0 & \ldots & 0 & 0 \\ c_2 & -\lambda_0 & 0 & \ldots & 0 & 0 \\ 0 & c_3 & -\lambda_0 & \ldots & 0 & 0 \\ \cdots & \cdots & \cdots & \cdots & \cdots & \cdots \\ 0 & 0 & 0 & \ldots & c_{N-1} & -\lambda_0 \end{vmatrix} +$$

$$+(-1)^{2N} c_N \begin{vmatrix} c_1 & -\lambda_0 & 0 & \ldots & 0 \\ 0 & c_2 & -\lambda_0 & \ldots & 0 \\ 0 & 0 & c_3 & \ldots & 0 \\ \cdots & \cdots & \cdots & \cdots & \cdots \\ 0 & 0 & 0 & \ldots & c_n \end{vmatrix}.$$

The first determinant has only zeros above the principal diagonal and the second one, below the principal diagonal. Therefore

$$\Delta = (-1)^{N+1}(-\lambda_0)^N + (-1)^{2N} c_N c_1 \ldots c_{N-1} = (-1)^{2N+1}\lambda_0^N + \lambda = \lambda - \lambda = 0,$$

since $\lambda_0^N = \lambda$.

Since the first subdeterminant of order $N-1$ is (by assumption) different from zero, we solve system (3.17) by cancelling the last equation and putting $\tilde{a}(\varepsilon) = \tilde{a}$, where a is an arbitrary complex number. We obtain

$$a(\varepsilon^2) = c_1 \tilde{a}/\lambda_0, \qquad \tilde{a}(\varepsilon^{m+1}) = c_m a(\varepsilon^m)/\lambda_0 \quad \text{for } m = 2, 3, \ldots, N-1.$$

Hence $\tilde{a}(\varepsilon^{m+1}) = c_m c_{m-1} \ldots c_1 a/\lambda_0^m$, where $\tilde{a}$ is an arbitrary complex number and $m = 1, 2, \ldots, N-1$. We have thus determined $\tilde{a}(\varepsilon^{m+1}) = \sum_{k=0}^{N-1} \tilde{a}_k \varepsilon^{(m+1)k}$. We shall now determine the constants $\tilde{a}_k$. We obtain the following system of equations:

$$\sum_{k=0}^{N-1} \tilde{a}_k \varepsilon^{(m+1)k} = \frac{c_m c_{m-1} \ldots c_1}{\lambda_0^m} \tilde{a} \qquad (m = 1, 2, \ldots, N).$$

Observing that by definition $\tilde{a} = \tilde{a}(\varepsilon) = \sum_{k=0}^{N-1} \tilde{a}_k \varepsilon^k = \sum_{k=0}^{N-1} \tilde{a}_k \varepsilon^{(N+1)k}$ and $\lambda_0^N = c_N c_{N-1} \ldots c_1 = \lambda$, we can write this system as follows:

$$\sum_{k=0}^{N-1} \tilde{a}_k \varepsilon^{(m+1)k} = \frac{c_m c_{m-1} \ldots c_1}{\lambda_0^m} \tilde{a} \qquad (m = 1, 2, \ldots, N). \tag{3.18}$$

We have thus obtained a system of N linear non-homogeneous equations in the N unknowns $a_0, \ldots, a_{N-1}$. The determinant V of system (3.18) is the Vandermonde determinant of the distinct numbers $\varepsilon^2, \varepsilon^3, \ldots, \varepsilon^{N+1} = \varepsilon$. Hence $V = \prod_{1 \leqslant k, m \leqslant N;\, k \neq m} (\varepsilon^k - \varepsilon^m) \neq 0$. Denote by $V_{k,m}$ the subdeterminant of V obtained by cancelling the $(k+1)$th column and the mth row. The unique solution of system (3.18) is then of the form

$$\tilde{a}_k = \frac{\tilde{a}}{V} \sum_{m=1}^{N-1} \frac{c_1 \ldots c_m}{\lambda_0^m} V_{k,m} \qquad (k = 0, 1, \ldots, N-1).$$

Since $\tilde{a}$ and $\tilde{z}_0 \in Z_{D-\lambda_0 I}$ are arbitrary, we write $z_0 = \tilde{a}\tilde{z}_0/V$ and this gives the required form of the set Z_A in the case under consideration.

We now prove the general theorem without any restriction on $\dim Z_{D-\lambda_0 I}$.

THEOREM 3.2.

$$Z_A = \Big\{ z \in X \colon z = \sum_{k=0}^{N-1} d_k S^k z_0;\, z_0 \in Z_{D-\lambda_0 I},\, d_k = \sum_{m=1}^{N} \lambda_0^{-m} c_1 \ldots c_m V_{k,m} \Big\},$$

where $V_{k,m}$ are determined as in Proposition 3.1.

PROOF. Since $Z_A \subset Z_{D^N - \lambda I}$ (Lemma 3.2), we infer that $z \in Z_A$ is of the form $z = \sum_{k=0}^{N-1} S^k z_k$, $z_k \in Z_{D-\lambda_0 I}$ (Theorem 3.1). Writing $d(S) = a(S)\,[b(S)]^{-1}$, we conclude that the equation $Az = [a(S) - b(S)D]z = 0$ is equivalent to the equation

$$[D - d(S)]z = 0. \tag{3.19}$$

We write $d(S) = \sum_{m=0}^{N-1} d_m S^m$. Since $z \in Z_{D^N - \lambda I}$, we have

$$Dz = D \sum_{k=0}^{N-1} S^k z = \sum_{k=0}^{N-1} DS^k z_k = \sum_{k=0}^{N-1} \varepsilon^k S^k D z_k = \sum_{k=0}^{N-1} \varepsilon^k \lambda_0 S^k z_k.$$

Hence, if $z \in Z$, then

$$0 = [D - d(S)]z = \sum_{k=0}^{N-1} \varepsilon^k \lambda_0 S^k z_k - \sum_{k=0}^{N-1} d(S) S^k z_k = \sum_{k=0}^{N-1} \varepsilon^k \lambda_0 S^k z_0 - \sum_{k,m=0}^{N-1} d_m S^{m+k} z_k.$$

Since $S^j z_k \in Z_{D-\lambda_j I}$ (Theorem 3.1), all the elements $S^j z_k$ are linearly independent. This implies that equation (3.18) is equivalent to the following system:

$$\begin{aligned}
&(d_0 - \lambda_0) z_0 + d_{N-1} z_1 + d_{N-2} z_2 + \dots + d_1 z_{N-1} = 0,\\
&d_1 S z_0 + (d_0 - \varepsilon \lambda_0) S z_1 + d_{N-1} S z_2 + \dots + d_2 S z_{N-1} = 0,\\
&\dots\dots\dots\dots\dots\dots\dots\dots\dots\dots\dots\dots\\
&d_{N-1} S^{N-1} z_0 + d_{N-2} S^{N-1} z_1 + d_{N-3} S^{N-1} z_2 + \dots + (d_0 - \varepsilon^{N-1} \lambda_0) S^{N-1} z_{N-1} = 0.
\end{aligned}$$

Applying the operator S^{N-k} $(k = 0, 1, \dots, N-1)$ to both sides of the kth equation and using the identity $S^N = I$, we obtain the following system of equations:

$$\begin{aligned}
&(d_0 - \lambda_0) z_0 + d_{N-1} z_1 \quad + \dots + d_1 z_{N-1} \quad = 0,\\
&d_1 z_0 + (d_0 - \varepsilon \lambda_0) z_1 \quad + \dots + d_2 z_{N-1} \quad = 0,\\
&\dots\dots\dots\dots\dots\dots\dots\dots\dots\dots\\
&d_{N-1} z_0 + d_{N-2} z_1 + \dots + (d_0 - \varepsilon^{N-1} \lambda_0) z_{N-1} = 0.
\end{aligned} \tag{3.20}$$

We observe that the matrix M of system (3.20) does not depend on the dimension of the space $Z_{D-\lambda_0 I}$. This implies that the rank of the matrix M is also independent of the dimension of the space $Z_{D-\lambda_0 I}$. An immediate consequence of Proposition 3.1 is that *rank* $M = N-1$ in the case where $\dim Z_{D-\lambda_0 I}$ equals 1. Hence we must have *rank* $M = N-1$ in the general case. Further considerations follow the same lines as Proposition 3.1.

4. NON-HOMOGENEOUS EQUATIONS WITH A PERMUTING OPERATOR

The notation and assumptions of the preceding Section continue unchanged. We now determine the general form of the solutions of the non-homogeneous equation $Ax = y$, $y \in X$.

LEMMA 4.1. *Let* $d(S) = a(S)[b(S)]^{-1}$. *Then* $\prod_{m=0}^{N-1} d(\varepsilon^m S) = \lambda I$.

PROOF. Property II.6.2 implies that

$$d(\varepsilon^m S) = \sum_{j=1}^{N} d(\varepsilon^{m+j}) P_j = \sum_{j=1}^{N} a(\varepsilon^{m+j})[b(\varepsilon^{m+j})]^{-1} P_j \quad (m = 1, 2, \ldots, N-1).$$

Property II.2.5 implies

$$\prod_{m=0}^{N-1} d(\varepsilon^m S) = \sum_{j=1}^{N} \Big[\prod_{m=0}^{n-1} d(\varepsilon^{m+j})\Big] P_j = \sum_{j=1}^{N} \Big[\prod_{m=0}^{N-1} \frac{a(\varepsilon^{m+j})}{b(\varepsilon^{m+j})}\Big] P_j.$$

Consider the coefficients in the last sum. For $j = 1$ we have

$$\prod_{m=0}^{N-1} \frac{a(\varepsilon^{m+1})}{b(\varepsilon^{m+1})} = c_1 c_2 \ldots c_N = \lambda.$$

For an arbitrary $1 < j \leqslant N$ we obtain a product of N numbers $a(\varepsilon^{m+j})/b(\varepsilon^{m+j})$ for N different values $m+j$. Using the equality $\varepsilon^{N+k} = \varepsilon^k$ for $k = 1, 2, \ldots, N$ we obtain the product of the same numbers $c_1, \ldots, c_N$ but in a different order for each j. Hence

$$\prod_{m=0}^{N-1} \frac{a(\varepsilon^{m+j})}{b(\varepsilon^{m+j})} = \lambda \quad \text{for } j = 1, 2, \ldots, N$$

and Corollary II.2.1 implies the required formula.

LEMMA 4.2. *Let* $\tilde{A} = D^{N-1} + T$, *where*

$$T = d(\varepsilon^{N-1} S) D^{N-2} + \ldots + d(\varepsilon^{N-1} S) \ldots d(\varepsilon^2 S) D + d(\varepsilon^{N-1} S) \ldots 3(\varepsilon S).$$

Then

$$[D - d(S)]\tilde{A} = \tilde{A}[D - d(S)] = D^N - \lambda I. \tag{4.1}$$

PROOF. Let $x \in X$ be fixed arbitrarily and let $u = [d(S) - D]x$. Then $Dx = d(S)x - u$. Applying powers of D to both sides of this equation and applying Corollary 3.1 we obtain successively:

$$\begin{aligned} D^2 x &= D[d(S)x - u] = d(\varepsilon S) Dx - Du = d(\varepsilon S)[d(S)x - u - Du] \\ &= d(\varepsilon S) d(S) x - d(\varepsilon S) u - Du, \\ D^3 x &= D[d(\varepsilon S) d(S) x - d(\varepsilon S) u - Du] = d(\varepsilon^2 S) d(\varepsilon S) Dx - d(\varepsilon^2 S) Du - D^2 u \\ &= d(\varepsilon^2 S) d(\varepsilon S)[d(S) x - u] - d(\varepsilon^2 S) Du - Du^2 \\ &= d(\varepsilon^2 S) d(\varepsilon S) d(S) x - d(\varepsilon^2 S) d(\varepsilon S) u - d(\varepsilon^2 S) Du - Du^2, \\ D^N x &= d(\varepsilon^{N-1} S) \ldots d(S) x - [d(\varepsilon^{N-1} S) \ldots d(\varepsilon S) + d(\varepsilon^{N-1} S) \ldots d(\varepsilon^2 S) D + \\ &\quad + \ldots + d(\varepsilon^{N-1} S) D^{N-2} + D^{N-1} u]. \end{aligned}$$

Lemma 4.1 implies $D^N x = \lambda x - \tilde{A}u$. But $u = [d(S)-D]x$. Hence $(D^N - \lambda I)x = -\tilde{A}[d(S)-D]x = \tilde{A}[D-d(S)]x$. Since x was chosen arbitrarily, we find $D^N - \lambda I = \tilde{A}[D-d(S)]$.

To prove the first part of Formula (4.1) we show that

$$d(S)\tilde{A} = DT + \lambda I. \tag{4.2}$$

Indeed, by Lemma 4.1 and Property 3.7,

$$\begin{aligned} d(S)\tilde{A} &= d(S)D^{N-1} + d(S)T = d(\varepsilon^N S)D^{N-1} + d(\varepsilon^N S)T \\ &= Dd(\varepsilon^{N-1}S)D^{N-2} + d(\varepsilon^N S)D[d(\varepsilon^{N-2}S)D^{N-3} + \ldots + d(\varepsilon^{N-2}S)\ldots d(\varepsilon S)] + \\ &\quad + d(\varepsilon^N S)d(\varepsilon^{N-1}S)\ldots d(\varepsilon S) \\ &= D[d(\varepsilon^{N-1}S)D^{N-2} + \ldots + d(\varepsilon^{N-1}S)\ldots d(\varepsilon S)] + d(\varepsilon^{N-1}S)\ldots d(\varepsilon S)d(S) \\ &= DT + \lambda I. \end{aligned}$$

On the other hand, by definition

$$D\tilde{A} = D^N + DT. \tag{4.3}$$

Hence, from (4.2) and (4.3) we obtain

$$[D-d(S)]\tilde{A} = D^N + DT - (DT + \lambda I) = D^N - \lambda I.$$

Lemma 4.3. *Let $R_A = -\tilde{A}[b(S)]^{-1}$, where $\tilde{A}$ is determined as in Lemma 4.2. Then*

$$R_A A = A R_A = D^N - \lambda I.$$

Proof. Since

$$D - d(S) = D - a(S)[b(S)]^{-1} = -[b(S)]^{-1}[a(S) - b(S)D] = -[b(S)]^{-1}A,$$

we find

$$D^N - \lambda I = \tilde{A}[D-d(S)] = \tilde{A}[-b(S)]^{-1}A = -\tilde{A}[b(S)]^{-1}A = R_A A.$$

On the other hand,

$$D^N - \lambda I = [D-d(S)]\tilde{A} = -[b(S)]^{-1}A\tilde{A}.$$

Hence $-A\tilde{A} = b(S)(D^N - \lambda I)$. Now Property 3.4 implies that $(D^N - \lambda I)$ commutes with S, and therefore $b(S)$ commutes with $D^N - \lambda I$ and $-A\tilde{A} = (D^N - \lambda I)b(S)$. Thus

$$D^N - \lambda I = A\tilde{A}[b(S)]^{-1} = A\{-\tilde{A}[b(S)]^{-1}\} = AR_A.$$

This lemma immediately implies

Proposition 4.1. *The operators A and R_A are commutative.*

LEMMA 4.4. *If $\tilde{x}$ is a solution of the equation $(D^N - \lambda I)\tilde{x} = y$, then $x = R_A\tilde{x}$ is a solution of the equation $Ax = y$.*

Indeed, by Lemma 4.3,

$$Ax = AR_A\tilde{x} = (D^N - \lambda I)\tilde{x} = y.$$

We can now formulate the main theorem:

THEOREM 4.1. *Let S be an involution of order N acting in a linear space X (over the complex field) and let D permute S. Let $A = a(S) - b(S)D$, where $a(t)$ and $b(t)$ are polynomials with constant complex coefficients, such that $a(\varepsilon^j) \neq 0$, $b(\varepsilon^j) \neq 0$ for $j = 1, 2, \ldots, N$, where $\varepsilon = e^{2\pi i/N}$. Then every solution of the equation $Ax = y$, $y \in X$, is of the form*

$$x = R_A\tilde{x} + \sum_{k=1}^{N-1} d_k S^k z_0,$$

where:

z_0 *is a solution of the equation* $(D - \lambda_0 I)z_0 = 0$,

$\tilde{x}$ *is a solution of the equation* $(D^N - \lambda I)\tilde{x} = y$,

$$R_A = -D^{N-1} - \frac{1}{N}\sum_{k=0}^{N-1}\sum_{j=1}^{N} \varepsilon^{-kj} S^k [c_{N+j-1} D^{N-2} + c_{N+j-1} c_{+N+j-2} D^{N-3} +$$

$$+ \ldots + c_{N+j-1} \ldots c_{j+1} I][b(S)]^{-1},$$

$$c_{N+m} = c_m = a(\varepsilon^m)/b(\varepsilon^m) \quad \text{for } m = 1, 2, \ldots, N; \; \lambda = c_1 c_2 \ldots c_N;$$

$$d_k = \sum_{m=1}^{N-1} \lambda_0^{-m} V_{k,m} c_1 \ldots c_m \qquad (k = 0, 1, \ldots, N-1),$$

$$\lambda_0 = \sqrt[N]{|\lambda|}\, e^{\varphi i/N}, \qquad \varphi = \operatorname{Arg} \lambda \; (0 \leqslant \varphi < 2\pi),$$

$V_{k,m}$ is the subdeterminant obtained by cancelling the $(k+1)$th column and the mth row of the Vandermonde determinant V of the numbers $\varepsilon^2, \varepsilon^3, \ldots, \varepsilon^N, \varepsilon$ $(m = 1, 2, \ldots, N;\; k = 0, 1, \ldots, N-1)$.

PROOF. Since from Property II.2.5 we have

$$d(S) = [b(S)]^{-1} a(S) = \sum_{j=1}^{N} \frac{a(\varepsilon^j)}{b(\varepsilon^j)} P_j = \sum_{j=1}^{N} c_j P_j$$

and for $m = 1, 2, \dots, N$

$$d(\varepsilon^m S) = \sum_{j=1}^{N} \frac{a(\varepsilon^{j+m})}{b(\varepsilon^{j+m})} P_j = \sum_{j=1}^{N} c_{j+m} P_j, \quad \text{where } c_{N+k} = c_k \ (k = 1, 2, \dots, N),$$

we find from Property II.2.5 and Property 3.1

$$\begin{aligned} R_A &= -[D^{N-1} + d(\varepsilon^{N-1}S)\, D^{N-2} + \dots + d(\varepsilon^{N-1}S) \dots d(\varepsilon S)]\,[b(S)]^{-1} \\ &= -\Big\{D^{N-1} + \sum_{j=1}^{N} (c_{N+j-1} P_j D^{N-2} + \dots + c_{N+j-1} \dots c_{j+1} P_j)[b(S)]^{-1}\Big\} \\ &= -\Big\{D^{N-1} + \sum_{j=1}^{N} P_j (c_{n+j-1} D^{N-2} + \dots + c_{N+j-1} \dots c_{j+1} I)[b(S)]^{-1}\Big\} \\ &= -D^{N-1} - \frac{1}{N} \sum_{j=1}^{N} \sum_{k=0}^{N-1} \varepsilon^{-kj} S^k (c_{N+j-1} D^{N-2} + \dots + \\ &\quad + c_{N+j-1} \dots c_{j+1} I)[b(S)]^{-1}. \end{aligned}$$

Since we already have the form of R_A, we immediately obtain our theorem from Theorem 3.3, and Lemmas 4.2, 4.3, 4.4 together with the linearity of the operator A.

So far we have assumed that $a(\varepsilon^j) \neq 0 \neq b(\varepsilon^j)$ for $j = 1, 2, \dots, N$. We now drop this assumption, and consider some of the most typical cases. As in Lemma 3.1 and Formula (3.13), the equation

$$Ax = y \tag{4.4}$$

can be rewritten as an equivalent system of equations

$$\begin{aligned} &a(\varepsilon^m) x_m - b(\varepsilon^m) D x_{(m+1)} = y_{(m)} \quad \text{for } m = 1, 2, \dots, N-1, \\ &a(\varepsilon^N) x_{(N)} - b(\varepsilon^N) D x_{(1)} = y_{(N)}, \end{aligned} \tag{4.5}$$

where $x_{(m)} = P_m x$ and $y_{(m)} = P_m y$. Of course, if $a(\varepsilon^m) = b(\varepsilon^m) = 0$ for $m = 1, 2, \dots, N$, then $A = 0$ (Corollary 3.1).

1° If $b(\varepsilon^m) = 0$ for $m = 1, 2, \dots, N$, then the solution of (4.4) is given by Theorem II.5.2.

2° If $a(\varepsilon^m) = 0$ for $m = 1, 2, \dots, N$, then we solve equation (4.4) with respect to the unknown Dx. We thus reduce our problem (as in 1°) to the equation $Dx = y_0$, where

$$y_0 = -\sum_{m:\, b(\varepsilon^m) \neq 0} \frac{1}{b(\varepsilon^m)} P_m y - \sum_{m:\, b(\varepsilon^m) = 0} z_{(m)},$$

and the $z_{(m)} \in X_m$ are arbitrary, under the necessary and sufficient condition that $P_m y = 0$ for all m such that $b(\varepsilon^m) = 0$.

3° Suppose now that $a(\varepsilon^m) \neq 0$ for all m and $b(\varepsilon^m) = 0$ for at least one m.

Consider, for instance, the case $b(\varepsilon^N) = 0$. From the last equation of (4.5) we obtain $x_{(N)} = [a(\varepsilon^N)]^{-1} y_{(N)}$, and solving the system (4.5) successively, we have

$$x_{(N)} = [a(\varepsilon^N)]^{-1} y_{(N)},$$

$$x_{(m)} = \frac{1}{a(\varepsilon)} y_{(m)} + \frac{b(\varepsilon^m)}{a(\varepsilon^m)} D x_{(m+1)} \qquad (m = 1, 2, \ldots, N-1).$$

Hence, by Properties 3.2 and 3.3,

$$\begin{aligned} x_{(m)} &= \frac{1}{a(\varepsilon^m)} \left[y_{(m)} + \frac{b(\varepsilon^m)}{a(\varepsilon^{m+1})} D y_{(m+1)} + \ldots + \frac{b(\varepsilon^m) \ldots b(\varepsilon^{N-1})}{a(\varepsilon^{m+1}) \ldots a(\varepsilon^N)} D^{N-m} y_{(N)} \right] \\ &= \frac{1}{a(\varepsilon^m)} \left[P_m + \frac{b(\varepsilon^m)}{a(\varepsilon^{m+1})} D P_{m+1} + \ldots + \frac{b(\varepsilon^m) \ldots b(\varepsilon^{N-1})}{a(\varepsilon^{m+1}) \ldots a(\varepsilon^N)} D^{N-m} P_N \right] y \\ &= \frac{1}{a(\varepsilon^m)} P_m \left[I + \frac{b(\varepsilon^m)}{a(\varepsilon^{m+1})} D + \ldots + \frac{b(\varepsilon^m) \ldots (b\varepsilon^{N-1})}{a(\varepsilon^{m+1}) \ldots a(\varepsilon^N)} D^{N-m} \right] y \end{aligned}$$

and

$$x = \sum_{m=1}^{N} x_{(m)} = \sum_{m=1}^{N} \frac{1}{a(\varepsilon^m)} P_m \left[I + \frac{b(\varepsilon^m)}{a(\varepsilon^{m+1})} + \frac{b(\varepsilon^m) \ldots b(\varepsilon^{N-1})}{a(\varepsilon^{m+1}) \ldots a(\varepsilon^N)} D^{N-m} \right] y.$$

In a similar way we can determine the solution of (4.4) if $b(\varepsilon^m) = 0$ for an $m \neq N$.

4° Suppose that $b(\varepsilon^m) \neq 0$ for all m and $a(\varepsilon^m) = 0$ for at least one m. As before, we consider the case $a(\varepsilon^N) = 0$. We then find $x_{(1)}$ from the equation $D x_{(1)} = -\frac{1}{b(\varepsilon^N)} y_{(N)}$ obtained from the last equation of (4.5). Having found $x_{(1)}$ we successively solve the equations

$$D x_{m+1} = \frac{-1}{b(\varepsilon^m)} y_{(m)} + \frac{a(\varepsilon^m)}{b(\varepsilon^m)} x_{(m)} \qquad (m = 1, 2, \ldots, N-1).$$

Similarly, we can solve equation (4.4) if $a(\varepsilon^m) = 0$ for an $m \neq N$.

CHAPTER IV

Algebraic Derivative and Abstract Differential Equations

In this Chapter we will deal with equations containing an algebraic derivative (also: algebraic derivatives) and algebraic operators.

The notion of abstract derivative has been used by many authors in different ways and from different points of view (see for instance: the survey articles of Averbuh and Smoljanov [1], Mikusiński ([1]–[6]), Schatte [1], also Bellert [1], Bittner [1], Kahane [1], Kaplansky [1], Lord [1], Johnson and Sinclair [1], Sikorski [1], Zariski and Samuel [1], Singer and Wermer [1].

The main problem is that in the classical case the derivative is not defined on the whole space of continuous functions, and this leads to very complicated constructions. Our approach is different, and is based on two classical facts, namely:

1° that the Volterra linear integral equation (of the second kind) always has a unique solution in the space of continuous functions (Section V of Chapter I) and 2° that the derivative of an integral with respect to its upper limit is the integrated function in the same space.

Using these facts, in a simple way, we will obtain the general form of the solutions of linear differential equations with constant and variable coefficients (see: the present author [12], [13], Mażbic-Kulma [2], [3]).

1. ALGEBRAIC DERIVATIVE IN LINEAR SPACES

Let X be a linear space over the complex scalar field[1]. We say that a linear operator D transforming X into itself is an *algebraic derivative* if there is a linear operator R transforming X into itself and such that

[1] Without any change of definitions and proofs we can in fact allow X to be a linear space over any algebraically closed field.

1° $D_R = X$ and $RX \subset D_D$,

2° $DR = I$,

3° the operator $I - \lambda R$ is invertible for every scalar λ.

The operator R is called the *algebraic integral*. The kernel Z_D of the algebraic derivative D is called the *space of constants* (with respect to the derivative D).

We first discuss some examples. The functions considered may be real or complex-valued and defined on the closed interval $[0, 1]$ merely for simplicity. We note here that a real function should always be understood to be an element of the corresponding linear space of complex functions.

EXAMPLE 1.1. Let X be the space $C[0, 1]$ of all functions continuous for $0 \leqslant t \leqslant 1$ (regarded as a linear space without any topology). Let $D = d/dt$, $(Rx)(t) = \int_0^t x(s)ds$ for $x \in C[0, 1]$, $0 \leqslant t \leqslant 1$. This is the most classical case. Indeed, every continuous function is integrable, and any integral of a continuous function has a derivative with respect to the upper limit of the integral which is again a continuous function. Hence $D_R = X$, $RX \subset D_D$, and $DR = I$. Condition 3° is also satisfied, which follows from Example 5.4 of Chapter I. Moreover, it is well known that in this case Z_D consists of all the constant functions. Hence $\alpha_D = \dim Z_D = 1$.

EXAMPLE 1.2. Let X be the space of continuous functions $x(t, s)$ defined for $0 \leqslant t, s \leqslant 1$. Let

$$(Dx)(t, s) = \frac{\partial x(t, s)}{\partial t} \quad \text{and} \quad (Rx)(t, s) = \int_0^t x(\tau, s)d\tau.$$

It is easily verified that conditions 1°, 2°, 3° are satisfied and that in this case $Z_D = \{x(t, s): x(t, s) = x(s) \in C[0, 1]\}$, hence $\alpha_D = +\infty$.

EXAMPLE 1.3. Let $0 \leqslant a \leqslant 1$ be fixed. Let $X = \{x \in C^\infty[0, 1]: x^{(k)}(a) = 0 \text{ for } k = 1, 2, \ldots\}$, where $C^\infty[0, 1]$ denotes the space of all infinitely differentiable functions defined on $[0, 1]$. It is clear that X is a linear space. Let $D = d/dt$ and $(Rx)(t) = \int_a^t x(s)ds$. We have to check that the operator R maps the space X into itself and that $I - \lambda R$ is invertible for all λ. Let $x \in X$, so that the derivative $x^{(k)}(t)$ is defined and continuous for all $t \in [0, 1]$ and $x^{(k)}(t) = 0$ $(k = 0, 1, 2, \ldots)$. Let $y(t) = (Rx)(t) = \int_a^t x(s)ds$. So that $y(a) = 0$. Since $y'(t) = x(t)$ we have $y'(a) = x(a) = 0$. In the same way, we find for an arbitrary k that $y^{(k+1)}(a) = x^{(k)}(a) = 0$. Hence $RX \subset X$. The theorem concerning the differentiation of an integral with respect to its upper limit implies that $RX \subset D_D$ and $DR = I$. If $x(t)$ is a solution of the equation

$$x(t) - \lambda \int_a^t x(s)ds = y(t), \quad y \in X, \tag{1.1}$$

then $x(a) = y(a) = 0$. Differentiating both sides of equation (1.1) $(k+1)$-times we obtain $x^{(k+1)}(t) - \lambda x^{(k)}(t) = y^{(k+1)}(t)$ and

$$x^{(k+1)}(a) = \lambda x^{(k)}(a) + y^{(k+1)}(a) = \lambda x^{(k)}(a) = \ldots = \lambda^{k+1} x(a) = 0.$$

Hence $x \in X$. Since all solutions of the equation $Dx = 0$ are of the form $x(t) = \text{const}$, the condition $x(a) = 0$ implies that $x(t) \equiv 0$ for $x \in Z_D$. Hence $Z_D = \{0\}$ and $\alpha_D = 0$. This fact implies that any equation $P(D)x = y$, where $y \in X$ and $P(t)$ is an arbitrary polynomial with constant complex coefficients, has a unique solution in the space X (Corollary 2.8).

EXAMPLE 1.4. Let X be the space $C^\infty[0, 1]$ of all infinitely differentiable functions on the interval $[0, 1]$. In this linear space we consider the *Hermite derivative*, defined by Antosik and Mikusiński [1] [1] in the following way

$$Dx = tx - \frac{dx}{dt}$$

and we show that it is an algebraic derivative. Indeed, let

$$(\tilde{R}x)(t) = \int_0^t x(s)ds \quad \text{and} \quad R = -\tilde{R}(I - t\tilde{R})^{-1}.$$

The validity of the definition of the operator R follows from the fact that the Volterra integral equation

$$x(t) - t\int_0^t x(s)ds = y(t), \qquad y \in X$$

has a unique solution in X (Example 5.2 in Chapter I). This implies that the operator $I - t\tilde{R}$ is invertible in X. It is clear that $RX \subset D_D = X$, because $RX = X$, and also that $D_R = X$. Furthermore, since $D\tilde{R} = I$, we have

$$\begin{aligned} DR &= D[-\tilde{R}(I - t\tilde{R})^{-1}] = (-t\tilde{R} + D\tilde{R})(I - t\tilde{R})^{-1} = (-t\tilde{R} + I)(I - t\tilde{R})^{-1} \\ &= (I - t\tilde{R})(I - t\tilde{R})^{-1} = I. \end{aligned}$$

Writing

$$I - \lambda R = I - \lambda[-\tilde{R}(I - t\tilde{R})]^{-1} = [I - (t - \lambda)\tilde{R}](I - t\tilde{R})^{-1}$$

we see that $I - \lambda R$ is invertible for every scalar λ for the same reasons as the operator $I - t\tilde{R}$. Since all solutions of the equation $Dx \equiv tx - \frac{dx}{dt} = 0$ are of the form: $x(t) = Ce^{t^2/2}$, here we find that the space Z_D of constants is one-dimensional: $\alpha_D = 1$.

EXAMPLE 1.5. Let X be the space $C(\boldsymbol{R})$ of functions continuous throughout the real line which satisfy the condition $x(t) = 0$ for t not less than an $a > 0$. Let

$$(Dx)(t) = \begin{cases} t\dfrac{dx(t)}{dt} & \text{for } t \geqslant a > 0, \\ 0 & \text{otherwise,} \end{cases}$$

and let

$$(Rx)(t) = \begin{cases} \displaystyle\int_a^t \frac{x(s)}{s} ds & \text{for } t \geqslant a > 0, \\ 0 & \text{otherwise.} \end{cases}$$

[1] Antosik and Mikusiński have studied this derivative in another space.

We see that $D_R = X$ and that

$$(DRx)(t) = \begin{cases} t\dfrac{d}{dt}\displaystyle\int_a^t \dfrac{x(s)}{s}\,ds = t\,\dfrac{x(t)}{t} = x(t) & \text{for } t \geqslant a > 0, \\ 0 & \text{otherwise.} \end{cases}$$

Suppose now that the equation $(I-\lambda R)x = 0$ $(t \geqslant a)$ has a non-trivial solution $x(t)$. Since $x(t) = \lambda(Rx)(t)$, we have by definition $Dx = \lambda DRx = \lambda x$, hence $tx' = \lambda x$, which implies $x(t) = Ct^{\lambda}$, where C is an arbitrary constant. However, $x(a) = 0$, whence $C = 0$ and $x(t) \equiv 0$ for $t \geqslant a$. For $t \leqslant a$ we have $Rx = 0$, and hence we conclude from the equation $x = \lambda Rx$ that $x = 0$. The operator $I-\lambda R$ is therefore invertible for every scalar λ.

We now give some properties of algebraic derivatives which will be very useful in the sequel.

PROPERTY 1.1. *$D^k R^k = I$ for $k = 0, 1, \ldots$*

Indeed, for $k = 1$, $DR = I$ by definition. If we assume that $D^k R^k = I$ for some arbitrary k, then $D^{k+1}R^{k+1} = D(D^k R^k)R = DR = I$.

PROPERTY 1.2.

(1.2) $\quad (D-\lambda I)^k (I-\lambda R)^{-k} = D^k$ *for every scalar λ and $k = 1, 2, \ldots$*

Indeed, $D-\lambda I = D-\lambda DR = D(I-\lambda R)$, which implies $(D-\lambda I)(I-\lambda R)^{-1} = D$. Now suppose that Formula (1.2) is true for some arbitrary k. We find

$$\begin{aligned}(D-\lambda I)^{k+1}(I-\lambda R)^{-(k+1)} &= (D-\lambda I)[(D-\lambda I)^k (I-\lambda R)^{-k}](I-\lambda R)^{-1} \\ &= (D-\lambda I)^k D^k (I-\lambda R)^{-1} \\ &= D^k (D-\lambda I)(I-\lambda R)^{-1} = D^{k+1},\end{aligned}$$

which was to be proved.

This and Property 1.1 immediately imply

PROPERTY 1.3. *$(D-\lambda I)^k (I-\lambda R)^{-k} R^k = I$ for every scalar λ and for $k = 1, 2, \ldots$*

Let

$$P(t) = \prod_{k=1}^{n} (t-t_k)^{r_k}, \qquad \text{where } N = r_1 + \ldots + r_n = \deg P(t),$$

so that the t_k are roots of the polynomial $P(t)$ (with complex coefficients) and the r_k are their multiplicities (we assume $t_k \neq t_k$ for $j \neq k$). We write

$$P(t, s) = \prod_{k=1}^{n} (t-t_k s)^{r_k}.$$

PROPERTY 1.4. *$P(D, I)[P(I, R)]^{-1}R^N = I$, where $N = \deg P(t)$.*

Indeed from Properties 1.3 and 1.1

$$\begin{aligned} P(D, I)[P(I, R)]^{-1}R^N &= \prod_{k=1}^{n} (D-t_k I)^{r_k} \Big[\prod_{k=1}^{n} (I-t_k R)\Big]^{-1} R^N \\ &= (D-t_1 I)^{r_1} \dots (D-t_n I)^{r_n}(I-t_n R)^{-r_n} \dots (I-t_1 R)^{-r_1} R^N \\ &= (D-t_1 I)^{r_1} \dots (D-t_{n-1} I)^{r_{n-1}} D^{r_n}(I-t_{n-1} R)^{-r_{n-1}} \dots \\ &\qquad \dots (I-t_1 R)^{-r_1} R^N \\ &= D^{r_n}(D-t_1 I)^{r_1} \dots (D-t_{n-1} I)^{r_{n-1}}(I-t_{n-1} R)^{-r_{n-1}} \dots \\ &\qquad \dots (I-t_1 R)^{-r_1} R^N \\ &= D^{r_n} \dots D^{r_1} R^N = D^{r_1 \dots + r_n} R^N = D^N R^N = I. \end{aligned}$$

PROPERTY 1.5. *If D is an algebraic derivative in a linear space X, then every polynomial $P(D)$ with scalar coefficients is also an algebraic derivative, and the corresponding integral is*

$$\tilde{R} = [P(I, R)]^{-1} R^N, \qquad \textit{where } N = \deg P(t),$$

and R is the algebraic integral corresponding to D.

Indeed, suppose that $DR = I$ and that $I-\lambda R$ is invertible for every scalar λ. Then from Property 1.4 $P(D)\tilde{R} = I$ and for all scalars λ the operator

$$\begin{aligned} I-\lambda\tilde{R} &= I-\lambda[P(I, R)]^{-1}R^N = [P(I, R)]^{-1}[P(I, R)-\lambda R^N] \\ &= [P(I, R)]^{-1}[p_0 R^N + p_1 R^{N-1} + \dots + I - \lambda R^N] \\ &= [P(I, R)]^{-1}[(p_0-\lambda) R^N + p_1 R^{N-1} + \dots + I] \end{aligned}$$

is invertible, since it follows from the assumption that $I-\lambda R$ is invertible for all scalars λ that every polynomial in R with scalar coefficients and with non-zero "free term" is invertible.

PROPERTY 1.6. *If $D_1, \dots, D_m$ are algebraic derivatives in a linear space X such that for the corresponding algebraic integrals $R_1, \dots, R_m$ and for every scalar λ the operator $I-\lambda R_m \dots R_1$ is invertible, then the operator $D = D_1 \dots D_m$ is an algebraic derivative with the algebraic integral $R = R_m \dots R_1$.*

Indeed, by hypothesis,

$$\begin{aligned} DR &= (D_1 \dots D_{m-1})(D_m R_m)(R_{m-1} \dots R_1) \\ &= D_1 \dots D_{m-1} R_{m-1} \dots R_1 = \dots = D_1 R_1 = I \end{aligned}$$

and $I-\lambda R = I-\lambda R_m \dots R_1$ is invertible for every scalar λ.

PROPERTY 1.7. *If* $D_1, \ldots, D_m$ *are algebraic derivatives in a linear space* X, *such that for the corresponding algebraic integrals* $R_1, \ldots, R_m$ *and for every scalar* λ *the operator* $I-\lambda R_m^{k_m} \ldots R_1^{k_1}$ *is invertible, then the operator* $D = D_1^{k_1} \ldots D_m^{k_m}$ *is also an algebraic derivative with the algebraic integral* $R = R_m^{k_m} \ldots R_1^{k_1}$.

Indeed, Property 1.5 implies that D^{k_j} is an algebraic derivative with an algebraic integral R^{k_j} $(j = 1, 2, \ldots, m)$, and so our theorem follows immediately from Property 1.6.

More generally, we can obtain from Properties 1.5 and 1.7

PROPERTY 1.8. *If* $D_1, \ldots, D_m$ *are algebraic derivatives in a linear space* X *such that for the corresponding algebraic integrals* $R_1, \ldots, R_m$ *and for every scalar* λ *the operator* $I-\lambda R$ *is invertible, where* $R = [P_m(I, R_m)]^{-1}R^{N_m} \ldots [P_1(I, R_1)]^{-1}R^{N_1}$, $N_j = \deg P_j(t)$ $(j = 1, 2, \ldots, m)$, *then* $D = P_1(D_1) \ldots P_m(D_m)$ *is an algebraic derivative with the integral* R, *the* $P_j(t)$ *are here arbitrary polynomials with scalar coefficients.*

2. DIFFERENTIAL EQUATIONS WITH CONSTANT COEFFICIENTS IN LINEAR SPACES

In this Section we study differential equations with scalar coefficients.

THEOREM 2.1. *For all scalars* λ

$$Z_{D-\lambda I} = \{(I-\lambda R)^{-1}z\colon z \in Z_D\} = (I-\lambda R)^{-1}Z_D.$$

PROOF. Let $x = (I-\lambda R)^{-1}z$, where $z \in Z_D$. Then x is a solution of the equation

$$(D-\lambda I)x = 0. \tag{2.1}$$

Indeed, Property 1.2 implies that $(D-\lambda I)x = (D-\lambda I)(I-\lambda R)^{-1}z = Dz = 0$, since $z \in Z_D$. On the other hand, let x be a solution of (2.1) and let $u = (I-\lambda R)x$. Then $Du = D(I-\lambda R)x = (D-\lambda DR)x = (D-\lambda I)x = 0$. Hence $u \in Z_D$. But $x = (I-\lambda R)^{-1}u$. This gives the required form of the set $Z_{D-\lambda I}$.

The above immediately implies

COROLLARY 2.1. $\alpha_{D-\lambda I} = \alpha_D$ *for all* λ.

THEOREM 2.2. *The general form of the solutions of the equation*

$$(D-\lambda I)x = y, \quad y \in X, \tag{2.2}$$

is given by

$$x = (I-\lambda R)^{-1}(Ry+z), \quad \text{where } z \in Z_D.$$

PROOF. Note that $\tilde{x} = (I-\lambda R)^{-1}Ry$ is a solution of equation (2.2). Indeed, Property 1.3 implies that $(D-\lambda I)\tilde{x} = (D-\lambda I)(I-\lambda R)^{-1}Ry = y$. The linearity of the operator $D-\lambda I$ implies that all other solutions are of the form $\tilde{x}+\tilde{z}$, where $\tilde{z} \in Z_{D-\lambda I}$. This and Theorem 2.1 give the required formula for all solutions of equation (2.2).

COROLLARY 2.2. *Every solution of the equation $Dx = y$ is the form $x = Ry+z$, where $z \in Z_D$.*

Write

$$e_\lambda(z) = (I-\lambda R)^{-1}z, \quad z \in Z_D.$$

It follows from the definition and from Theorem 2.1 that $e_\lambda(z) \in Z_{D-\lambda I}$ for every $z \in Z_D$ and every scalar λ and that the set $E_\lambda = Z_{D-\lambda I}$ is the eigenspace of the operator D corresponding to the eigenvalue λ. The elements $e_\lambda(z)$ play, in a sense, the role of exponential functions. From the definition it also follows that $e_\lambda(z_1), \dots, e_\lambda(z_m)$ are linearly independent if and only if $z_1, \dots, z_m \in Z_D$ are linearly independent.

COROLLARY 2.3. *There exists an x such that $Dx = x$, namely $x = e_1(z)$, where $z \in Z_D$, provided, that $Z_D \neq \{0\}$.*

Obviously

$$\dim\{x \in X : Dx = x\} = \dim\{e_1(z) : z \in Z_D\} = \alpha_{D-\lambda I} = \alpha_D.$$

THEOREM 2.3. *For every m, if $z_k \in Z_D$, $z_k \neq 0$ $(k = 0, 1, \dots, m)$, then the elements $z_0, Rz_1, \dots, R^m z_m$ are linearly independent.*

PROOF by induction. Let $m = 1$, and suppose that z_0 and Rz_1 are linearly dependent. Then there are scalars a_0, a_1 non-vanishing simultaneously and such that $a_0 z_0 + a_1 Rz_1 = 0$. Hence $a_0 Dz_0 + a_1 DRz_1 = 0$, which implies $a_0 Dz_0 + a_1 z_1 = 0$. But $Dz_0 = 0$ by assumption, and hence $a_1 z_1 = 0$. By assumption $z_1 \neq 0$, which implies $a_1 = 0$. But, if $a_1 Rz_1 = 0$, then $a_0 z_0 = 0$. Since $z_0 \neq 0$, we not only have $a_1 = 0$ but also $a_0 = 0$, which is a contradiction. Therefore z_0 and Rz_1 are linearly independent.

Let $m > 1$ be arbitrary and suppose that $z_0, Rz_1, \dots, Rz_m$ are linearly independent. Let $R^{m+1}z_{m+1}$ be linearly dependent on the elements $z_0, Rz_1, \dots, R^m z_m$, i.e. suppose there are scalars $a_0, \dots, a_m, a_{m+1}$, not all zero, such that

$$(2.3) \qquad a_0 z_0 + \dots + a_m R^m z_m + a_{m+1} R^{m+1} z_{m+1} = 0, \quad \text{where } a_{m+1} \neq 0.$$

Let $\tilde{z}_0 = a_0 z_0 + \dots + a_m z_m R^m$ and $\tilde{z}_1 = R^m z_{m+1}$. By the inductive hypothesis $\tilde{z}_0 \neq 0$. Also $z_1 \neq 0$. Indeed, the operator $I-\lambda R$ is invertible for every scalar λ, which implies that $Rx = 0$ if and only if $x = 0$. Hence $Z_R = \{0\}$ and

$Z_{R^m} = \{0\}$ also. Since $z_{m+1} \neq 0$, we have $\tilde{z}_1 \in R^m z_{m+1} \neq 0$.

Equality (2.3) can be written as $\tilde{z}_0 + a_{m+1} R\tilde{z}_1 = 0$, which implies $D^{m+1}\tilde{z}_0 + a_{m+1} D^{m+1} R\tilde{z}_1 = 0$. But

$$D^{m+1}\tilde{z}_0 = D^{m+1}(a_0 z_0 + \ldots + a_m R^m z_m) = a_0 D^{m+1} z_0 + \ldots + a_m D^{m+1} R^m z_m$$
$$= a_0 D^{m+1} z_0 + \ldots + a_m D z_m = 0,$$

from Property 1.1. Since $D^{m+1} R\tilde{z}_1 = D^{m+1} R^{m+1} z_{m+1} = z_{m+1}$, we conclude that $a_{m+1}\tilde{z}_1 = 0$, which implies $a_{m+1} = 0$. It follows that $\tilde{z}_0 = 0$, which contradicts our inductive hypothesis. Hence $z_0, \ldots, R^m z_m, R^{m+1} z_{m+1}$ are linearly independent.

THEOREM 2.4. *If* $m \leqslant \dim Z_D$ *and* $z_1, \ldots, z_m \in Z_D$ *are linearly independent, then for every* k *the elements* $R^k z_1, \ldots, R^k z_m$ *are linearly independent.*

PROOF. Suppose that $R^k z_1, \ldots, R^k z_m$ are linearly dependent, i.e. that there are scalars $a_0, \ldots, a_m$, not all zero, such that $a_1 R^k z_1 + \ldots + a_m R^k z_m = 0$. Then also $a_1 D^k R^k z_1 + \ldots + a_m D^k R^k z_m = 0$. This and Property 1.1 together imply that $a_1 z_1 + \ldots + a_m z_m = 0$, which is a contradiction.

THEOREM 2.5. $Z_{D^k} = \left\{\sum_{m=0}^{k-1} R^m z_m : z_m \in Z_D\right\}$ *for* $k = 1, 2, \ldots$

PROOF. If $u = \sum_{m=0}^{k-1} R^m z_m$, where $z_m \in Z_D$, then from Property 1.1

$$D^k u = D^k(R^{k-1} z_{k-1} + \ldots + R z_1 + z_0) = D^k R^{k-1} z_{k-1} + \ldots + D^k R z_1 + D^k z_0$$
$$= D z_{k-1} + \ldots + D^{k-1} z_1 + D^k z_0 = 0.$$

On the other hand, suppose that u is a solution of the equation $D^k u = 0$ for $k > 1$ and write: $v_m = D^m u$ for $m = 0, 1, \ldots, k-1$. By definition $v_m = D v_{m-1}$ $(m = 1, 2, \ldots, k-1)$. Since $v_1 = Du$, Corollary 2.2 implies that $u = R v_1 + z_0$, where $z_0 \in Z_D$. But $v_2 = D v_1$, whence this Corollary implies $v_1 = R v_2 + z_1$, where $z_1 \in Z_D$. Therefore $u = R v_1 + z_0 = R(R v_2 + z_1) + z_0 = R^2 v_2 + R z_1 + z_0$, where $z_0, z_1 \in Z_D$. Arguing successively in the same way, we finally conclude, that

$$u = R^{k-1} v_{k-1} + R^{k-2} z_{k-2} + \ldots + R z_1 + z_0, \quad \text{where } z_0, \ldots, z_{k-2} \in Z_D.$$

But $D v_{k-1} = D(D^{k-1} u) = D^k u = 0$. Hence $v_{k-1} \in Z_D$. Writing $z_{k-1} = v_{k-1}$, we find

$$u = R^{k-1} z_{k-1} + \ldots + R z_1 + z_0, \quad \text{where } z_0, \ldots, z_{k-1} \in Z_D,$$

which shows that the set Z_{D^k} is of the required form.

THEOREM 2.6. *For every scalar* λ *and for* $k = 1, 2, \dots$

$$Z_{(D-\lambda I)^k} = \left\{(I-\lambda R)^{-k}\sum_{m=0}^{k-1} R^m z_m : z_m \in Z_D\right\} = (I-\lambda R)^{-k} Z_{D^k}.$$

PROOF. Let $x = (I-\lambda R)^{-k}\sum_{m=0}^{k-1} R^m z_m$, where $z_m \in Z_D$. From Property 1.2 and from Theorem 2.5 it follows that

$$(D-\lambda I)^k x = (D-\lambda I)^k (I-\lambda R)^{-k}\sum_{m=0}^{k-1} R^m z_m = D^k \sum_{m=0}^{k-1} R^m z_m = 0.$$

Conversely, let x be a solution of the equation $(D-\lambda I)^k x = 0$ and let $u = (I-\lambda R)^k x$. Then $x = (I-\lambda R)^{-k} u$ and from Property 1.2

$$D^k u = (D-\lambda I)^k (I-\lambda R)^{-k} u = (D-\lambda I)^k x = 0.$$

Hence Theorem 2.5 implies that $u = \sum_{m=0}^{k-1} R^m z_m$, where $z_m \in Z_D$, and $x = (I-\lambda R)^{-k}\sum_{m=0}^{k-1} R^m z_m$, which was to be proved.

This immediately implies

COROLLARY 2.4. $\alpha_{(D-\lambda I)^k} = \alpha_{D^k} = k\alpha_D$.

THEOREM 2.7. *The general form of the solutions of the equation*

$$(D-\lambda I)^k x = y, \quad y \in X, \tag{2.4}$$

where λ *is an arbitrary scalar and* $k = 1, 2, \dots,$ *is*

$$x = (I-\lambda R)^{-k} R^k y + \sum_{m=0}^{k-1} R^m z_m, \quad \textit{where } z_0, \dots, z_{k-1} \in Z_D \textit{ are arbitrary.}$$

Let $P(t) = p_0 + p_1 t + \dots + p_N t^N$ be an arbitrary polynomial with scalar coefficients. We assume $p_N = 1$. Hence $P(t) = \prod_{k=1}^{n} (t-t_k)^{r_k}$, where $r_1 + \dots + r_n = N = \deg P(t)$ and the t_k are the roots of $P(t)$. As in Section 1, we write $P(t, s) = \prod_{k=1}^{n} (t-t_k s)^{r_k}$.

THEOREM 2.8. $\alpha_{P(D)} = \alpha_D \cdot \deg P(t)$.

PROOF. The derivative D is an algebraic operator of the order $N = \deg P(t)$ on the space $Z_{P(D)}$, since $P(D)z = 0$ for $z \in Z_{P(D)}$. From Theorem II.4.1 we therefore conclude that the space $Z_{P(D)}$ is the direct sum

$$Z_{P(D)} = Z_{(D-t_1 I)^{r_1}} \oplus \ldots \oplus Z_{(D-t_n I)^{r_n}}.$$

This and Corollary 2.4 together imply that

$$\begin{aligned}\alpha_{P(D)} &= \dim Z_{P(D)} = \dim Z_{(D-t_1 I)^{r_1}} + \ldots + \dim Z_{(D-t_n I)^{r_n}} \\ &= r_1 \alpha_D + \ldots + r_n \alpha_D = (r_1 + \ldots + r_n)\alpha_D = \alpha_D \cdot \deg P(t).\end{aligned}$$

COROLLARY 2.6. *If* $\alpha_D = 1$, *then* $\alpha_{P(D)} = \deg P(t)$.

COROLLARY 2.7. *If $P(t)$ and $Q(t)$ are arbitrary polynomials with scalar coefficients, then*

$$\alpha_{P(D)Q(D)} = \alpha_{P(D)} + \alpha_{Q(D)}.$$

Indeed,

$$\begin{aligned}\alpha_{P(D)Q(D)} &= \deg\big(P(t)Q(t)\big)\cdot\alpha_D = [\deg P(t) + \deg Q(t)]\cdot\alpha_D \\ &= \alpha_D \deg P(t) + \alpha_D \deg Q(t) = \alpha_{P(D)} + \alpha_{Q(D)}.\end{aligned}$$

From Theorem 2.6 and from the proof of Theorem 2.8 we immediately infer

THEOREM 2.9.

$$\begin{aligned}Z_{P(D)} &= \Big\{z \in X : z = \sum_{k=1}^{n} (I - t_k R)^{-r_k} \sum_{m=0}^{r_k - 1} R^m z_{m,k},\ z_{m,k} \in Z_D\Big\} \\ &= \bigoplus_{k=1}^{n} (I - t_k R)^{-r_k} Z_{D^{r_k}}.\end{aligned}$$

THEOREM 2.10. *The general form of the solutions of the equation $P(D)x = y$, $y \in X$, is*

$$x = [P(I, R)]^{-1} R^N y + \sum_{k=1}^{n} (I - t_k R)^{-r_k} \sum_{m=0}^{r_k - 1} R^m z_{m,k}, \quad \text{where } z_{m,k} \in Z_D$$

are arbitrary and $N = r_1 + \ldots + r_n = \deg P(t)$.

For the proof it suffices to check that $x = [P(I, R)]^{-1} R^N y$ is a solution of the equation $P(D)x = y$. Indeed, Property 2.4 implies

$$P(D)x = P(D, I)x = P(D, I)[P(I, R)]^{-1} R^N y = y.$$

COROLLARY 2.8. *If $\alpha_D = 0$, then the equation $P(D)x = y$ has a unique solution $x = [P(I, R)]^{-1} R^N y$.*

Indeed, Theorem 2.8 implies that in this case $\alpha_{P(D)} = 0$, and therefore $Z_{P(D)} = \{0\}$.

EXAMPLE 2.1. We here examine the equation

$$x^{(N)}+p_{N-1}x^{N-1}+\dots+p_0x = y,$$

where y is a continuous function defined on the interval $[0, 1]$. The algebraic derivative and algebraic integral are determined as in Example 1.1. Let the polynomial $P(t) = t^N+p_{N-1}t^{N-1}+ + \dots +p_0 = \prod_{k=1}^{N}(t-t_k)$ have simple roots only. According to Example 5.4 of Chapter I, the resolvent of the Volterra integral equation $(I-t_kR)v = u$ is $I-t_kR_{t_k}$, where $(R_\lambda u)(t) = \int_0^t e^{\lambda(t-s)}u(s)ds$ for $\lambda \neq 0$ and $R_0 = I$. Hence

$$[P(I, R)]^{-1} = \prod_{k=1}^{N}(I-t_kR)^{-1} = \prod_{k=1}^{N}(I-t_kR_{t_k}).$$

Note that integrating by parts, we have $(R^ku)(t) = \int_0^t \frac{(t-s)^k}{k!}u(s)ds$ for all $u \in X$ and $k = 1, 2, \dots$ and that Z_D is the space of constants. Then for $z = c_k \in Z_D$

$$z_k(t) = (R_{t_k}z)(t) = c_k\int_0^t e^{t_k(t-s)}ds = \begin{cases} c_k & \text{for } t_k = 0, \\ \dfrac{c_k}{t_k}(1-e^{t_kt}) & \text{for } t_k \neq 0. \end{cases}$$

Hence

$$x = \Big[\prod_{k=1}^{N}(I-t_kR_k)\Big]\tilde{y}+\sum_{k=1}^{N}z_k(t), \quad \text{where } \tilde{y}(t) = (R^Ny)(t) = \int_0^t \frac{(t-s)^N}{N!}y(s)ds.$$

This formula allows us to find any solution of the equation in question by integration only The condition that the t_k are simple roots is not essential, it follows from the general theory.

3. DIFFERENTIAL EQUATIONS WITH ALGEBRAIC OPERATORS IN LINEAR SPACES

At the beginning of this section we consider differential equations having as coefficients polynomials in an algebraic operator commuting with the algebraic derivative[1]. When the algebraic operator has simple roots only, the method of solving such equations is fairly easy and we therefore start with this case.

Let D be an algebraic derivative in a linear space X over the complex scalar field[2] with the algebraic integral R. Let S be an algebraic operator acting in X and such that

1° $DS-SD = 0$,

---

[1] The results of this section were obtained by Mażbic-Kulma [3].

[2] As before, we can equally well consider a linear space over any algebraically closed field.

2° the characteristic polynomial $P(t) = p_0+p_1+ \ldots +p_N t^N$, $p_N = 1$, of the operator S has simple roots only, i.e. $P(t) = \prod_{k=1}^{N} (t-t_k)$.

THEOREM 3.1. *The general form of the solutions of the equation*

$$Ax \equiv \sum_{m=0}^{M} \sum_{k=0}^{N-1} a_{mk} S^k D^m x = y, \quad y \in X \tag{3.1}$$

(where S and D satisfy conditions 1° *and* 2°*) is the following*

$$x = \sum_{k=1}^{N} [Q_k(I, R)]^{-1} \{R^M ([a_{\mu_k}(t_k)]^{-1} y_{(k)}) + z_k\}, \tag{3.2}$$

where

(1) $$z_k \in Z_{Q_k(D)}, \quad y_{(k)} = \Big[\prod_{j=1, j\neq k}^{N} (t_j-t_k)^{-1}(S-t_k I)\Big] y \quad (k = 1, 2, \ldots, N),$$

(2) $$Q_k(t, s) = \frac{1}{a_{\mu_k}(t_k)} \sum_{m=\mu_k}^{M} a_m(t_k) t^m s^{M-m}, \quad Q_k(t) = Q_k(t, 1),$$

(3) $$a_m(t) = \sum_{k=0}^{N-1} a_{mk} t^k \quad (m = 0, 1, \ldots, M),$$

(4) μ_k *is the least index* m *such that, given a fixed* k, $a_m(t_k) \neq 0$ $(m = 0, 1, \ldots, M;\ k = 1, 2, \ldots, N)$.

PROOF. Since the operator S has simple roots only, Formula II.2.10 and Property II.2.4 imply that

$$S = \sum_{k=1}^{N} t_k P_k \quad \text{and} \quad a_m(S) = \sum_{k=1}^{N} a_m(t_k) P_k \quad (m = 0, 1, \ldots, M),$$

where the P_k are the projectors defined by Formulae II.2.3 and II.1.2. Since S commutes with D, the operators $P_1, \ldots, P_N$ also commute with D and with an arbitrary polynomial in D with scalar coefficients. Hence, using the notation of (3), and (2), (4) and writing $x_{(k)} = P_k x$, we have for every $x \in X$:

$$Ax = \sum_{m=0}^{M} a_m(S) D^m x = \sum_{m=0}^{M} \sum_{k=1}^{N} a_m(t_k) P_k D^m x = \sum_{m=\mu_k}^{M} \sum_{k=1}^{N} a_m(t_k) D^m x_{(k)}$$

$$= \sum_{k=1}^{N} a_{\mu_k}(t_k) Q_k(D) x_{(k)}.$$

The space $P_k X$ is invariant with respect to the operators $Q_k(D)$, and Theorem II.5.1 therefore implies that equation (3.1) is equivalent to the k independent equations

(3.3) $$a_{\mu_k}(t_k)Q_k(D)x_{(k)} = y_{(k)} \quad (k = 1, 2, \ldots, N).$$

Since $a_{\mu_k}(t_k) \neq 0$, we have

$$Q_k(D)x_{(k)} \neq \frac{1}{a_{\mu_k}(t_k)} y_{(k)} \quad (k = 1, 2, \ldots, N).$$

From Theorem 2.10 it follows that

$$x_{(k)} = [Q_k(I, R)]^{-1}\left[R^M \frac{y_{(k)}}{a_{\mu_k}(t_k)} + z_k\right], \quad \text{where } z_k \in Z_{Q_k(D)}.$$

But $x = \sum_{k=1}^{N} x_{(k)}$, and the proof is complete.

REMARK 3.1. Theorem remains true if the coefficients a_{mk} are now linear operators commuting with S and D such that $a_{\mu_k}(t_k)$ is invertible for $k = 1, 2, \ldots$ $\ldots, N$. The proof is exactly the same as that just given.

REMARK 3.2. Theorem 3.1 is also valid when S is a multi-involution, m, k, M, N are the respective multi-indices and $D^k = D_1^{k_1} \ldots D_q^{k_q}$, where $D_1, \ldots, D_q$ are algebraic derivatives in the space X such that for the corresponding algebraic integrals and for all scalars λ the operator $I - \lambda R_q \ldots R_1$ is invertible (compare Section 7 in Chapter II and with Property 1.7). The proof is the same as the proof of Theorem 3.1.

When we turn to a differential equation with an algebraic operator having multiple characteristic roots, we cannot apply the method used above. In this case we will only obtain some partial results.

Thus let D be an algebraic derivative on a linear space X (over the complexes) with the algebraic integral R. Let A be an algebraic operator acting in X which commutes with $D: AD - DA = 0$. We write $P(t, s) = \prod_{k=1}^{n} (t - t_k s)^{r_k} = p_0 s^N + \ldots + p_N t^N$ ($p_N = 1$; $N = r_1 + \ldots + r_n$), where $P(t) = P(t, 1)$ is the characteristic polynomial of the operator A.

PROPOSITION 3.1. *Every solution of the equation*

(3.4) $$Dx = Ax$$

belonging to $\mathbf{D}_{D^N}$ *belongs to* $Z_{P(D)}$, *whence it is of the form*

$$x = \sum_{k=1}^{n} (I - t_k R)^{-r_k} \sum_{m=0}^{r_k - 1} R^m z_{m,k}, \quad \text{where } z_{m,k} \in Z_D. \tag{3.5}$$

Indeed, if there is a solution $x \in \mathbf{D}_{D^N}$ of equation (3.4), then by assumption

$$D^2 x = D(Dx) = D(Ax) = A(Dx) = A^2 x$$

and by simple induction $D^k x = A^k x$ for $k = 1, 2, \ldots, N$. Therefore

$$P(D)x = (p_0 I + p_1 D + \ldots + p_N D^N)x = (p_0 I + p_1 D + \ldots + p_N D^N)x$$
$$= P(A)x = 0,$$

and Theorem 2.9 yields the required formula.

PROPOSITION 3.2. *Every solution, belonging to* $\mathbf{D}_{D^N}$, *of the equation*

$$Dx = Ax + y, \quad \text{where } y \in \mathbf{D}_{D^{N-1}}, \tag{3.5'}$$

is of the form

$$x = [P(I, R)]^{-1} R^N A_1 y + z, \quad \text{where } z \in Z_{P(D)}, \quad A_1 = \sum_{k=1}^{N} p_k \sum_{m=0}^{k-1} A^{k-m} D^m. \tag{3.6}$$

PROOF. Just as in the proof of the preceding proposition, we obtain

$$D^2 x = D(Ax + y) = ADx + Dy = A(Ax + y) + Dy = A^2 x + Ay + Dy,$$

$$D^k x = A^k x + A^{k-1} y + A^{k-2} Dy + \ldots + D^{k-1} y = A^k x + \sum_{m=0}^{k-1} A^{k-1-m} D^m y$$

$$(k = 1, \ldots, N).$$

Hence

$$P(D)x = (p_0 I + p_1 D + \ldots + p_N D^N)x = (p_0 I + p_1 A + \ldots + p_N A^N)x + p_1 y +$$
$$+ \ldots + p_N \sum_{m=0}^{N-1} A^{N-1-m} D^m y = P(A)x + A_1 y = A_1 y,$$

as $P(A) = 0$ on X by assumption. Applying Theorem 2.10 and Property 3.1 to the equation $P(D)x = y_1$, where $y_1 = A_1 y$, we obtain Formula (3.6).

COROLLARY 3.1. *Every solution, belonging to* $\mathbf{D}_{D^{N(M-1)}}$ *of the equation* $(D-$ $-A)^M x = y$, *where* $y \in \mathbf{D}_{D^{(N-1)(M-1)}}$ *is of the form*

$$x = (D-A)^{M-1} [P(I, R)]^{-1} R^N A_1 y + (D-A)^{M-1} z, \quad \text{where } z \in Z_{P(D)} \tag{3.7}$$

and A_1 *is defined as in Formula* (3.6).

Indeed, if we write $x_k = (D-A)^{M-k}x$ $(k = 1, 2, \ldots, M)$, then x_1 satisfies the equation $(D-A)x_1 = y$ and $x_k = (D-A)^j x_{k-j}$ for $j = k+1 = 2, \ldots, M$. Hence $x = x_M = (D-A)^{M-1}x_1$. Solving the equation $(D-A)x_1 = y$, we obtain (Proposition 3.2) the required formula.

In a similar way we can solve every equation which can be written in the form

$$(D_1-A_1)^{M_1} \ldots (D_L-A_L)^{M_L}x = y, \quad y \in X,$$

where the operators $A_1, \ldots, A_L$ and $D_1, \ldots, D_L$ satisfy, respectively, the same conditions as the operators A and D in Propositions 3.1, 3.2 and Corollary 3.1. The proof follows the same lines as the proof of Corollary 3.1. Applying these results and Properties 1.5, 1.6, 1.7, 1.8, we obtain a large class of equations all of which can be solved in this manner.

EXAMPLE 3.1. Let X be the space of all continuous vector valued functions $x(t) = x_1(t), \ldots$ $\ldots, x_q(t)$ defined on the interval $[0, 1]$ and let $Dx = (Dx_1, \ldots, Dx_q)$, $Rx = (Rx_1, \ldots, Rx_q)$, where $D = d/dt$ and $(Rx)(t) = \int_0^t x(s)ds$ (compare Examples 1.1 and 2.1). It is easy to check that D is an algebraic derivative on X with the algebraic integral R. Let A be a scalar matrix of dimension $q \times q$. Such an A is an algebraic operator commuting with D. The equation $Dx = Ax+y$, $y \in X$ is simply a system of ordinary linear differential equations of order 1 with constant coefficients.

The equation

$$(AD+B)x = y, \quad y \in X \tag{3.8}$$

with an algebraic derivative acting in X can be solved by means of Proposition 3.2. We consider several different cases.

THEOREM 3.2. *Let D be an algebraic derivative on a linear space X with the algebraic integral R. Suppose that the linear operators A and B acting in X commute with D, that A is invertible and that $A^{-1}B$ is algebraic. If $A^{-1}y \in \mathbf{D}_{D^{N-1}}$, then every solution of equation* (3.8) *belonging to $\mathbf{D}_{D^N}$ is of the form*

$$x = -[P(I, R)]^{-1}R^N A_1 A^{-1}y + z, \quad \textit{where } z \in Z_{P(D)},$$

$P(t, s) = p_0 s^N + \ldots + p_N t^N$, $P(t) = P(t, 1)$ is the characteristic polynomial of the operator $A^{-1}B$ and A_1 is given by Formula (3.6).

PROOF. We have assumed that $AD-DA = 0$, $BD-DB = 0$. Therefore $A^{-1}D-$ $-DA^{-1} = A^{-1}(D-ADA^{-1}) = A^{-1}(DA-AD)A^{-1} = 0$, which implies

$$(A^{-1}B)D-D(A^{-1}B) = A^{-1}BD-A^{-1}DB = A^{-1}(BD-DB) = 0.$$

Hence the algebraic operator $A^{-1}B$ commutes with D. Acting on both sides of equation (3.8) with the operator A^{-1} we obtain an equivalent equation

$$Dx = -A^{-1}Bx+A^{-1}y.$$

To solve this equation we can now, apply Proposition 3.2, since all the conditions are satisfied. This gives the required formula.

THEOREM 3.3. *Let D be an algebraic derivative on a linear space X with the algebraic integral R. Suppose that the operators A and B acting in X commute with D and that A and $Q_1(A)B$ are algebraic, where* $Q(t,s) = \sum_{k=0}^{L} q_k t^k s^{L-k}$, $q_0 \neq 0$, $q_L \neq 0$, $Q(t) = Q(t, 1)$ *is the characteristic polynomial of the operator A and* $Q_1(t) = \sum_{k=1}^{L} q_k t^{k-1}$. *If* $\frac{1}{q_0} Q_1(A) y \in \boldsymbol{D}_{D^N}$, *then every solution of equation* (3.8) *is of the form*

$$x = \frac{1}{q_0} [P(I, R)]^{-1} A_1 Q_1(A) y + z, \quad \text{where } z \in Z_{P(D)},$$

$P(t)$ *is the characteristic polynomial of the operator* $\frac{1}{q_0} Q_1(A)B$ *and* $A_1 = \sum_{k=1}^{N} p_k \sum_{m=0}^{k-1} A^{k-1-m} D^m$.

PROOF. Writing equation (3.8) in the form $ADx = y - Bx$ and acting on both sides of the last equation with the operators $A, A^2, \ldots, A^{L-1}$ successively we obtain, just as in the proof of Proposition 3.1

$$\begin{aligned} 0 &= Q(A)Dx - (q_0 I + q_1 A + \ldots + q_L A^L) Dx \\ &= q_0 Dx + (q_1 I + \ldots + q_L A^{L-1})(y - Bx) = q_0 Dx + Q_1(A)(y - Bx). \end{aligned}$$

Since $q_0 \neq 0$ we obtain the following equation:

$$Dx + \frac{1}{q_0} Q_1(A) Bx = \frac{1}{q_0} Q_1(A) y.$$

Since A and B commute with D by assumption, $Q_1(A)B$ also commutes with D. All the conditions of Proposition 3.2 are satisfied, and applying this proposition, we now obtain our theorem.

We can also obtain similar generalizations of Theorems 3.2 and 3.3 as it has already been done in the case of the equation $Dx = Ax + y$.

EXAMPLE 3.2. Let X be the space of all continuous m-dimensional vector valued functions $x(t) = (x_1(t), \ldots, x_m(t))$ defined for $t = (t_1, \ldots, t_q) \in \boldsymbol{R}^q$. Let

$$Dx_j = \frac{\partial^{k_1}}{\partial t_1^{k_1}} \cdots \frac{\partial^{k_q}}{\partial t_q^{k_q}}, \quad (R_j x)(t) = \int_0^{t_j} x(t_1, \ldots, t_{j-1}, s, t_{j+1}, \ldots, t_q) ds$$

$(j = 1, 2, \ldots, m)$ and $Dx = (Dx_1, \ldots, Dx_m)$, $Rx = (Rx_1, \ldots, Rx_m)$ follows from Example 1.2 and Property 1.7, that the operator D is an algebraic derivative with the algebraic integra-

$R = R_q^{k_q} \dots R_1^{k_1}$, therefore Theorem 3.3 give us a formula for the solutions of a system of partial differential equations with constant coefficients:

$$\sum_{i=1}^{m} [a_{ij} D x_i + b_{ij} x_i] = y_j \quad (j = 1, 2, \dots, m),$$

which can be written in the form $(AD+B)x = y$, where A and B are scalar matrices.

We now consider differential equations with an involution of order N in the case where the algebraic derivative permutes this involution. From Theorems III.4.1 and 2.10 we obtain the following

THEOREM 3.5. *Let S be an involution of order N acting in a linear space X, and let an algebraic derivative D permute S. Let $A = a(S) - b(S)D = \sum_{k=0}^{N-1} (a_k I - b_k S)D$, where a_k, b_k are complexes constant, $a(\varepsilon^j) \neq 0 \neq b(\varepsilon^j)$ for $j = 1, 2, \dots, N$ and $\varepsilon = e^{2\pi i/N}$. Then every solution of the equation $Ax = y$ is of the form*

$$x = R_A\Big[(I - \lambda R^N)^{-1} R^N y + \sum_{k=1}^{N} (I - \lambda_k R)^{-1} z_k\Big] + \sum_{k=0}^{N-1} d_k S^k (I - \lambda_0 R)^{-1} z_0,$$

where

R is the algebraic integral corresponding to the algebraic derivative D,

$$R_A = -D^{N-1} - \frac{1}{N} \sum_{k=0}^{N-1} \sum_{j=1}^{N} \varepsilon^{-kj} S^k [c_{N+j-1} D^{N-2} + \dots + c_{N+j-1} \dots c_{j+1} I][b(S)]^{-1},$$

$z_0, z_1, \dots, z_{N-1} \in Z_D$ are arbitrary,

$$\lambda = \prod_{m=1}^{N} c_m; \qquad c_{N+m} = c_m = \frac{a(\varepsilon^m)}{b(\varepsilon^m)} \quad \text{for } m = 1, 2, \dots, N,$$

$$d_k = \sum_{m=1}^{N} \lambda_0^{-m} c_1 \dots c_m V_{k,m} \quad (k = 0, 1, \dots, N-1),$$

$$\lambda_k = \sqrt[N]{|\lambda|}\, e^{(2\pi k + \varphi)i/N} = \lambda_0 \varepsilon^k \quad (k = 0, 1, \dots, N-1), \quad \varphi = \operatorname{Arg}\lambda \ (0 \leqslant \varphi < 2\pi).$$

$V_{k,m}$ is the subdeterminant obtained by cancelling the $(k+1)$-column and the mth row of the Vandermonde determinant of the numbers $\varepsilon^2, \varepsilon^3, \dots, \varepsilon^N, \varepsilon$ ($k = 0, 1, \dots, N-1$; $m = 1, 2, \dots, N$).

For $N = 2$ we obtain, similarly, from Theorems III.1.3 and 2.10:

THEOREM 3.6. *Let $A = (a_0 I + b_0 S) + (a_1 I + b_1 S)D$, where S is an involution acting in a linear space X and D is an algebraic derivative acting in X and anti-*

commuting with S. If $a_0^2 - b_0^2 \neq a_1^2 - b_1^2$, then any solution of the equation $Ax = y$ is of the form

$$x = R_A[(I-\lambda R^2)^{-1}R^2 y + (I-\sqrt{\lambda}R)^{-1}z_1 + (I+\sqrt{\lambda}R)z_2] + [(a_0 - a_1\sqrt{\lambda})I - \\ -(b_0+b_1\sqrt{\lambda})S](I-\lambda_0 R)^{-1}z_0,$$

where $z_0, z_1, z_2 \in Z_D$, $R_A = -(a_1^2-b_1^2)^{-1}[(a_0 I - b_0 S)-(a_1 I + b_1 S)D]$ and R is the algebraic integral corresponding to the algebraic derivative D.

In the same way we obtain solutions when we drop the assumption $a(\varepsilon^j) \neq 0 \neq b(\varepsilon^m)$ for $m = 1, 2, \ldots, N$ (compare Chapter III).

EXAMPLE 3.3. Let $X = \left\{x(t) = \sum_{k=0}^{\infty} a_k t^k : \sum_{k=0}^{\infty} |a_k| < +\infty\right\}$, where t is either a real or a complex variable. Let $(Dx)(t) = \sum_{k=1}^{\infty} k a_k t^{k-1}$, where $D_D = \left\{x \in X : \sum_{k=0}^{\infty} |a_k| < +\infty \text{ and } \sum_{k=1}^{\infty} k|a_k| + \infty\right\}$, and let $(Rx)(t) = \sum_{k=0}^{\infty} \frac{a_k}{k+1} t^{k+1}$. It is easy to check that D is an algebraic derivative with the integral R. Let $(Sx)(t) = \sum_{k=0}^{\infty} a_k \varepsilon^k t^k$, where $\varepsilon = e^{2\pi i/N}$. Since $(S^N x)(t) = \sum_{k=0}^{\infty} a_k \varepsilon^{Nk} t^k = x(t)$, S is an involution of order N on X. Moreover,

$$(SDx)(t) = \sum_{k=1}^{\infty} k a_k \varepsilon^k t^{k-1} = \sum_{k=1}^{\infty} k a_k \varepsilon^{k-1} t^{k-1} = (DSx)(t).$$

Hence D permutes S, which enables us to apply the last two theorems.

For applications see also: Chapter VII, Ordinary Differential Equations with Reflection and Rotation of Argument.

4. ALGEBRAIC DERIVATIVE AND DIFFERENTIAL EQUATIONS IN A LINEAR RING

The purpose of this section is to solve linear differential equations with coefficients in a linear ring. This will give a formula for the solutions of differential equations with variable coefficients.

Let X be a commutative linear ring (over complex scalars [1]). We modify the definition of an algebraic derivative in an obvious way:

[1] Without any change in the definitions and proofs we can consider an arbitrary field of scalars as long as we do not consider scalar coefficients. If we include the case of scalar coefficients, it is necessary to assume that the field of scalar is algebraically closed.

If for a linear operator D acting in X there is a linear operator R acting in X and such that

1° $RX \subset D_D$ and $D_R = X$,

2° $DR = I$,

3° the operator $I+Rp$ is invertible for any $p \in X$, where by Rp we always mean the superposition of the operator R and the operator of multiplication by an element $p \in X$,

4° $D(xy) = xDy+yDx$ for all $x, y \in X$,

then D is said to be an *algebraic derivative* on X and R is the corresponding *algebraic integral.*

In particular, if the ring X has a unit e, then condition 3° implies that the operator $I-\lambda R$ is invertible for every scalar λ. Indeed, for every $x \in X$ we have $(I-\lambda R)x = x-R(\lambda x) = (I-Rp)x$, where $p = \lambda e$.

It is easy to check that the algebraic derivatives from Examples 1.1, 1.2, 1.3 are also algebraic derivatives in the sense of the last definition in the respective spaces considered as linear ring with respect to the usual multiplication of functions (see also Example 5.3 of Chapter I for the proof that the operator $I-R_p$ is invertible for every $p \in X$).

In Example 1.5 condition 4° is also satisfied, for

$$[D(xy)](t) = t\frac{dx(t)y(t)}{dt} = t\left[\frac{dx(t)}{dt}y(t)+\frac{dy(t)}{dt}x(t)\right]$$

$$= t\frac{dx(t)}{dt}y(t)+t\frac{dy(t)}{dt}x(t) = (Dx)(t)y(t)+(Dy)(t)x(t).$$

Arguing as before, we show that the operator $I-Rp$ is invertible for every $p \in X$.

We can also easily verify that the derivatives from Examples 1.4 and 1.6 do not satisfy condition 4°, and thus are not algebraic if we consider the respective spaces as linear rings. We now give the most classical example of an algebraic derivative in a linear ring, namely that considered by Mikusiński [6].

EXAMPLE 4.1. Let $L(0, +\infty)$ be the set of all Lebesgue integrable functions on the half-axis $(0, +\infty)$ with multiplication of two functions defined to be *convolution*[1]:

$$(x * y)(t) = \int_0^t x(t-s)y(s)ds \quad \text{for } x, y \in L(0, +\infty).$$

It is well known that $L(0, +\infty)$ is a linear commutative ring with this multiplication. Let $(Dx)(t) = -tx(t)$ for $x \in L(0, +\infty)$. The linearity of D is immediate.

[1] Some properties of the convolution will be given in Chapter XV, Section 1.

Condition 4° is also easily verified. Indeed,

$$[D(x*y)](t) = -t\int_0^t x(t-s)y(s)ds = \int_0^t (-t+s)x(t-s)y(s)ds + \int_0^t x(t-s)(-s)y(s)ds$$
$$= [(Dx)*y](t) + [x*(Dy)](t).$$

Let $X = \{x \in L(0, +\infty): x(t) \equiv 0 \text{ for } 0 \leqslant t < a, \text{ for some } a > 0\}$. This is also a commutative linear ring. Let $(Rx)(t) = -\int_0^t \frac{x(s)}{s} ds$ for $x \in X$. Arguing as in Example 1.5 we can show that D is an algebraic derivative on X with the algebraic algebraic integral R.

PROPOSITION 4.1. *If X has a unit, then $D(\lambda e) = 0$ for every scalar λ.*

Indeed, for all $x \in \boldsymbol{D}_D$ we have $Dx = D(ex) = (De)x + eDx = (De)x + Dx$. Hence $(De)x = 0$ and the arbitrariness of x now implies that $De = 0$. Therefore $D(\lambda e) = \lambda De = 0$ for all λ.

PROPOSITION 4.2. *If X has a unit, then $Dx^n = nx^{n-1}Dx$ for $n = 1, 2, \ldots$*

This follows from simple induction.

PROPOSITION 4.3. *If X has a unit, then there is a $g \in X$ such that $Dg = e$, namely $g = Re$.*

Indeed, from the definition $Dg = DRe = e$.

The following property, proposition, lemma and theorems were obtained by Mażbic-Kulma [2] and [3].

PROPERTY 4.4. *$(D+pI)(I+Rp)^{-1} = D$ for every $p \in X$.*

Indeed, since

$$D+pI = D+Ip = D+DRp = D(I+Rp) \tag{4.1}$$

we find

$$(D+pI)(I+Rp)^{-1} = D(I+Rp)(I+Rp)^{-1} = D.$$

PROPOSITION 4.5. *Every solution of the equation*

$$(D+pI)x = y, \quad \text{where } p, y \in X \text{ are given,} \tag{4.2}$$

is of the form

$$x = (I+Rp)^{-1}(Ry+z), \quad \text{where } z \in Z_D. \tag{4.3}$$

Indeed, if x is of the form (4.3), then by Property 4.4

$$(D+pI)x = (D+pI)(I+Rp)^{-1}(Ry+z) = D(Ry+z) = DRy+Dz = y,$$

since $Dz = 0$. If x and $\tilde{x}$ are two solutions of equation (4.2), then $\tilde{z} = x - \tilde{x}$ is a solution of the equation $(D+pI)\tilde{z} = 0$. Hence from equation (4.1) we have $D(I+Rp)\tilde{z} = 0$. Hence $(I+Rp)\tilde{z} = z \in Z_D$, which implies that $x - \tilde{x} = \tilde{z} = (I+Rp)^{-1}z$, where $z \in Z_D$. This shows that all solutions of equation (4.2) are of the form (4.3).

To solve an equation of order n we adopt the usual notation, for convenience:

(4.4) $Dp = p', \quad D^2 p = p'', \quad \ldots, \quad D^n p = p^{(n)}$ for $n = 0, 1, 2, \ldots$ and $p \in \mathbf{D}_D$,

and we prove the following

LEMMA 4.1. *For arbitrary $a, x \in \mathbf{D}_{D^k}$ and for all positive integers k*

$$aD^k = \sum_{j=0}^{k} (-1)^j \binom{k}{j} D^{k-j}(a^{(j)}x). \tag{4.5}$$

PROOF by induction. For $k = 1$ Formula (4.5) holds, because $aDx = D(ax) - a'x$. We now assume that (4.5) is true for an arbitrary k and we show that

$$aD^{k+1}x = \sum_{j=0}^{k+1} (-1)^j \binom{k+1}{j} D^{k+1-j}(a^{(j)}x).$$

Since

$$aD^{k+1}x = aD(D^k x) = D(aD^k x) - a'D^k x; \quad a'D^k x = \sum_{j=0}^{k} (-1)^j \binom{k}{j} D^{k-j}(a^{(j+1)}x),$$

the induction hypothesis implies that

$$\begin{aligned} aD^{k+1} &= D\Big[\sum_{j=0}^{k} (-1)^j \binom{k}{j} D^{k-j}(a^{(j)}x)\Big] - \sum_{j=0}^{k} (-1)^j \binom{k}{j} D^{k-j}(a^{(j+1)}x) \\ &= \sum_{j=0}^{k} (-1)^j \binom{k}{j} D^{k+1-j}(a^{(j)}x) + \sum_{j=0}^{k} (-1)^{j+1} \binom{k}{j} D^{k-j}(a^{(j+1)}x) \\ &= \sum_{j=0}^{k} (-1)^j \binom{k}{j} D^{k+1-j}(a^{(j)}x) + \sum_{j=1}^{k+1} (-1)^j \binom{k}{j-1} D^{k+1-j}(a^{(j)}x) \\ &= D^{k+1}(ax) + \sum_{j=1}^{k} (-1)^j \Big[\binom{k}{j} + \binom{k}{j-1}\Big] D^{k+1-j}(a^{(j)}x) + (-1)^{j+1} a^{(k+1)}x \\ &= \sum_{j=0}^{k+1} (-1)^j \binom{k+1}{j} D^{k+1-j}(a^{(j)}x), \end{aligned}$$

which was to be proved.

THEOREM 4.1. *The general form of the solutions of the equation*

$$D^n x + p_1 D^{n-1} x + \dots + p_n x = y, \quad \text{where } y \in X, p_j \in X, \tag{4.6}$$

is

$$x = \Big(I + \sum_{j=1}^{n} R^j \tilde{p}_j\Big)^{-1} \Big(R^n y + \sum_{m=0}^{n-1} R^m z_m\Big), \quad \text{where } z_0, \dots, z_{n-1} \in Z_D \text{ are arbitrary,}$$

provided that

(a) $p_j \in D_{D^{n-j}}$ *for* $j = 1, 2, \dots, n$,

(b) *the operator* $I + \sum_{j=1}^{n} R^j \tilde{p}_j$ *is invertible, where*

$$\tilde{p}_j = \sum_{m=0}^{j-1} (-1)^m \binom{n+m-j}{m} p_{j-m}^{m} \quad (j = 1, 2, \dots, n). \tag{4.7}$$

PROOF. Applying Lemma 4.1 to the left-hand side of equation (4.6), we obtain

$$D^n x + \sum_{i=0}^{n-1} (-1)^i \binom{n-1}{i} D^{n-1-i}(p_1^{(i)} x) + \sum_{i=0}^{n-2} (-1)^i \binom{n-2}{i} D^{n-2-i}(p_2^{(i)} x) +$$
$$+ \dots + \sum_{i=0}^{1} (-1)^i \binom{1}{i} D^{1-i}(p_{n-1}^{(i)} x + p_n x) = y.$$

If we arrange the left-hand side of this equation in descending powers of D, we obtain

$$D^n x + D^{n-1}\binom{n-1}{0} p_1 x + D^{n-2}\binom{n-2}{0} p_2 x - \binom{n-1}{1} p_1' x + \dots +$$
$$+ p_n x + (-1)^1 p_{n-1}' x + \dots + (-1)^{n-1} p_1^{(n-1)}$$
$$= D^n x + D^n R(p_1 x) + \dots + D^n R^n p_n x + \dots + (-1)^{n-1} p^{(n-1)} x.$$

Using the notation of (4.7), we obtain equation (4.6) in the following form

$$D^n \Big(I + \sum_{j=1}^{n} R^j \tilde{p}_j\Big) x = y.$$

Hence Theorem 2.7 implies that

$$\Big(I + \sum_{j=1}^{n} R^j \tilde{p}_j\Big) x = R^n y + \sum_{m=0}^{n-1} R^m z_m, \quad \text{where } z_0, \dots, z_{n-1} \in Z_D \text{ are arbitrary.}$$

From assumption (b) we now conclude that the required formula holds.

EXAMPLE 4.2. Consider the space $C[0, 1]$ and the following ordinary differential equation:

$$x'' + px + qx = y, \quad \text{where } p, q, p', y \in C[0, 1].$$

Since in this case Z_D is the space of all constants, we have from Theorem 4.1

$$x = [1+Rp+R^2(q-p')]^{-1}[R^2y+c_1t+c_2], \quad \text{where } c_1, c_2 \text{ are arbitrary constants,}$$

and $(Rx)(t) = \int_0^t x(s)ds$. Since $(R^2u)(t) = \int_0^t\int_0^s u(\tau)d\tau = \int_0^t (t-s)x(s)ds$, we find solving the corresponding Volterra equation (compare Section 5 of Chapter I),

$$x(t) = \int_0^t (t-s)y(s)ds+c_1t+c_2+\int_0^t \mathfrak{N}(t,s)\Big[\int_0^s (s-u)y(u)du+c_1s+c_2\Big]ds, \tag{4.8}$$

where c_1, c_2 are arbitrary constants and $\mathfrak{N}(t,s) = \sum_{n=0}^{\infty} N_n(t,s)$, $N_0(t,s) = p(s)+(t-s)[q(s)-$ $-p'(s)]$, $N_n(t,s) = \int_s^t N_0(t,u)N_{n-1}(u,s)du$ $(n = 1, 2, \ldots)$. It is easy to check that $x(t)$ as defined by Formula (4.8) is twice continuously differentiable.

We note here that in a linear ring (with non-trivial multiplication) no algebraic derivative satysfies Properties 1.5, 1.7, 1.8, as follows from

THEOREM 4.2. *Let X be a commutative linear ring with non-trivial multiplication. If D is an algebraic derivative in X, then D^2 does not satisfy condition 4°, and hence it is not an algebraic derivative on X.*

PROOF. By assumption there are $x, y \in X$ such that $xy \neq 0$. Suppose that both D and D^2 are algebraic derivatives on X. Then for all $x, y \in X$

$$D^2(xy) = xD^2y+yD^2x \quad \text{and} \quad D(xy) = xDy+yDx.$$

Hence

$$D^2(xy) = D(D(xy)) = D[xDy+yDx] = xD^2y+(Dx)(Dy)+(Dy)(Dx)+yD^2x$$
$$= xD^2y+yD^2x+2(Dx)(Dy) = D^2(xy)-2(Dx)(Dy).$$

This implies that

$$(Dx)(Dy) = 0 \quad \text{for all } x, y \in X. \tag{4.9}$$

Let u and v be arbitrary elements of X and let $u_1 = Ru$ and $v_1 = Rv$. Formula (4.9) implies that

$$uv = (DRu)(DRv) = (Du_1)(Dv_1) = 0 \quad \text{for all } u, v \in X,$$

which contradicts our assumption. Hence D^2 is not an algebraic derivative in X.

In the same way, we can show by simple induction, that in a linear ring no power of an algebraic derivative is itself an algebraic derivative.

5. DIFFERENTIAL EQUATIONS WITH ALGEBRAIC OPERATORS IN A LINEAR RING

For differential equations with algebraic operators in a linear ring we can prove similar theorems to these obtained in Section 3. The theorems presented here were obtained by Mażbic-Kulma [3].

Let D be an algebraic derivative in a linear ring over complex scalars[1] with the algebraic integral R. Let S be an algebraic operator defined on X, and such that

1° $DS-SD = 0$,

2° the characteristic polynomial $P(t) = p_0+p_1t+ \dots +p_Nt^N$ $(p_N = 1)$ has simple roots only, i.e. $P(t) = \prod_{k=1}^{N} (t-t_k)$.

THEOREM 5.1. *Let* $a_m(t) = \sum_{j=0}^{N-1} a_{mj}t^j$, *where* $a_{mj} \in X$ *and* $(Sa_{mj}-a_{mj}S)x = 0$ *for all* $x \in X$ $(m = 0, 1, \dots, n;\ j = 0, \dots, N-1)$. *Then the general form of the solutions of the equation*

$$(5.1)\qquad Ax \equiv \sum_{m=0}^{M} a_m(S)D^m x = y, \quad y \in X$$

(where S *satisfies conditions* 1° *and* 2°*) is*

$$(5.2)\qquad x = \sum_{k=1}^{N}\Big[\sum_{j=0}^{M} R^j\tilde{b}_{kj}\Big]^{-1}[R^M\tilde{y}_k+z_k],$$

where

(1) $$z_k = Z_{Q_k(D)}, \qquad \tilde{y} = [a_{\mu_k}(t_k)]^{-1}\Big[\prod_{j=1,\, j\neq k}^{N} (t_k-t_j)^{-1}(S-t_jI)\Big]y,$$

(2) $$Q_k(t) = \frac{1}{a_{\mu_k}(t_k)}\sum_{m=\mu_k}^{M} a_m(t_k)t^m,$$

(3) μ_k *is the least index such that for a given fixed* k $a_m(t_k)$ *is invertible* $(m = 0, 1, \dots, M;\ k = 1, 2, \dots, N)$,

(4) $$\tilde{b}_{kj} = \sum_{m=0}^{j}(-1)^m\binom{n+j+m}{m}b^{(m)}_{k,n-j+m}, \qquad b_{mk} = a_k(t_m)$$
$$(m = 1, 2, \dots, M; \quad j = 0, 1, \dots, N),$$

(5) *operators* $\sum_{j=0}^{M} R^j\tilde{b}_{kj}$ *are invertible for* $k = 0, 1, \dots, N$.

---

[1] As before, we can consider a linear space over any algebraically closed field.

The proof follows the same lines as the proof of Theorem 3.1, since $(Sa_{kj}-a_{kj}S)x = 0$ for all x and $k = 0, 1, \ldots, N$; $j = 0, \ldots, M$. We merely use Theorem 4.1 instead of Theorem 2.10.

REMARK 5.1. The condition $(Sa_{kj}-a_{kj}S)x = 0$ is satisfied in particular if S is multiplicative and all the a_{kj} are invariant with respect to S, i.e. if $S(xy) = (Sx)(Sy)$ for all $x, y \in X$ and if $Sa_{kj} = a_{kj}$. Indeed, if these conditions are satisfied, then for all $x \in X$

$$\begin{aligned}(Sa_{kj}-a_{kj}S)x &= S(a_{kj}x)-a_{kj}Sx = (Sa_{kj})(Sx)-a_{kj}(Sx) \\ &= (Sa_{kj}-a_{kj})(Sx) = 0.\end{aligned}$$

Suppose that in equation (5.1) we have $N = 0$. This means, that we consider an equation of the form

$$\Big(\sum_{m=0}^{N-1} a_j S^j\Big)x = y, \quad y \in X. \tag{5.3}$$

We can solve this by the method of Theorem 5.3 of Chapter II, to obtain the following

PROPOSITION 5.1. *If $a_j \in X$, $(Sa_j-a_jS)x = 0$ for all $x \in X$ $(j = 0, 1, \ldots, N-1)$, and the $a(t_m)$ are invertible for $m = 1, 2, \ldots, N$, where $a(t) = \sum_{j=1}^{N-1} a_j t^j$, then equation* (5.3) *has the unique solution*

$$x = \sum_{m=0}^{N-1} [a(t_m)]^{-1} y_{(m)}, \quad \text{where } y_{(m)} = \Big[\prod_{j=1, j\neq m}^{N} (t_m-t_j)^{-1}(S-t_jI)\Big]y.$$

A theorem analogous to Theorem 5.1 also holds for multi-involutions.

CHAPTER V

Periodic Solutions of Differential-Difference Equations

In this chapter, in order to make clear what is meant by "a solution of a differential equation" we will adopt the following convention although this is not essential for our methods:

A function $x(t)$ determined for all real t is called a *solution of a differential equation* of order p if $x(t)$ possesses a $(p-1)$th derivative which is absolutely continuous and if $x(t)$ satisfies the given equation almost everywhere. We will also assume that all given functions are measurable and locally integrable.

A *differential-difference equation* is an equation such that together with the unknown function $x(t)$ and its derivatives the terms $x(t-h_1), \ldots, x(t-h_m)$ and their derivatives also appear. We consider only constant *deviations* $h_1, \ldots, h_m$. If all $h_1, \ldots, h_m > 0$, we say that the argument is *delayed*, if all $h_1, \ldots, h_m < 0$, it is *advanced*. In other cases we say that the equation is of the *neutral type*.

Differential-difference equations appear in various problems of engineering, physics and economics. An example of a physical system with delay is shown in Fig. 5.

Since we are interested here in periodic solutions, no initial conditions will be postulated.

1. ORDINARY LINEAR DIFFERENTIAL-DIFFERENCE EQUATIONS WITH CONSTANT COEFFICIENTS

Consider the following ordinary linear-differential-difference equation with constant coefficients:

$$Ax \equiv \sum_{k=0}^{n} \sum_{j=0}^{m} a_{kj} x^{(k)}(t-\omega_j) = y(t), \tag{1.1}$$

where $\omega_0 = 0, \omega_1, \ldots, \omega_m$ are real constants, the complex valued function $y(t)$ defined on the real line is periodic with period ω_{m+1} and $x^{(k)}$ denotes the kth derivative of the function $x(t)$: $x^{(k)} = d^k x/dt^k$.

If a complex-valued or real-valued function $y(t)$ is periodic with period p, we call it briefly a *p-periodic function.*

Suppose that all the numbers $\omega_1, \ldots, \omega_{m+1}$ are *commensurable*. This means that there are integers $\eta_1, \ldots, \eta_{m+1}$ and a real $r \neq 0$ such that

$$\omega_j = \eta_j r \quad \text{for } j = 1, 2, \ldots, m+1.$$

For the sake of convenience we assume additionally that $\eta_0 = 0$, so that $\omega_0 = \eta_0 r$. Let N be a common multiple of the positive integers $|\eta_1|, \ldots, |\eta_{m+1}|$, not necessarily the least one, and let

$$\omega = Nr.$$

Consider the set E_ω of all complex valued ω-periodic functions defined on the real line. It is easy to check that E_ω is a linear space (over the field of complex scalars). In the space E_ω we study the *shift operator*:

$$(Sx)(t) = x(t-r).$$

The operator S is obviously linear and transforms the space E_ω onto itself. Moreover, it is an involution of order N on E_ω. Indeed, by simple induction we have

$$(S^k x)(t) = x(t-kr) \quad (k = 0, 1, 2, \ldots). \tag{1.2}$$

Hence, for any function $x \in E_\omega$

$$(S^k x)(t) = x(t-Nr) = x(t-\omega) = x(t)$$

and

$$S^N = I \quad \text{on} \quad E_\omega, \tag{1.3}$$

where N is the least integer satisfying (1.3).

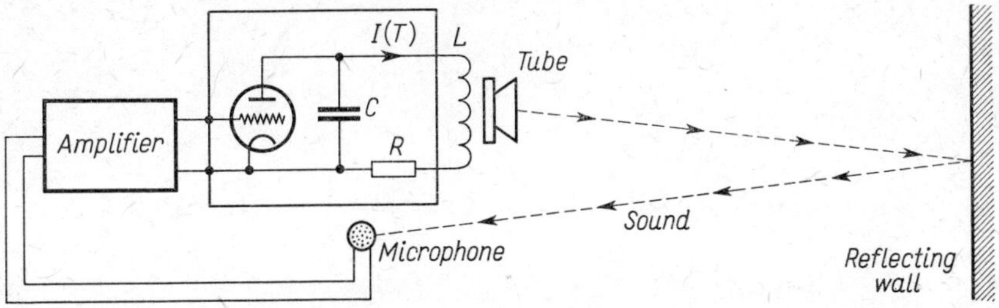

Fig. 5. Example of a physical system with delay.
The system is described by the Minorski [1] equation:

$$LCI''(T)+RCI'(T)+I(T)+AkLI'\left(T-\frac{2\delta}{c}\right) = Bk^3LI'^3\left(T-\frac{2\delta}{c}\right),$$

where R — resistance, L — induction, C — capacity, $I(T)$ — current, T — time, c — velocity of sound, δ — distance from reflecting wall, k — amplification constant, A, B, — constants characterizing tube

Formula (1.2) also holds for negative integers. Indeed, for $k = 1, \ldots, N-1$ we have $N-k > 0$ and

$$(S^{-k}x)(t) = (S^{N-k}x)(t) = x\big(t-(N-k)r\big) = x(t-Nr+kr) = x(t-\omega+kr)$$
$$= x(t+kr) = x\big(t-(-k)r\big).$$

Observe now that S, being a shift operator, commutes with derivation. Indeed,

$$S\left[\frac{d}{dt}x(t)\right] = x'(t-r) = \left[\frac{d}{ds}x(s)\right]_{s=t-r} = \frac{d}{dt}x(t-r) = \frac{d}{dt}(Sx)(t).$$

Hence, by Theorem II.6.2, we can determine all solutions of equation (1.1) which belong to the space E_ω, i.e. all ω-periodic solutions of equation (1.1). In fact, the space E_ω is the direct sum of the spaces $E_{(j)} = P_j E_\omega$, where

$$x_{(j)}(t) = (P_j x)(t) = \frac{1}{N}\sum_{k=0}^{N-1} \varepsilon^{-kj}(S^k x)(t-kr) \quad (j = 1, 2, \ldots, N)$$

and $\varepsilon = e^{2\pi i/N}$.

Using this notation and the fact that S commutes with derivation, we can rewrite the operator A defined in (1.1) in the following form:

$$(Ax)(t) = \sum_{k=0}^{n}\sum_{j=0}^{m} a_{kj}x^{(k)}(t-\omega_j) = \sum_{k=0}^{n}\sum_{j=0}^{m} a_{kj}x^{(k)}(t-\eta_j r)$$
$$= \sum_{k=0}^{n}\sum_{j=0}^{m} a_{kj}(S^{\eta_j}x^{(k)})(t) = \sum_{k=0}^{n}\sum_{j=0}^{m} a_{kj}\Big[S^{\eta_j}\sum_{\nu=1}^{N} x^{(k)}_{(\nu)}\Big](t)$$
$$= \sum_{k=0}^{n}\sum_{j=0}^{m} a_{kj}\Big[\sum_{\nu=1}^{N} \varepsilon^{\nu\eta_j}x^{(k)}_{(\nu)}(t)\Big] = \sum_{\nu=1}^{N}\Big[\sum_{k=0}^{n}\sum_{j=0}^{m} a_{kj}\varepsilon^{\nu\eta_j}\Big]x^{(k)}_{(\nu)}(t)$$
$$= \sum_{\nu=1}^{N}\Big[\sum_{k=0}^{n}\sum_{j=0}^{m} a_{kj}\varepsilon^{\nu\eta_j}\Big](P_\nu x^{(k)})(t)$$

since (Formula II.(4.1)) $x^{(k)} = \sum_{\nu=1}^{N} x^{(k)}_{(\nu)}$ $(k = 0, 1, \ldots, n;\ \nu = 1, 2, \ldots, N)$ and (Formula II.(6.3))

$$S^k x_{(\nu)} = \varepsilon^{k\nu}x_{(\nu)} \quad (k = 0, \pm 1, \ldots;\ \nu = 1, 2, \ldots, N).$$

If we now write

$$(1.4) \qquad b_{k\nu} = \sum_{j=0}^{m} a_{kj}\varepsilon^{\nu\eta_j} \quad (k = 0, 1, \ldots, n;\ \nu = 1, 2, \ldots, N)$$

and

(1.5) $$A_\nu u = \sum_{k=0}^{n} b_{k\nu} u^{(k)} \qquad (\nu = 1, 2, \ldots, N),$$

then the operator A appears in the following form

(1.6) $$A = \sum_{\nu=1}^{N} A_\nu P_\nu$$

and

$$Ax = \sum_{\nu=1}^{N} A_\nu x_{(\nu)} \qquad \text{for all } x \in E_\omega.$$

Each of the operators A_ν commutes with S. Hence, by Theorem II.6.2, the equation $Ax = y$ is equivalent to the N independent equations:

(1.7) $$A_\nu x_{(\nu)} = y_{(\nu)} \qquad (\nu = 1, 2, \ldots, N),$$

where $y_{(\nu)}$ is a well-known function:

$$y_{(\nu)}(t) = (P_\nu y)(t) = \frac{1}{N} \sum_{k=0}^{N-1} \varepsilon^{-k\nu} y(t-kr) \qquad (\nu = 1, 2, \ldots, N).$$

Now every equation of system (1.7) is an ordinary differential equation with constant coefficients *without shifts* of the argument. It is therefore easy to check whether or n there exists an ω-periodic solution of the νth equation (1.7). Theorem II.6.2 implies that equation (1.1) has an ω-periodic solution if and only if all equations (1.7) have ω-periodic solutions (including solutions identically equal to zero).

Suppose that every equation $A_\nu u = y_{(\nu)}$ $(\nu = 1, 2, \ldots, N)$ has an ω-periodic solution x_ν. Then

$$x = \sum_{\nu=1}^{N} x_\nu$$

is an ω-periodic solution of equation (1.1) (Theorem II.6.2). Finally we obtain the following

THEOREM 1.1. *Let the complex valued function $y(t)$ defined on the real line be ω_{m+1}-periodic, where ω_{m+1} is commensurable with the commensurable reals $\omega_1, \ldots$ $\ldots, \omega_m$. Then equation* (1.1)

$$\sum_{k=0}^{n} \sum_{j=0}^{m} a_{kj} x^{(k)}(t-\omega_j) = y(t), \qquad \omega_0 = 0$$

has ω-periodic solutions if and only if all the ordinary differential equations

(1.8) $$\sum_{k=0}^{n} b_{k\nu} x_\nu^{(k)}(t) = y_{(\nu)}(t) \qquad (\nu = 1, 2, \ldots, N)$$

have ω-periodic solutions, where

$$y_{(\nu)}(t) = \frac{1}{N}\sum_{k=0}^{N-1} \varepsilon^{-k\nu} y(t-kr)$$

$$b_{k\nu} = \sum_{j=0}^{m} a_{kj}\,\varepsilon^{\nu\eta_j} \qquad (\nu = 1, 2, \ldots, N;\ k = 0, 1, \ldots, n),$$

$$\varepsilon = e^{2\pi i/N}, \quad \omega_j = \eta_j r \quad \textit{for } j = 1, 2, \ldots, m+1,\ \eta_0 = 0,\ r \neq 0,$$

η_j are integer and N is a common multiple of the positive integers $|\eta_1|, \ldots, |\eta_{m+1}|$ (not necessarily the least one). The solutions of (1.1) *are of the form*

$$x(t) = \sum_{\nu=1}^{N} x_\nu(t),$$

where x_ν is an ω-periodic solution of the νth equation (1.8).

REMARK 1.1. If the given function $y(t)$ is identically equal to zero, i.e. if equation (1.1) is homogeneous, then $y(t)$ can be regarded as a periodic function with an arbitrary period. The easiest way to see this is to take for N a common multiple of the numbers $\eta_1, \ldots, \eta_m$, since we can put $\omega_{m+1} = \omega_j$ for an arbitrarily fixed j ($j = 1, 2, \ldots, m$). On the other hand the homogeneous equation (1.1) always has an ω-periodic solution, which may be identically equal to zero.

REMARK 1.2. Theorem 1.1 is also true without any essential change in the proof if $x(t)$ and $y(t)$ are vector valued functions. To fact, we can assume that these functions defined on the real line take values in any Banach space (over complex scalars).

REMARK 1.3. Suppose that the numbers $\omega_1, \ldots, \omega_m$ in equation (1.1) are as before. We drop the assumption of the periodicity of the function $y(t)$. Suppose now that this function is defined on a finite interval, for instance in $[0, T]$. So that we are now looking for solutions of equation (1.1) in the interval $[0, T]$. We can use the previously described method in the following manner: Let us choose N sufficiently large that $\omega = Nr \geqslant T$. We extend $y(t)$ to the interval $[0, \omega]$ in an arbitrary way and we define

$$\tilde{y}(t) = y_\omega(t) \quad \text{for } t \in [0, \omega],$$
$$\tilde{y}(t+k\omega) = \tilde{y}(t) \quad \text{for } t \in [0, \omega] \text{ and } k = \pm 1, \pm 2, \ldots,$$

where $y_\omega(t)$ denotes the extension of the function $y(t)$ to the interval $[0, \omega]$. Furthermore, we find ω-periodic solutions $x(t)$ of the equation $A\tilde{x} = \tilde{y}$, where the operator A is defined using of the left-hand side of equation (1.1). The function $x = \tilde{x}|_{[0,T]}$, which is the restriction of $\tilde{x}$ to the interval $[0, T]$, is a solution of equation (1.1) (Fig. 6). Unfortunately this method imposes restrictions on the set of initial functions.

We now give some examples.

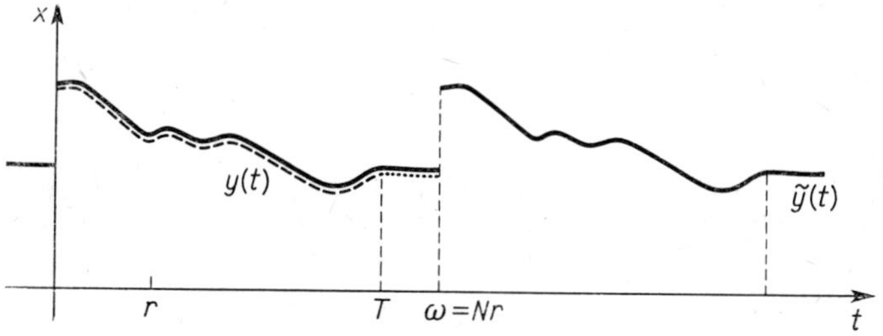

Fig. 6. Solutions of equation (1.1) on a finite interval

EXAMPLE 1.1. Consider the equation

$$ax(t)+bx(t-\pi)+cx''(t)+dx''(t-\pi) = 0 \tag{1.9}$$

with arbitrary real coefficients $a, b, c \neq 0$, $d \neq 0$. We are looking for 2π-periodic solutions of this equation. In our case $r = \pi$, $N = 2$, $\varepsilon = e^{2\pi i/2} = -1$ and

$$b_{01} = a-b, \quad b_{02} = a+b, \quad b_{11} = b_{12} = 0, \quad b_{21} = c-d, \quad b_{22} = c+d.$$

Equations (1.8) will be of the form:

$$(a-b)x_1+(c-d)x_1'' = 0, \tag{1.10}$$

$$(a+b)x_2+(c+d)x_2'' = 0. \tag{1.11}$$

It is easy to see that 2π-periodic solutions of equations (1.10) and (1.11) exist if and only if there are integers p and q such that

$$a-b = p^2(c-d), \quad a+b = q^2(c+d), \tag{1.12}$$

i.e. such that

$$a = c(p^2+q^2), \quad b = d(q^2-p^2). \tag{1.13}$$

Under these conditions we find the 2π-periodic solutions of equations (1.10) and (1.11):

$$x_1(t) = A\sin pt+B\cos pt, \quad x_2(t) = C\sin qt+D\cos qt,$$

where A, B, C, D are arbitrary reals. The 2π-periodic solutions of equation (1.9) will be of the form

$$x(t) = A\sin pt+B\cos pt+C\sin qt+D\cos qt,$$

where A, B, C, D are arbitrary real constants and p, q are integers satisfying condition (1.12).

EXAMPLE 1.2. Let the following difference equation be given:

$$\sum_{j=0}^{m} a_j x\left(t-\frac{2\pi}{T}j\right) = y(t), \tag{1.14}$$

where $y(t)$ is a $\frac{2\pi}{T}$-periodic vector-function and T is a given real. Let N be a common multiple of numbers $1, 2, \ldots, m$, for instance let $N = m!$. We are looking for the ω-periodic solutions of equation (1.14), where $\omega = \frac{2}{T}N$.

It follows from Theorem 1.1 that (1.14) is equivalent to the system of N independent equations

$$b_\nu x_\nu(t) = y_{(\nu)}(t) \quad (\nu = 1, 2, \ldots, N), \tag{1.15}$$

where $b_\nu = \sum_{j=0}^{m} a_j \varepsilon^{\nu j}$, $\varepsilon = e^{2\pi i/N}$ and, since $y(t)$ is $\frac{2\pi}{T}$-periodic

$$y_{(\nu)}(t) = \frac{1}{N}\sum_{k=0}^{N-1} \varepsilon^{-k\nu} y\left(t - \frac{2\pi k}{T}\right) = \left[\frac{1}{N}\sum_{k=0}^{N-1} \varepsilon^{-k}\right] y(t) = \begin{cases} 0 & \text{for } \nu = 1, 2, \ldots, N-1, \\ y(t) & \text{for } \nu = N \end{cases}$$

(compare Section I of Chapter II). Theorem II.5.2 implies that a necessary and sufficient condition for system (1.15) to have a solution in the space of ω-periodic functions is that

$$y_{(\nu)} = 0 \quad \text{when } b_\nu = 0.$$

But $y_{(\nu)} = 0$ for $\nu = 1, 2, \ldots, N-1$. Hence a necessary and sufficient condition for equation (1.15) to have a solution is

$$\sum_{j=0}^{m} a_j \varepsilon^{\nu j} = 0 \quad \text{for } \nu = 1, 2, \ldots, N-1 \text{ and } a_1 + \ldots + a_N \neq 0. \tag{1.16}$$

Under this condition, by Theorem II.5.2 the solution is of the form

$$x(t) = \frac{1}{a_1 + \ldots + a_N} y(t) + \sum_{\nu=1}^{N-1} z_{(\nu)}(t),$$

where

$$z_{(\nu)}(t) = \frac{1}{N}\sum_{k=0}^{N-1} \varepsilon^{-k\nu} z\left(t - \frac{2\pi k}{T}\right),$$

and $z(t)$ is an arbitrary ω-periodic function.

It follows from our previous studies that the conditions on the numbers b_ν depend only on the period of the function $y(t)$.

2. ORDINARY LINEAR DIFFERENTIAL-DIFFERENCE EQUATIONS WITH PERIODIC COEFFICIENTS

In this section we consider the following ordinary linear differential-difference equation with periodic coefficients:

$$Ax \equiv \sum_{k=0}^{n}\sum_{j=0}^{m} a_{kj}(t) x^{(k)}(t - \omega_j) = y(t), \quad \omega_0 = 0. \tag{2.1}$$

All the notation of the preceding section continues unchanged.

If the coefficients $a_{kj}(t)$ are of period r:

$$a_{kj}(t+r) = a_{kj}(t) \quad (k = 0, 1, \ldots, n;\ j = 0, 1, \ldots, m), \tag{2.2}$$

then the problem of determinating the periodic solutions of equation (2.1) is very simple. In fact, the operator of multiplication by an r-periodic function $a(t)$ commutes with the shift operator defined by Formula (1.2):

$$[(Sa-aS)x](t) = a(t-r)x(t-r)-a(t)x(t-r) = [a(t)-a(t)]x(t-r) = 0.$$

Hence, by Theorem II.6.2, we obtain in the same way as in Section 1 the following

THEOREM 2.1. *Let a complex valued function $y(t)$ defined on the real line be ω_{m+1}-periodic, where ω_{m+1} is a real number commensurable with the commensurable real numbers $\omega_1, \ldots, \omega_m$. Then equation* (2.1) *with r-periodic coefficients $a_{kj}(t)$ has ω-periodic solutions if and only if all the ordinary differential equations*

$$\sum_{k=0}^{n} b_{k\nu}(t)x_{(k)}(t) = y_{(\nu)}(t) \quad (\nu = 1, 2, \ldots, N) \tag{2.3}$$

have ω-periodic solutions, where

$$y_{(\nu)}(t) = \frac{1}{N}\sum_{k=0}^{N-1} \varepsilon^{-k\nu}y(t-kr)$$

$$(\nu = 1, 2, \ldots, N;\ k = 0, 1, \ldots, n),$$

$$b_{k\nu}(t) = \sum_{j=0}^{m} \varepsilon^{\nu\eta_j}a_{kj}(t)$$

$$\varepsilon = e^{2\pi i/N}, \quad \omega_j = \eta_j r \quad \text{for } j = 0, 1, \ldots, m+1,\ \eta_0 = 0,\ r \neq 0,$$

η_j are integers, and N is a common multiple of the positive integers $|\eta_1|, \ldots, |\eta_{m+1}|$ (not necessarily the least one). The solutions of equation (2.1) are of the form

$$x(t) = \sum_{\nu=1}^{N} x_\nu(t),$$

where x_ν is an ω-periodic solution of the νth equation (2.3).

Of course, Remarks 1.1 and 1.2 are valid here also.

EXAMPLE 2.1. We examine the possible existence of 2π-periodic solutions of the equation

$$2x(t) - \tfrac{1}{2}\cos 2t x''(t) - \tfrac{1}{2}x''(t-\pi) = 0. \tag{2.4}$$

Here $r = \pi$, $N = 2$, $\varepsilon = e^{2\pi i/2} = -1$. The coefficient $\frac{1}{2}\cos 2t$ is a π-periodic functions as $\cos 2(t+\pi) = \cos(2t+2\pi) = \cos 2t$. We also find

$$b_{01} = b_{02} = 2, \quad b_{11} = b_{12} = 0,$$
$$b_{21} = -\tfrac{1}{2}\cos 2t + \tfrac{1}{2} = \sin^2 t,$$
$$b_{22}(t) = -\tfrac{1}{2}\cos 2t - \tfrac{1}{2} = -\cos^2 t.$$

Equations (2.3) are of the form

$$2x_1 - \cos^2 t x_1'' = 0, \tag{2.5}$$

$$2x_2 + \sin^2 t x_2'' = 0. \tag{2.6}$$

Equation (2.5) has a 2π-periodic solution $x_1(t) = \tan t$ and the 2π-periodic solution of equation (2.6) is $x_2(t) = \cot t$. Since both equations are homogeneous, the functions $A\tan t$ and $B\cot t$, where A and B are arbitrary reals, are also 2π-periodic solutions of equations (2.5) and (2.6) respectively.

It follows from Theorem 2.1 that equation (2.4) has 2π-periodic solutions of the form

$$x(t) = A\tan t + B\cot t,$$

where A and B are arbitrary reals.

When the coefficients of equation (2.1) are periodic with periods not equal to r (but still commensurable with r) we are faced with a completely different situation, because such coefficients do not commute with the operator S. We therefore solve this equation using another method.

Without loss of generality we can assume that all the $a_{kj}(t)$ are of period ω:

$$a_{kj}(t+\omega) = a_{kj}(t) \qquad (k = 0, 1, \ldots, n;\ j = 0, 1, \ldots, m)$$

since otherwise we only need to take N sufficiently large. In our further considerations we need the following

PROPOSITION 2.1. *A function $u(t)$ belongs to the space $E_{(\nu)} = P_\nu E_\omega$ (where ν is fixed, $\nu = 1, 2, \ldots, N$) if and only if*

$$u(t) = e^{-2\pi i\nu t/\omega} v(t), \tag{2.7}$$

where $v(t)$ is an r-periodic function.

Indeed if $u \in E_{(\nu)}$, then $u(t-r) = (Su)(t) = \varepsilon^\nu u(t)$ and

$$v(t-r) = e^{2\pi i\nu(t-r)/\omega} u(t-r) = e^{2\pi i\nu t/\omega} e^{-2\pi i\nu r/Nr} u(t-r)$$
$$= e^{2\pi i\nu t/\omega} \varepsilon^{-\nu} \varepsilon^\nu u(t) = e^{2\pi i\nu t/\omega} u(t) = v(t).$$

Conversely, if $v(t-r) = v(t)$, then

$$u(t-r) = e^{-2\pi i\nu(t-r)/\omega} v(t-r) = e^{-2\pi i\nu t/\omega} e^{2\pi i\nu r/Nr} v(t)$$
$$= e^{-2\pi i\nu t/\omega} \varepsilon^\nu v(t) = \varepsilon^\nu u(t)$$

which proves that $u \in E_{(\nu)}$.

Now write

$$\tilde{x}_\nu(t) = x_{(\nu)}(t)e^{2\pi i\nu t/\omega} \qquad (\nu = 1, 2, \ldots, N). \tag{2.8}$$

To simplify our calculations, we consider not equation (2.1) but a corresponding system of differential-difference equations of order one:

$$Ax \equiv \sum_{j=0}^{m} [a_{0j}(t)x(t-\omega_j)+a_{1j}(t)x'(t-\omega_j)] = y(t). \tag{2.9}$$

Here x, y denote n-dimensional vector valued functions, $x'(t) = (x_1'(t), \ldots, x_n'(t))$ and a_{0j}, a_{1j} are ω-periodic matrix valued functions. These square matrices are determined in a well-known manner by the coefficients of equation (2.1). Our basic space is the space E_ω^n which consists of all n-dimensional ω-periodic vector valued functions. As before, we introduce the shift operator $(Sx)(t) = x(t-r)$ and we decompose the space E_ω^n into the corresponding direct sum. Using Proposition 2.1, we find for every $x \in E_\omega^n$

$$x'(t) = \sum_{\nu=1}^{N} x'_{(\nu)}(t) = \sum_{\nu=1}^{N} \frac{d}{dt}[e^{-2\pi i\nu t/\omega}\tilde{x}_\nu(t)]$$

$$= \sum_{\nu=1}^{N} e^{-2\pi i\nu t/\omega}\left[x_\nu'(t) - \frac{2\pi i\nu}{\omega}\tilde{x}_\nu(t)\right],$$

where $\tilde{x}(t) = e^{2\pi i\nu t/\omega}x_\nu\ (t)$ are r-periodic vector valued functions. This implies that for μ chosen arbitrarily from among the integers $1, 2, \ldots, N$ we have

$$P_\mu Ax = \frac{1}{N}\sum_{k=0}^{N-1}\sum_{j=0}^{m} \varepsilon^{-\mu k}[a_{0j}(t-kr)x(t-kr-\omega_j)+a_{1j}(t-kr)x'(t-kr-\omega_j)]$$

$$= \frac{1}{N}\sum_{k=0}^{N-1}\sum_{j=0}^{m-1} \varepsilon^{-\mu k}[a_{0j}(t-kr)x(t-kr-\eta_j r)+a_{1j}(t-kr)x'(t-kr-\eta_j r)]$$

$$= \frac{1}{N}\sum_{k=0}^{N-1}\sum_{j=0}^{m} \varepsilon^{-\mu k}\sum_{\nu=1}^{N} e^{-2\pi i\nu(t-kr-\eta_j r)/\omega}\Big\{a_{0j}(t-kr)\tilde{x}_\nu(t-kr-\eta_j r)+$$

$$+a_{1j}(t-kr)\left[\tilde{x}_\nu'(t-kr-\eta_j r)-\frac{2\pi i\nu}{\omega}\tilde{x}_\nu(t-kr-\eta_j r)\right]\Big\}$$

$$= \frac{1}{N}\sum_{k=0}^{N-1}\sum_{j=0}^{m} \varepsilon^{-\mu k}\sum_{\nu=1}^{N} e^{2\pi i\nu(k+\eta_j)r/Nr}e^{-2\pi i\nu t/\omega}\Big\{\Big[a_{0j}(t-kr)-$$

$$-\frac{2\pi i\nu}{\omega}\boldsymbol{a}_{1j}(t-kr)\Big]\tilde{\boldsymbol{x}}_\nu(t)+\boldsymbol{a}_{1j}(t-kr)\tilde{\boldsymbol{x}}'_\nu(t)\Big\}$$

$$=\frac{1}{N}\sum_{k=0}^{N-1}\sum_{j=0}^{m}\sum_{\nu=1}^{N}\varepsilon^{-(\mu-\nu)k}\varepsilon^{\nu\eta_j}e^{-2\pi i\nu t/\omega}\Big\{\Big[\boldsymbol{a}_{0j}(t-kr)-$$

$$-\frac{2\pi i\nu}{\omega}\boldsymbol{a}_{1j}(t-kr)\Big]\tilde{\boldsymbol{x}}_\nu(t)+\boldsymbol{a}_{1j}(t-kr)\tilde{\boldsymbol{x}}'_\nu(t)\Big\}$$

$$=\sum_{j=0}^{m}\sum_{\nu=1}^{N}\varepsilon^{\nu\eta_j}e^{-2\pi i\nu t/\omega}\Big\{\frac{1}{N}\sum_{k=0}^{N-1}\Big[\boldsymbol{a}_{0j}(t-kr)-$$

$$-\frac{2\pi i\nu}{\omega}\boldsymbol{a}_{1j}(t-kr)\Big]\varepsilon^{-(\mu-\nu)k}\tilde{\boldsymbol{x}}_\nu(t)+\frac{1}{N}\sum_{k=0}^{N-1}\varepsilon^{-(\mu-\nu)k}\boldsymbol{a}_{1j}(t-kr)\tilde{\boldsymbol{x}}'_\nu(t)\Big\}$$

$$=\sum_{j=0}^{m}\sum_{\nu=1}^{N}\varepsilon^{\nu\eta_j}e^{-2\pi i\nu t/\omega}\Big\{\Big[P_{\mu-\nu}\Big(\boldsymbol{a}_{0j}-\frac{2\pi i\nu}{\omega}\boldsymbol{a}_{1j}\Big)\Big](t)\tilde{\boldsymbol{x}}_\nu(t)+(P_{\mu-\nu}\boldsymbol{a}_{1j})(t)\tilde{\boldsymbol{x}}'_\nu(t)\Big\},$$

where the indices $\mu-\nu$ of the projectors in the last expression have to be taken modulo N. Let

$$\text{(2.10)}\qquad \begin{aligned}\boldsymbol{d}_{0\nu}(t)&=\sum_{j=0}^{m}\varepsilon^{\nu\eta_j}\Big[\boldsymbol{a}_{0j}(t)-\frac{2\pi i\nu}{\omega}\boldsymbol{a}_{1j}(t)\Big]\\ \boldsymbol{d}_{1\nu}(t)&=\sum_{j=0}^{m}\varepsilon^{\nu\eta_j}\boldsymbol{a}_{1j}(t)\end{aligned}\qquad (\nu=1,2,\ldots,N).$$

We rewrite the operator $P_\mu A$ as follows:

$$P_\mu A\boldsymbol{x}=\sum_{\nu=1}^{N}e^{-2\pi i\nu t/\omega}\{[P_{\mu-\nu}\boldsymbol{d}_{0\nu}](t)\tilde{\boldsymbol{x}}_\nu(t)+[P_{\mu-\nu}\boldsymbol{d}_{1\nu}](t)\tilde{\boldsymbol{x}}'_\nu(t)\}.$$

Proposition 2.1 implies that the matrix valued functions

$$\boldsymbol{\alpha}_{p\nu\mu}(t)=e^{2\pi i(\mu-\nu)t/\omega}[P_{\mu-\nu}\boldsymbol{d}_{p\nu}](t)\qquad(p=0,1)$$

are r-periodic. This means that acting on both sides of equation (2.9) with the operator $e^{2\pi i\mu t/\omega}P_\mu$ and writing as in (2.8)

$$\tilde{\boldsymbol{y}}_\mu(t)=\boldsymbol{y}_{(\mu)}(t)e^{2\pi i\mu t/\omega}\qquad(\mu=1,2,\ldots,N)$$

we obtain the following system of N ordinary differential equations with N unknown r-periodic vector valued functions $\tilde{\boldsymbol{x}}_\nu$:

$$(2.11)\qquad \sum_{\nu=1}^{N} [\boldsymbol{\alpha}_{0\nu\mu}\tilde{\boldsymbol{x}}_\nu(t)+\boldsymbol{\alpha}_{1\nu\mu}\tilde{\boldsymbol{x}}_\nu'(t)] = \tilde{\boldsymbol{y}}_\mu(t) \qquad (\mu = 1, 2, \ldots, N).$$

The differential operator $A_{\mu\nu}$ defined by the formula

$$A_{\mu\nu}\boldsymbol{u} = \boldsymbol{\alpha}_{0\nu\mu}\boldsymbol{u}+\boldsymbol{\alpha}_{1\nu\mu}\boldsymbol{u}' \qquad (\mu, \nu = 1, 2, \ldots, N)$$

has r-periodic coefficients, whence it commutes with S and with $P_1, \ldots, P_N$. Moreover, if $\boldsymbol{u}$ is an r-periodic vector valued function, then the function $A_{\mu\nu}\boldsymbol{u}$ is also r-periodic, as well as the function $\tilde{\boldsymbol{y}}_\mu$. This shows that a solution of system (2.11), if it exists, is a system of Nr-periodic vector valued functions. By Formula (2.8) this solution is of the form

$$\boldsymbol{x}(t) = \sum_{\nu=1}^{N} e^{-2\pi i\nu t/\omega}\tilde{\boldsymbol{x}}_\nu(t),$$

where $(\tilde{\boldsymbol{x}}_1, \ldots, \tilde{\boldsymbol{x}}_N)$ is an r-periodic solution of system (2.11), provided such a solution exists. Finally we obtain

THEOREM 2.2. *Suppose that the n-dimensional complex vector valued function $\boldsymbol{y}(t)$ defined on the real line is ω_{m+1}-periodic, where ω_{m+1} is commensurable with the commensurable reals $\omega_1, \ldots, \omega_m$. Let $\omega_j = \eta_j r$ for $j = 0, 1, \ldots, m+1$ and $\eta_0 = 0$, $r \neq 0$. Let $\omega = Nr$, where N is a common multiple of the positive integers $|\eta_1|, \ldots, |\eta_{m+1}|$. Let the coefficients $\boldsymbol{a}_{0j}(t)$, $\boldsymbol{a}_{1j}(t)$ be ω-periodic square matrix valued functions of dimension $n\times n$*[1]*. Then equation (2.9) has ω-periodic solutions if and only if system (2.11) of N ordinary linear differential equations has an r-periodic solution, where*

$$\boldsymbol{\alpha}_{0\nu\mu}(t) = \frac{1}{N}e^{2\pi i(\mu-\nu)t/\omega}\sum_{k=0}^{N-1}\sum_{j=0}^{m}\varepsilon^{\nu\eta_j}\varepsilon^{-(\mu-\nu)k}\left[\boldsymbol{a}_{0j}(t-kr)-\frac{2\pi i\nu}{\omega}\boldsymbol{a}_1(t-kr)\right],$$

$$\boldsymbol{\alpha}_{1\nu\mu}(t) = \frac{1}{N}e^{2\pi i(\mu-\nu)t/\omega}\sum_{k=0}^{N-1}\sum_{j=0}^{m}\varepsilon^{\nu\eta_j}\varepsilon^{-(\mu-\nu)k}\boldsymbol{a}_{1j}(t-kr),$$

$$\tilde{\boldsymbol{y}}_\mu(t) = \frac{1}{N}e^{2\pi i\mu t/\omega}\sum_{k=0}^{N-1}\varepsilon^{-\mu k}\boldsymbol{y}(t-kr) \qquad (\mu, \nu = 1, 2, \ldots, N),\ \varepsilon = e^{2\pi i/N}.$$

---

[1] This may be assumed without any loss of generality if all the $\boldsymbol{a}_{0j}$, $\boldsymbol{a}_{1j}$ are periodic with periodics commensurable with $\omega_1, \ldots, \omega_{m+1}$, since we can always choose N large enough.

The solution is of the form

$$x(t) = \sum_{\nu=1}^{N} e^{-2\pi i\nu t/\omega}\tilde{x}_\nu(t),$$

where $(\tilde{x}_1, \ldots, \tilde{x}_N)$ *is an r-periodic solution of system* (2.11), *provided such a solution exists.*

REMARK 2.1. System (2.9) with $m = 1$ has an ω-periodic solution for every y if and only if the corresponding homogeneous system ($y \equiv 0$) has only zero as an ω-periodic solution. Moreover, if that system has a bounded solution, then there also exists an ω-periodic solution (Halanay [1], [2]). On the other hand, system (2.9) with an arbitrary number of deviations is equivalent to system (2.11), which can be written as

$$\tilde{x}' = \tilde{A}\tilde{x}+\tilde{y}, \tag{2.12}$$

provided that $\det \boldsymbol{\alpha}_{1\nu\mu}(t) \neq 0$ for $\mu, \nu = 1, 2, \ldots, N$, where $\tilde{y} = (\tilde{y}_1, \ldots, \tilde{y}_N)$ is the given nN-dimensional r-periodic vector valued function and $\tilde{A} = ([\boldsymbol{\alpha}_{1\nu\mu}(t)]^{-1}\boldsymbol{\alpha}_{0\nu\mu}(t))_{\mu,\nu=1,2,\ldots,N}$ is an $nN\times nN$-dimensional r-periodic matrix valued function. The existence of r-periodic solutions of system (2.12) then depends on the matrix $\tilde{A}$. Consequently the existence of ω-periodic solutions of system (2.9) also depends on the matrix $\tilde{A}$.

EXAMPLE 2.2. We now examine the possible existence of 2π-periodic solutions of the equation

$$\cos t x(t)-x(t-\pi)+x'(t)-\cos t x'(t-\pi) = 3+\cos 2t. \tag{2.13}$$

The coefficients are 2π-periodic, and in our notation $r = \pi$, $N = 2$, $\omega = 2\pi$, $\varepsilon = -1$, $\eta_j = j = 0$ or 1, $a_{00} = \cos t$, $a_{01} = -1$, $a_{10} = 1$, $a_{11} = -\cos t$. Since

$$(P_1 x)(t) = \frac{x(t)-x(t-\pi)}{2}, \quad (P_2 x)(t) = \frac{x(t)+x(t-\pi)}{2},$$

we have $P_1 \cos t = \cos t$, $P_2 \cos 2t = \cos 2t$, $P_1 \cos 2t = P_2 \cos t = 0$ and, if $f(t) \equiv$ const, then $P_1 f = 0$, $P_2 f = f$. This and Theorem 2.2 together imply that equation (2.13) is equivalent to the following system:

$$\begin{aligned} &(1-i)\tilde{x}_1+\tilde{x}_1'-e^{-it}\cos t[-(1+2i)\tilde{x}_2+\tilde{x}_2'] = 0, \\ &e^{it}\cos t[(1-i)\tilde{x}_1+\tilde{x}_1']+[-(1+2i)\tilde{x}_2+\tilde{x}_2'] = e^{2it}(3+\cos 2t). \end{aligned} \tag{2.14}$$

If we introduce new unknowns

$$u = (1-i)\tilde{x}_1+\tilde{x}_1, \quad v = -(1+2i)\tilde{x}_2+\tilde{x}_2, \tag{2.15}$$

we obtain a new system of algebraic equations

$$u-e^{-it}\cos t v = 0, \quad e^{it}\cos t u+v = e^{2it}(3+\cos 2t), \tag{2.16}$$

with the determinant $\Delta = 1+\cos^2 t = \frac{1}{2}(3+\cos 2t) \neq 0$. Solving system (2.16) we obtain $u = 2e^{it}\cos t$, $v = 2e^{it}$, which enables us to solve equations (2.14). The first of these equations has a π-periodic solution $\tilde{x}_1(t) = e^{it}(\cos t+\sin t)$ and a π-periodic solution of the second one is $\tilde{x}_2(t) = -2e^{2it}$. Thus the 2π-periodic solution of the given equation (2.13) is

$$x(t) = e^{-it}\tilde{x}_1(t)+e^{-2it}\tilde{x}_2(t) = \cos t+\sin t-2.$$

3. PERTURBATIONS OF DEVIATIONS PRESERVING PERIODIC SOLUTIONS OF LINEAR DIFFERENTIAL-DIFFERENCE EQUATIONS

In the first two sections we have considered linear differential-difference equations with deviations commensurable with the periods of the coefficients and of the right-hand side function. The following questions arise: 1° when do linear differential-difference equations with non-commensurable deviations have periodic solutions and how are they determined? 2° Is it possible to approximate ω-periodic solutions of such equations by ω-periodic solutions of differential-difference equations with deviations "near", in a sense, to the previous ones but commensurable? It will be shown that the answers to both questions are positive, but with some additional restrictions regarding the form of the differential-difference equations.

Denote by C_ω^n the space of all ω-periodic n-times continuously differentiable complex valued functions defined on the real line with the norm

$$\|x\|_n = \sum_{k=0}^{n} \sup_{0\leqslant t\leqslant\omega} |x^{(k)}(t)| \quad (n = 0, 1, \ldots), \tag{3.1}$$

C_ω^0 is a Banach space. We also write $C_\omega = C_\omega^0$.

Consider the following linear differential-difference equation

$$\frac{dx}{dt} + \sum_{j=0}^{m} a_j(t)x(t-h_j) = y(t) \quad (h_0 = 0), \tag{3.2}$$

where the functions $a_1(t), \ldots, a_m(t)$, $y(t)$ are ω-periodic complex valued functions defined for all $t \in \boldsymbol{R}$. We are looking for solutions of equation (3.2) belonging to C_ω^1. Without loss of generality we can assume $\omega \neq 0$[1].

We have the following two theorems, which will be proved at the end of this section.

THEOREM 3.1. *If the homogeneous equation* (3.2) (*i.e. this equation with* $y \equiv 0$) *has only the ω-periodic solution zero, then equation* (3.2) *has a unique solution* $x_h \in C_\omega^1$ *for every* $y \in C_\omega$.

[1] Note that C_ω^1 is a subspace of C_ω.

THEOREM 3.2. *If equation* (3.2) *has a unique solution* $x_h \in C^1_\omega$, *then for arbitrary reals* $h'_1, \ldots, h'_m$ *such that the values* $|h'_j - h_j|$ *are sufficiently small* $(j = 1, \ldots, m)$ *the equation*

$$\frac{dx}{dt} + \sum_{j=0}^{m} a_j(t)x(t-h'_j) = y(t) \quad (h'_0 = 0) \tag{3.2'}$$

has a unique solution $x_{h'} \in C^1_\omega$. *Moreover, if* $h'_j \to h_j$ *for* $j = 1, 2, \ldots, m$, *then* $x_{h'}$ *tends uniformly to* x_h.

REMARK 3.1. Theorems 3.1 and 3.2 remain valid if instead of scalar functions we consider vector-valued functions of an arbitrary dimension k (in this case the coefficients $a_j(t)$ are taken to be $k \times k$ matrices). However, it will be shown that the proofs of these theorems do not work in the case of functions taking values in an infinitely dimensional Banach space (cf. Rolewicz [1]).

It will also be shown that these theorems are true for the equation

$$\frac{d^n t}{dt^n} + \sum_{k=0}^{n-1}\sum_{j=0}^{m} a_{kj}(t)x^{(k)}(t-h_j) = y(t) \quad (n > 1,\ h_0 = 0), \tag{3.3}$$

where we are looking for solutions belonging to C^n_ω.

We now can pass to our main corollary. Suppose the numbers $h_1, \ldots, h_m$ in equation (3.3) are not commensurable with ω nor possibly, commensurable with one another. Since we are looking for ω-periodic solutions, we can assume without loss of generality, that

$$0 < h_j \leqslant \omega \quad \text{for } j = 1, 2, \ldots, m. \tag{3.4}$$

For a given number $\delta > 0$ we can find numbers $\omega_1, \ldots, \omega_m$ commensurable with ω and with one another, such that $0 \leqslant \omega_j \leqslant \omega$ and

$$|\omega_j - h_j| < \delta \quad \text{for } j = 1, 2, \ldots, m. \tag{3.5}$$

Indeed, without loss of generality we can assume that $\delta < 1$. Let $N = \left[\frac{1}{\delta}\right]$[1]. For a fixed j there is an integer $\eta_j \leqslant N$ such that

$$\frac{\eta_j}{N}\omega - h_j < \delta \leqslant \frac{1}{N}.$$

Put

$$r = \frac{\omega}{N}, \quad \eta_{m+1} = N, \quad \eta_0 = 0, \quad \omega_j = \eta_j r \quad (j = 1, 2, \ldots, m).$$

---

[1] The symbol $[x]$, read "integer part of x", denotes the greatest integer $M \leqslant x$.

So that we have $m+1$ commensurable numbers $\omega_1, \ldots, \omega_{m+1}$ satisfying the required condition 3.5 and such that $0 \leqslant \omega_j \leqslant \omega_{m+1} \leqslant \omega$. Now consider the perturbed equation (3.3), where instead of the deviations h_j we put the numbers ω_j:

$$\frac{d^n x_\omega}{dt^n} + \sum_{j=0}^{m} a_j(t) x_\omega^{(k)}(t-\omega_j) = y(t). \tag{3.3'}$$

Using the method described in Section 2, we conclude that this equation is equivalent in the space C_ω to a system of N independent differential equations without deviations of argument (either (2.3) or (2.11) depending on the periods of the coefficients).

If each of these homogeneous equations has only zero as an ω-periodic solution, then all non-homogeneous equations have the unique ω-periodic solution. Hence equation (3.3′) has a unique ω-periodic solution $x_\omega(t)$ for every $y \in C_\omega$. It follows from Theorem 3.2 that for every $\varepsilon > 0$ there is a $\delta > 0$ such that for all numbers ω_j satisfying (3.5) (and (3.4), which is not essential but is merely a technical assumption) equation (3.3) has a unique ω-periodic solution $x \in C_\omega^1$ for every $y \in C_\omega$ such that

$$\|x - x^\omega\|_0 = \sup_{0 \leqslant t \leqslant \omega} |x(t) - x^\omega(t)| < \varepsilon.$$

This is which was to be proved.

Proof of theorem 3.1. We define an operator R on the space C by means of the equality:

$$(Rx)(t) = \int_0^t x(s)\,ds - \left(\frac{t}{\omega} + 1\right) \int_0^\omega x(s)\,ds \quad \text{for } x \in C^\omega. \tag{3.6}$$

It is clear that R maps continuous functions into differentiable functions. We now show that R transforms ω-periodic functions into ω-periodic functions. Indeed, let $x \in C_\omega$ be arbitrary. Then

$$(Rx)(t+\omega) - (Rx)(t)$$

$$= \int_0^{t+\omega} x(s)\,ds - \left(\frac{t+\omega}{\omega} + 1\right) \int_0^\omega x(s)\,ds - \left[\int_0^t x(s)\,ds - \left(\frac{t}{\omega} + 1\right) \int_0^\omega x(s)\,ds\right]$$

$$= \int_0^{t+\omega} x(s)\,ds - \int_0^\omega x(s)\,ds = \int_0^\omega x(s)\,ds - \int_0^\omega x(s)\,ds = 0.$$

Hence R transforms C_ω into C_ω^1. We now define the operator $\tilde{R}$ on the space C_ω^1 by the equality:

$$(\tilde{R}y)(t) = \frac{d}{dt}y(t) - \frac{1}{\omega}y(0). \tag{3.7}$$

It is easy to check that the operator $\tilde{R}$ transforms C_ω^1 into C_ω. Moreover, we have for $y \in C_\omega^1$

$$\begin{aligned}
(R\tilde{R}y)(t) &= \int_0^t \left[y'(s) - \frac{1}{\omega}y(0)\right] ds - \left(\frac{t}{\omega}+1\right)\int_0^\omega \left[y'(s) - \frac{1}{\omega}y(0)\right] ds \\
&= y(t) - y(0) - \frac{t}{\omega}y(0) - \left(\frac{t}{\omega}+1\right)\left[y(\omega) - y(0) - \frac{\omega}{\omega}y(0)\right] \\
&= y(t) + - \left(\frac{t}{\omega}+1\right)[y(0)+y(\omega)-2y(0)] = y(t) - \left(\frac{t}{\omega}+1\right)[y(0)-y(0)] \\
&= y(t)
\end{aligned}$$

for $y(\omega) = y(0)$, when $y \in C_\omega^1$. On the other hand we have for $x \in C_\omega$

$$\begin{aligned}
(\tilde{R}Rx)(t) = \frac{d}{dt}\left[\int_0^t x(s)\,ds - \left(\frac{t}{\omega}+1\right)\int_0^\omega x(s)\,ds\right] - \frac{1}{\omega}\Bigg[\int_0^t x(s)\,ds \,- \\
- \left(\frac{t}{\omega}+1\right)\int_0^\omega x(s)\,ds\Bigg]_{t=0} = x(t) - \frac{1}{\omega}x(s)\,ds + \frac{1}{\omega}\int_0^\omega x(s)\,ds = x(t).
\end{aligned}$$

Hence $\tilde{R} = R^{-1}$, which shows that R is a one-to-one mapping of C_ω onto C_ω^1.
Acting on both sides of equation (3.2) with the operator R, we obtain an equivalent equation:

$$\begin{aligned}
x(t) - x(0) + \sum_{j=0}^{m}\int_0^t a_j(s)x(s-h_j)\,ds - \left(\frac{t}{\omega}+1\right)\sum_{j=0}^{m}\int_0^\omega a_j(s)x(s-h_j)\,ds \\
= \int_0^t y(s)\,ds - \left(\frac{t}{\omega}+1\right)\int_0^\omega y(s)\,ds.
\end{aligned} \tag{3.8}$$

Since the operator R is invertible, either both of equations (3.2) and (3.8) have solutions or neither equation has one. In particular, the homogeneous equation (3.2) has only zero as a solution if and only if equation (3.8) has only zero as a solution (in the appropriate space). Equation (3.8) can also be written in the following way

$$(I+T_h)x = Ry, \tag{3.9}$$

where the operator T_h is defined by the formula

$$(T_h x)(t) = -x(0)+\sum_{j=0}^{m}\int_0^t a_j(s)x(s-h_j)\,ds-\left(\frac{t}{\omega}+1\right)\int_0^{\omega} a_j(s)x(s-h_j)\,ds.$$

The operator T_h is then a sum of two operators: one of them is an integral operator of Volterra type and the other one is finite dimensional (in fact two-dimensional). Thus T_h is a compact operator transforming C_ω into itself.

Hence by the Riesz theorem (Theorem I.4.2) if the equation $(I+T_h)x = 0$ has the unique solution zero, then the equation $(I+T_h)x = \tilde{y}$ has a unique solution for all $\tilde{y} \in C$, in particular for $\tilde{y} = Ry$. This completes the proof of Theorem 3.1.

Proof of Theorem 3.2. We retain all the notation of the preceding proof and we consider the equation

$$(I+T_{h'})x = Ry, \tag{3.9'}$$

where

$$(T_{h'} x)(t) = -x(0)+\sum_{j=0}^{m}\int_0^t a_j(s)x(s-h_j')\,ds-\left(\frac{t}{\omega}+1\right)\int_0^{\omega} a_j(s)x(s-h_j')\,ds.$$

We shall prove that

$$\lim_{h_j'\to h_j (j=1,2,\cdots,m)} ||T_{h'}-T_h|| = 0,$$

where by $||A||$ we mean the norm of an operator A acting in C_ω. Indeed, let $\varepsilon > 0$ be arbitrary and let $\delta > 0$ be chosen so that the following two conditions are satisfied:

1° $\delta < \varepsilon$, 2° $\omega m_j(\delta) < \varepsilon$,

where $m_j(|t|)$ denotes the modulus of continuity of the function $a_j(t)$.

Let $x \in C_\omega$ be arbitrary and let $d_j = h_j'-h_j$ $(j = 1, 2, \ldots, m)$. We will estimate the norm $||T_{h'} x-T_h x||$:

$$||T_{h'} x-T_h x|| = \sup_{0\leqslant t\leqslant\omega}\Bigg|\Bigg[-x(0)+\sum_{j=0}^{m}\int_0^t a_j(s)x(s-h_j')\,ds-$$

$$-\left(\frac{t}{\omega}+1\right)\int_0^{\omega} a_j(s)x(s-h_j')\,ds\Bigg]-\Bigg[-x(0)+\sum_{j=0}^{m}\int_0^t a_j(s)x(s-h_j)\,ds-$$

$$-\left(\frac{t}{\omega}+1\right)\sum_{j=0}^{m}\int_{0}^{\omega} a_j(s)x(s-h_j)\,ds\Bigg]\Bigg\|$$

$$= \sup_{0\leqslant t\leqslant\omega}\left|\sum_{j=0}^{m}\left\{\int_{0}^{t} a_j(s)x(s-h_j')\,ds-\right.\right.$$

$$\left.\left.- a_j(s)x(s-h_j)\,ds-\left(\frac{t}{\omega}+1\right)\int_{0}^{\omega} a_j(s)[x(s-h_j')-x(s-h_j)]\,ds\right\}\right|$$

$$\leqslant \sup_{0\leqslant t\leqslant\omega}\left\{\sum_{j=0}^{m}\left|\int_{-d_j}^{t-d_j} a(s+d_j)x(s-h_j)\,ds-\int_{0}^{t} a_j(s)x(s-h_j)\,ds\right|+\right.$$

$$\left.+2\sum_{j=0}^{m}\left|\int_{-d_j}^{\omega-d_j} a_j(s+d_j)x(s-h_j)-\int_{0}^{\omega} a_j(s)x(s-h_j)\,ds\right|\right\}.$$

But

$$\left|\int_{-d_j}^{t-d_j} a_j(s+d_j)x(s-h_j)\,ds-\int_{0}^{t} a_j(s)x(s-h_j)\,ds\right| \tag{3.9''}$$

$$\leqslant\left|\int_{-d_j}^{0} a(s+d_j)x(s-h_j)\,ds+\int_{0}^{t-d_j}[a_j(s+d_j)-a_j(s)]x(s-h_j)\,ds+\int_{t-d_j}^{t} a_j(s)x(s-h_j)\,ds\right|$$

$$\leqslant M_j\|x\|_0|d_j|+|t-d_j|\,m_j(|d_j|)\,\|x\|_0+M_j\|x\|_0|d_j|$$
$$= [2M_j|d_j|+|t-d_j|\,m_j(|d_j|)]\,\|x\|_0 \leqslant [2M_j|d_j|+2\omega m_j(|d|)]\,\|x\|_0,$$

where

$$M_j = \max_{0\leqslant t\leqslant\omega}\|a_j(t)\| \qquad (j=1,2,\ldots,m).$$

Furthermore, since a_j and x are ω-periodic, we find

$$\left|\int_{-d_j}^{\omega-d_j} a_j(s+d_j)x(s-h_j)\,ds-\int_{0}^{\omega} a_j(s)x(s-h_j)\,ds\right| \tag{3.10}$$

$$=\left|\int_{0}^{\omega} a_j(s+d_j)x(s-h_j)\,ds-\int_{0}^{\omega} a_j(s)x(s-h_j)\,ds\right|$$

$$=\left|\int_{0}^{\omega}[a_j(s+d_j)-a_j(s)]\,x(s-h_j)\,ds\right|\leqslant \omega m_j(|d_j|)\,\|x\|_0.$$

Finally, from (3.9″) and (3.10) we obtain

$$||T_{h'}x-T_hx|| \leqslant \sum_{j=0}^{m} [2M_j|d_j|+3\omega m_j(|d_j|)]\,||x||_0 .$$

Since, from our assumptions $|d_j| = |h_j'-h_j| < \delta < \varepsilon$ and $\omega m_j(|d_j|) \leqslant \omega m_j(\delta) < \varepsilon$, we conclude that $||T_{h'}x-T_hx|| < M\varepsilon$. Since x and ε are arbitrary, it follows that $T_{h'} \to T_h$ in the norm of the space C_ω.

Equation (3.9′): $(I+T_{h'})x = Ry$ is equivalent to the equation

$$[I+T_h+(T_{h'}-T_h)]x = Ry. \tag{3.11}$$

It follows from Theorem 3.1 that the operator $I+T_h$ is compact and one-to-one. The Riesz theorem (Theorem I.4.2) now implies that the operator $I+T_h$ is invertible and that the operator $(I+T_h)^{-1}$ is continuous. Acting on both sides of equation (3.11) with the operator $(I+T_h)^{-1}$, we obtain the equation

$$(I+B)x = (I+T_h)^{-1}Ry, \tag{3.12}$$

where the operator $B = (I+T_h)^{-1}(T_{h'}-T_h)$ has norm less than 1 for $|h_j'-h_j|$ sufficiently small ($j = 1, 2, \ldots, m$). Hence the Neumann theorem (Theorem I.4.4) implies that equation (3.12) has a unique solution $x_{h'} \in C_\omega$. This solution is of the form

$$x_{h'} = \sum_{j=0}^{\infty} B^j(I+T_h)^{-1}Ry.$$

Note that $x_h = (I+T_h)^{-1}Ry$, and so

$$||x_{h'}-x_h|| \leqslant \left|\left|\sum_{j=1}^{\infty} B^j(I+T_h)^{-1}Ry\right|\right| \leqslant ||Ry||\;||(I+T_h)^{-1}||\sum_{j=1}^{\infty} ||B||^j$$

$$\leqslant ||Ry||\;||(I+T_h)^{-1}||\,\frac{||B||}{1-||B||}\,.$$

By definition $||B|| \to 0$ when $h_j' \to h_j$ ($j = 1, 2, \ldots, m$). This implies that $||x_{h'}-x_h|| \to 0$, which establishes the continuity of the ω-periodic solutions of equation (3.2) with respect to deviations.

This proof does not work for functions with values in an infinite dimensional Banach space, since in this case the operator T is not compact and, moreover, neither is the operator $(Tx)(t) = x(0)$ (cf. Rolewicz [1] .

The proofs for equation (3.3) are exactly the same if we act on both sides of this equation with the operator R^n. We obtain an equivalent equation of the form $(I+T_h)x = R^n y$, where the operator T_h is compact. Further, the proofs follow the same lines as before.

4. PARTIAL LINEAR DIFFERENTIAL-DIFFERENCE EQUATIONS WITH CONSTANT AND PERIODIC COEFFICIENTS

Let $\boldsymbol{R}^q$ be a q-dimensional real Euclidean space. Let $\boldsymbol{t} = (t_1, \ldots, t_q) \in \boldsymbol{R}^q$, $q \neq 1$. As usual, we write

$$D^k x(\boldsymbol{t}) = \frac{\partial^{k_1+\cdots+k_q}}{\partial t_1^{k_1} \ldots t_q^{k_q}} x(\boldsymbol{t}),$$

where $\boldsymbol{k} = (k_1, \ldots, k_q)$ is a q-dimensional *multi-index* (cf. Section 7 of Chapter II). Let

$$\boldsymbol{h}_j = (h_{1j}, \ldots, h_{qj}), \quad j = 0, 1, \ldots, m$$

and consider the following partial linear differential-difference equation:

$$\sum_{(0)_q \leqslant \boldsymbol{k} \leqslant \boldsymbol{n}} \sum_{0 \leqslant j \leqslant m} A_{k,j}(\boldsymbol{t}) D^k x(\boldsymbol{t} - \boldsymbol{h}_j) = y(\boldsymbol{t}), \tag{4.1}$$

where $\boldsymbol{n}$ is a given multi-index. We assume that $h_{p,0} = 0$ for $p = 1, 2, \ldots, q$.

We say that a function $x(\boldsymbol{t})$ defined for $\boldsymbol{t} \in \boldsymbol{R}^q$ is *ω-periodic* if it is ω_p-periodic with respect to the pth variable t_p $(p = 1, 2, \ldots, q)$ and if $\boldsymbol{\omega} = (\omega_1, \ldots, \omega_q)$. The vector $\boldsymbol{\omega}$ is called the *period* (the *vector-period*) of the function x. If a function is $\boldsymbol{\omega}$-periodic, then for every multi-index $\boldsymbol{k} = (k_1, \ldots, k_q)$

$$x(\boldsymbol{t} - \boldsymbol{\omega}) = x(t_1 - k_1\omega_1, \ldots, t_q - k_q\omega_q) = x(t_1, \ldots, t_q) = x(\boldsymbol{t}).$$

Using Corollary II.7.1 on equations with multi-involutions, we obtain a generalization of Theorems 1.1 and 2.1:

THEOREM 4.1. *Let a (real-valued or complex-valued) function $y(\boldsymbol{t})$ defined for $\boldsymbol{t} \in \boldsymbol{R}^q$ be $\boldsymbol{h}_{m+1}$-periodic with the period $\boldsymbol{h}_{m+1} = (h_{1,m+1}, \ldots, h_{q,m+1})$, where for a fixed p the numbers $h_{p,m+1}$ are commensurable with the commensurable reals $h_{p,j}$ $(p = 1, \ldots, q,\ j = 1, 2, \ldots, m)$ i.e. there are such integers $\eta_{p,j}$ and such reals $r_p \neq 0$ such that $h_{p,j} = \eta_{p,j} r_p$. Let $\eta_{0,j} = 0$ and $\boldsymbol{\eta}_j = (\eta_{1,j}, \ldots, \eta_{q,j})$ for $j = 0$, $1, \ldots, m+1$, which implies that $\boldsymbol{h}_j = \boldsymbol{\eta}_j \boldsymbol{r}$, where $\boldsymbol{r} = (r_1, \ldots, r_q)$. Let N_p be a common multiple (not necessarily the least one) of the positive integers $|\eta_{p,1}|, \ldots, |\eta_{p,m+1}|$ and let $N = (N_1, \ldots, N_q)$, $\boldsymbol{\omega} = (\omega_1, \ldots, \omega_q)$, where $\omega_p = N_p r_p$. Let the (real or complex) functions $A_{k,j}(\boldsymbol{t})$ determined for $\boldsymbol{t} \in \boldsymbol{R}^q$ be $\boldsymbol{r}$-periodic or constant $(0 \leqslant j \leqslant m,\ (0)_q \leqslant \boldsymbol{k} \leqslant \boldsymbol{n})$. Then equation (4.7) has an $\boldsymbol{\omega}$-periodic solution if and only if all the partial differential equations*

$$A_\nu x \equiv \sum_{(0)_q \leqslant \boldsymbol{k} \leqslant \boldsymbol{n}} b_{k\nu}(\boldsymbol{t}) D^k x(\boldsymbol{t}) = y_{(\nu)}(\boldsymbol{t}), \quad (1)_q \leqslant \boldsymbol{\nu} \leqslant N \tag{4.2}$$

have ω-periodic solutions, where

$$b_{k\nu}(t) = \sum_{j=0}^{m} \varepsilon^{-\nu\eta_j} A_{k,j}(t), \qquad \varepsilon = (\varepsilon_1, \ldots, \varepsilon_q),\ \varepsilon_p = e^{2\pi i/N_p},$$

$$y_{(\nu)}(t) = \frac{1}{N_1 \ldots N_q} \sum_{(0)_q \leqslant k \leqslant N-(1)_q} \varepsilon^{-\nu k_j}(t-kr).$$

The number of equations (4.2) *is $N_0 = N_1 \ldots N_q$. Their solutions are of the form*

$$x(t) = \sum_{(1)_q \leqslant \nu \leqslant N} x_\nu(t),$$

where x_ν denotes an ω-periodic solution of the νth equation (4.2), *provided such a solution exists.*

PROOF. We consider the space E^q_ω of all ω-periodic real or complex valued functions $x(t)$ defined for $t \in R^q$ with the period ω described above. Let

$$(S_p x)(t) = x(t_1, \ldots, t_{p-1}, t_p - r_p, t_{p+1}, \ldots, t_q) \quad \text{for } x \in E^q_\omega \ (p = 1, 2, \ldots, q).$$

Each of the operators S_p is a linear operator mapping E^q_ω onto itself, and, moreover, is an involution of order N_p. Indeed,

$$(4.3) \qquad \begin{aligned} (S_p^{N_p} x)(t) &= x(t_1, \ldots, t_{p-1}, t_p - N_p r_p, t_{p+1}, \ldots, t_q) \\ &= x(t_1, \ldots, t_{p-1}, t_p - \omega_p, t_{p+1}, \ldots, t_q) \\ &= x(t_1, \ldots, t_p, \ldots, t_q) = x(t) \end{aligned}$$

and N_p is the smallest number satisfying (4.3). Let

$$(Sx)(t) = x(t-r) \quad \text{for } x \in E^q_\omega.$$

The operator S is a multi-involution of order $N = (N_1, \ldots, N_q)$ (compare Section 7 of Chapter II) since

$$\begin{aligned} (S^N x)(t) &= (S_1^{N_1} \ldots S_q^{N_q} x)(t) = x(t_1 - N_1 r_1, \ldots, t_q - N_q r_q) \\ &= x(t_1 - \omega_1, \ldots, t_q - \omega_q) = x(t). \end{aligned}$$

If $a(t)$ is an arbitrary (real-valued or complex-valued) r-periodic function defined on R^q, then the operator a of multiplication by this function acting in E^q_ω commutes with S:

$$(4.4) \qquad Sa - aS = 0 \quad \text{on } E^q_\omega.$$

Indeed,

$$\begin{aligned} [(Sa-aS)x](t) &= a(t-r)x(t-r) - a(t)x(t-r) = [a(t-r)-a(t)]x(t) \\ &= [a(t)-a(t)]x(t-r) = 0. \end{aligned}$$

The operator S, as a shift operator, commutes with the differential operators D^k, $k \leqslant n$ on the set of all n-times differentiable functions belonging to E^q_ω. Hence on the same set the superposition aD^k for $k \leqslant n$ also commutes with S:

$$aD^kS - SaD^k = aD^kS - aSD^k + aSD^k - SaD^k = a(D^kS - SD^k) + (aS - Sa)D^k = 0.$$

All the conditions of Corollary II.7.1 are satisfied. Hence equation (4.1) is equivalent in E^q_ω to $N_0 = N_1 \dots N_q$ partial differential equations (4.2). If this last system has an ω-periodic solution $(x_{(1)_q}, \dots, x_N)$, then equation (4.1) has an ω-periodic solution

$$x(t) = \sum_{(1)_q \leqslant v \leqslant N} x_v(t),$$

which was to be proved.

Our theorem remains valid if we consider vector valued functions defined on R^q, or, more generally, functions taking values in an arbitrary Banach space.

If the coefficients $A_{k,j}(t)$ are ω-periodic we can reiterate all the discussion leading to Theorem 2.2 and reformulate Proposition 2.1 in the following way:

PROPOSITION 4.1. *A function $u(t)$ belongs to $X_{(v)}$ for a v, $(1)_q \leqslant v \leqslant N$, if and only if*

$$u(t) = e^{-2\pi i v t/\omega} v(t), \quad t \in R^q,$$

where $v(t)$ is an r-periodic function and

$$e^{-2\pi i v t/\omega} = \exp\left(\sum_{p=1}^{q} \frac{v_p t_p}{\omega_p}\right).$$

This enables us to construct, as in Theorem 2.2, a finite system of partial differential equations without deviations equivalent to the given system.

5. NON-LINEAR DIFFERENTIAL-DIFFERENCE EQUATIONS

Suppose we are given a system of non-linear differential-difference equations

$$\frac{dx(t)}{dt} = G(t, x(t), x(t-\omega_1), \dots, x(t-\omega_m)), \tag{5.1}$$

where $x(t) = (x_1(t), \dots, x_n(t))$ is an unknown n-dimensional vector valued function and the given n-dimensional vector valued function $G(t, z_{1,0}, \dots, z_{1,m}, z_{m,0}, \dots \dots, z_{m,n})$ is defined for all real t and $z_{k,j}$ and is continuous [1]. We assume, as before, that all the numbers $\omega_1, \dots, \omega_m$ are commensurable, i.e. that $\omega_j = \eta_j r$, $r \neq 0$, where the η_j are integers and that $\omega_0 = \eta_0 = 0$. Let N be a common

[1] This assumption is not essential, it is enough to assume that G is a Borel function.

multiple of the positive integers $|\eta_1|, \ldots, |\eta_m|$ and let $\omega = Nr$. We are looking for ω-periodic solutions of equation (5.1). For this purpose we proceed in the same way as in the proof of Theorem 2.2. Namely, acting on both sides of equation (5.1) with the projectors P_ν and using Proposition (2.1) and the substitution (2.8), we see that system (5.1) can be rewritten in the form of nN ordinary differential equations without deviations:

$$(5.2)\qquad \dot{\tilde{x}}_\nu(t) = \frac{2\pi i}{\omega}\,\tilde{x}(t) + e^{2\pi i t/\omega} P_\nu[\tilde{G}(t, \tilde{x}_1(t), \ldots, \tilde{x}_N(t))] \qquad (\nu = 1, 2, \ldots, N),$$

where $\tilde{x}_1, \ldots, \tilde{x}_N$ are r-periodic vector-functions and

$$\tilde{G}(t, \tilde{x}_1, \ldots, \tilde{x}_N) = G\Big(t, \sum_{k=1}^{N} e^{-2\pi i k t/\omega}\tilde{x}_k, \ldots, \sum_{k=1}^{N} e^{-2\pi i k t/\omega} e^{2\pi k \eta_m}\tilde{x}_k\Big).$$

Since $x_1, \ldots, \tilde{x}_N$ are r-periodic functions, we have

$$(5.3)\qquad P_\nu[\tilde{G}(t, \tilde{x}_1(t), \ldots, \tilde{x}_N(t))] = \frac{1}{N}\sum_{k=0}^{N-1} \varepsilon^{-k\nu}\tilde{G}(t-kr, \tilde{x}_1(t-kr), \ldots, \tilde{x}_N(t-kr))$$

$$= \sum_{k=0}^{N-1} \frac{1}{N}\varepsilon^{-k\nu}\tilde{G}(t-kr, \tilde{x}_1(t), \ldots, \tilde{x}_N(t))$$

$$= P_\nu[G(t, \tilde{x}_1, \ldots\ldots, \tilde{x}_N)]|_{\tilde{x}_j = \tilde{x}_j(t)},$$

where on the right-hand side of equality (5.3) the operator P_ν acts on the variable t alone.

If we write

$$\tilde{G}_\nu(t, \tilde{x}_1, \ldots, \tilde{x}_N) = e^{2\pi i \nu t/\omega} P_\nu[\tilde{G}(t, \tilde{x}_1, \ldots, \tilde{x}_N)] \qquad (\nu = 1, 2, \ldots, N)$$

we obtain the following

THEOREM 5.1. *The n-dimensional system* (5.1) *has an ω-periodic solution if and only if the following nN-dimensional system* (*without deviations*)

$$(5.4)\qquad \frac{d\tilde{x}_\nu}{dt} = \frac{2\pi i\nu}{\omega}\tilde{x}_\nu + \tilde{G}_\nu(t, \tilde{x}_1, \ldots, \tilde{x}_N) \qquad (\nu = 1, 2, \ldots, N)$$

has an r-periodic solution. The solution is of the form

$$x(t) = \sum_{\nu=1}^{N} e^{-2\pi i \nu t/\omega}\tilde{x}_\nu(t),$$

where $(\tilde{x}_1, \ldots, \tilde{x}_N)$ *is an r-periodic solution of system* (5.4).

If the vector valued function $G(t, z_{1,0}, \ldots, z_{n,m})$ is ω-periodic with respect to the variable t, then the vector valued functions $\tilde{G}_\nu(t, \tilde{x}_1, \ldots, \tilde{x}_N)$ are r-periodic with respect to t.

Consider now a system dependent on a parameter μ (scalar or vector)

$$\frac{dx}{dt} = G(t, x(t), x(t-\omega_1), \ldots, x(t-\omega_m), \mu), \tag{5.5}$$

where ω_j and G satisfy the same condition as in system (5.1). We assume, moreover, that $G(t, z_{1,0}, \ldots, z_{s,m}, \mu)$ and $\dfrac{\partial G}{\partial z_{i,j}}$ $(i = 1, \ldots, n; j = 0, 1, \ldots, m)$ are continuous and ω-periodic with respect to the variable t.

Suppose that for $\mu = 0$ there is an ω-periodic solution $p(t)$ of system (5.5) and consider the *linearized system*

$$\frac{dy_i}{dt} = \sum_{j=0}^{m} \sum_{k=0}^{n} \left.\frac{\partial G_i}{\partial z_{k,j}}\right|_{(t,p(t),0)} y_i(t-\omega_j) \quad (i = 1, \ldots, n), \tag{5.6}$$

corresponding to system (5.5), where G_i is the ith coordinate of the vector G. The following theorem holds:

THEOREM 5.2. *If the linear system* (5.6) *only has one ω-periodic solution, namely* 0, *then for $|\mu|$ sufficiently small the system has a unique ω-periodic solution $q(t, \mu)$ depending continuously on (t, μ) and such that $q(t, 0) = p(t)$.*

Indeed, Theorems 5.1 and 2.2 imply that the existence of ω-periodic solutions of system (5.5) and (5.6) is equivalent to the existence of r-periodic solutions of the corresponding systems of ordinary differential equations.

Theorem 5.2 is thus a consequence of the classical theorem concerning perturbations of periodic ordinary differential equations (see Coddington and Levinson [1], Chapter XIV, § 1).

A disadvantage of the method given here is that an autonomous system is transformed into a non-autonomous system.

For non-linear differential-difference equations it would also be possible to prove theorems analogous to Theorems 3.1 and 3.2 on equations with perturbed deviations, but this would require much heavier mathematical tools. Some results were obtained recently by Ginzburg [1] and Stephan [1].

EXAMPLE 5.1. We are looking for non-trivial 2π-periodic solutions of the equation

$$x'(t) = x^2(t) - x^2(t-\pi). \tag{5.7}$$

Here $n = 1$, $r = \pi$, $\omega = 2\pi$, and $N = 2$. Since $G(t, z_1, z_2) = z_1^2 - z_2^2$ and $(P_1 x)(t) = \frac{1}{2}[x(t) - x(t-\pi)]$, $(P_2 x)(t) = \frac{1}{2}[x(t) + x(t-\pi)]$, we find

$$P_1[G(t, x(t), x(t-\pi))] = \frac{1}{2}[x^2(t) - x^2(t-\pi) - x^2(t-\pi) + x^2(t-2\pi)] = x^2(t) - x^2(t-\pi)$$
$$= G(t, x(t), x(t-\pi)),$$

and similarly $P_2[G(t, x(t), x(t-\pi))] = 0$. We have

$$x(t) = x_{(1)}(t) + x_{(2)}(t) = \tilde{x}_1(t)e^{-it} + \tilde{x}_2(t)e^{-2it},$$

where $\tilde{x}_1$ and $\tilde{x}_2$ are π-periodic functions. As in (5.4) we obtain the following system of differential equations:

$$(\tilde{x}_1' - i\tilde{x}_1)e^{-it} = (\tilde{x}_1 e^{-it} + \tilde{x}_2 e^{-2it})^2 - (-\tilde{x}_1 e^{-it} + \tilde{x}_2 e^{-2it})^2,$$
$$\tilde{x}_2 - 2i\tilde{x}_2 = 0.$$

After some simplifications we can rewrite the last system as follows

(5.8) $$\tilde{x}_2' - 2i\tilde{x}_2 = 0,$$

(5.9) $$\tilde{x}_1' - i\tilde{x}_1 = 4\tilde{x}_1\tilde{x}_2 e^{-2it}.$$

The general solution of equation (5.8) is $\tilde{x}_2(t) = C_2 e^{2it}$, where C_2 is an arbitrary constant. For all C_2 this is a π-periodic function. If we substitute for $\tilde{x}_2$ in the right-hand side of (5.9) we obtain the following differential equation:

$$\tilde{x}_1' - i\tilde{x}_1 = 4C_2\tilde{x}_1 \quad \text{or} \quad \tilde{x}_1' - (i + 4C_2)\tilde{x}_1 = 0.$$

The general form of the solutions of this last equation is

$$\tilde{x}_1(t) = C_1 e^{(i+C_2)t}, \quad \text{where } C_1 \text{ is an arbitrary constant.}$$

A π-periodic solutions exists if

(5.10) $$i + 4C_2 = 2ki, \quad k = 0, \pm 1, \pm 2, \ldots$$

Since C_2 is an arbitrary constant, we obtain π-periodic solutions of (5.9) by choosing C_2 in such a way that equality (5.10) is satisfied. From (5.10) we obtain

$$C_2 = \frac{2k-1}{4} i, \quad k = 0, \pm 1, \pm 2, \ldots$$

and $\tilde{x}_1(t) = C_1 e^{2kit}$. We finally obtain 2π-periodic solutions of the given equation in the form

$$x(t) = \tilde{x}_1(t)e^{-it} + \tilde{x}_2(t)e^{-2it} = C_1 e^{(2k-1it)} + \frac{2k-1}{4} i,$$

where C_1 is an arbitrary constant and k is an arbitrary integer.

EXAMPLE 5.2. Consider the equation

(5.11) $$\frac{dx}{dt} = x(t) + x^2(t) - x^2(t-\pi) + \mu h(t, x(t), x(t-\pi)),$$

where μ is a real parameter and the given function $h(t, z_1, z_2)$ satisfies the same conditions as the function G in Theorem 5.2. For $\mu = 0$ equation (5.11) has a 2π-periodic solution $p(t)$, namely $p(t) \equiv 0$. Since $G(t, z_1, z_2, \mu) = z_1 + z_1^2 - z_1 + \mu h(t, z_1, z_2)$, we find $\frac{\partial G}{\partial z_1} = 1 + 2z_1 +$

$+2z_1+\mu h'_{z_1}(t, z_1, z_2)$, $\dfrac{\partial G}{\partial z_2} = -2z_2+\mu h'_{z_2}(t, z_1, z_2)$. Then $\left.\dfrac{\partial G}{\partial z_1}\right|_{(t,\, p(t),\, 0)} = 1$, $\left.\dfrac{\partial G}{\partial z_2}\right|_{(t,\, p(t),\, 0)} = 0$ and the linearized system (5.6) assumes in our case the form: $dy/dt = y$. This equation has only one 2π-periodic solution: namely $y(t) \equiv 0$. All the conditions of Theorem 5.2 are satisfied, whence we conclude that for $|\mu|$ sufficiently small there exist 2μ-periodic solutions of equation (5.11).

REMARK 5.1. A function $x(t)$ of a real variable is called *half-periodic*[1] with period $d > 0$ if $x(t+d) = -x(t)$. The shift operator $(Sx)(t) = x(t+d)$ is an involution on the space of half-periodic functions. Indeed,

$$(S^2x)(t) = x(t+2d) = -x(t+d) = -[-x(t)] = x(t).$$

We can thus use all the methods described in this chapter for equations with given half-periodic functions.

REMARK 5.2. If equation (5.1) has a delayed argument, i.e. if all the $\omega_j > 0$, then every periodic solution of this equation is infinitely differentiable. This could be shown by the method of integration "step by step" and by the application of periodicity.

[1] See Neuman [1].

CHAPTER VI

Exponential-Periodic Solutions of Differential-Difference Equations

In this chapter the results of Chapter V will be extended to a larger class of functions. For this purpose we deal with algebraic operators. The convention introduced at the begining of Chapter V is retained.

1. ORDINARY LINEAR DIFFERENTIAL-DIFFERENCE EQUATIONS WITH CONSTANT AND PERIODIC COEFFICIENTS

As in Section 1 and 2 of Chapter V we study the following linear differential-difference equation

(1.1) $$Ax \equiv \sum_{k=0}^{n}\sum_{j=0}^{m} a_{kj}(t)x^{(k)}(t-\omega_j) = y(t) \qquad (\omega_0 = 0),$$

where all the numbers $\omega_1, \ldots, \omega_m$ are commensurable. This implies that there are integers η_j and a real $r \neq 0$ such that $\omega_j = \eta_j r$ for $j = 1, 2, \ldots, m$ (we assume $\eta_0 = 0$). In this section we assume that the coefficients are either constant or r-periodic. Let N be a common multiple (not necessarily the least one) and let $\omega = Nr$. As before, E_ω denotes the space of all complex valued ω-periodic functions defined on the real line. Let $E_\omega(\lambda_1, \ldots, \lambda_M)$ be the set of all functions of the form

(1.2) $$x(t) = \sum_{l=1}^{M} e^{\lambda_l t} y_l(t), \qquad \text{where } y_l(t) \in E_\omega,$$

and $\lambda_1, \ldots, \lambda_M$ are arbitrarily fixed complex (or possibly real) numbers. Without loss of generality we can assume that for $k \neq l$

(1.3) $$\lambda_k \neq \lambda_l + \frac{2\pi ij}{\omega} \qquad \text{for } j = 0, \pm 1, \pm 2, \ldots$$

Indeed if $\lambda_k = \lambda_l + 2\pi ij/\omega$ for some k, l and j, then the components with indices k and l can be rewritten as follows

$$e^{\lambda_k t} y_k(t) + e^{\lambda_l t} y_l(t) = e^{\lambda_l t} \tilde{y}_l(t),$$

where

$$\tilde{y}_l(t) = e^{2\pi ijt/\omega} y_k(t) + y_l(t)$$

is an ω-periodic function, as the functions $e^{2\pi ijt/\omega}$, $y_k(t)$, $y_l(t)$ are ω-periodic. Hence it is enough in this case to consider all functions of the form $x(t) = \sum_{l=1, l\neq k}^{M} e^{\lambda_l t} y_l(t)$, i.e. the space $E_\omega(\lambda_1, \dots, \lambda_{k-1}, \lambda_{k+1}, \dots, \lambda_M)$ instead of the space $E_\omega(\lambda_1, \dots, \lambda_M)$.

The class $E_\omega(\lambda_1, \dots, \lambda_M)$, called the *class of exponential-periodic functions*, is very useful. For example, the solutions of the Mathieu differential equation $x'' + \omega^2 x = \varepsilon \cos tx$, where ω and ε are parameters, are of the form (1.2) with $M = 2$, $\lambda_1 = \mu$, $\lambda_2 = -\mu$ and μ chosen suitably (see, for instance, Arscott [1]). Moreover, Ince [1] has proved that any linear homogeneous differential equation whose coefficients are π-periodic has at least one solution of the form $e^{\mu t}\varphi(t)$, where μ is chosen suitably and φ is a π-periodic function.

Many problems in physics and engineering are described by exponential-periodic functions. Two typical examples are shown in Fig. 7 and Fig. 8.

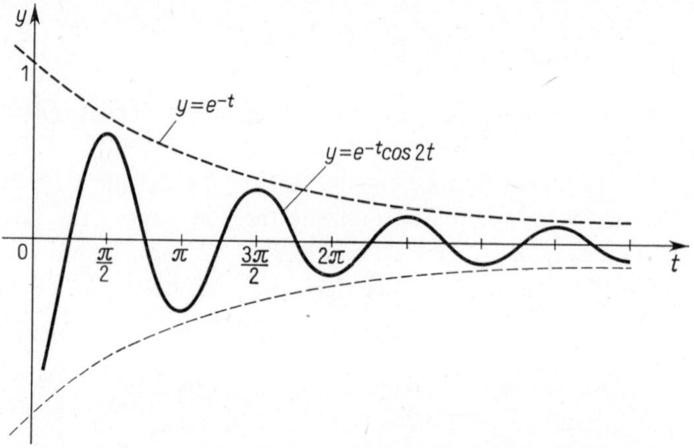

Fig. 7. An exponential-periodic function (damped oscillations)

It is clear that $E_\omega(\lambda_1, \dots, \lambda_M)$ is a linear space. When $\mathrm{re}\,\lambda_j = 0$ for $j = 1, 2, \dots, M$, the space $E_\omega(\lambda_1, \dots, \lambda_M)$ is a subspace of the space of almost periodic functions (see, for instance, Levitan [1]).

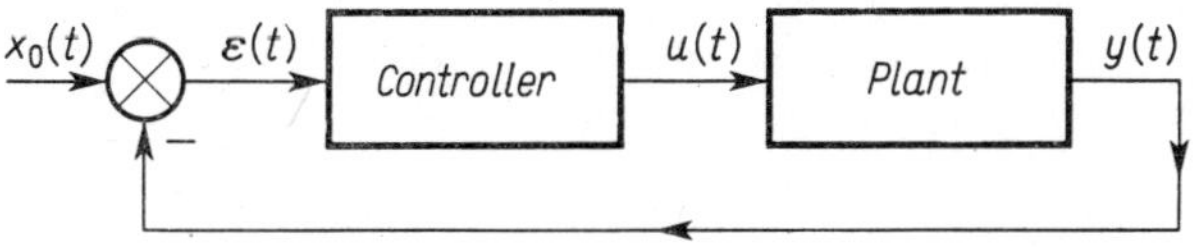

Controller:

$$u(t) = K_1\varepsilon(t)+K_2 \int_0^t \varepsilon(s)\,ds$$

Plant:

$$T\frac{dy(t)}{dt}+y(t) = u(t-\omega),$$

$$\varepsilon(t) = -y(t)$$

$$y(0) \neq 0$$

$$\Downarrow$$

$$-T\frac{d^2\varepsilon(t)}{dt^2}-\frac{d\varepsilon(t)}{dt}$$

$$= K_1\frac{d\varepsilon(t-\omega)}{dt}+K_2\varepsilon(t-\omega)$$

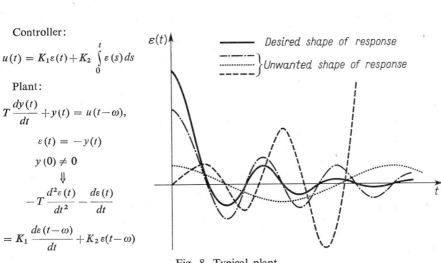

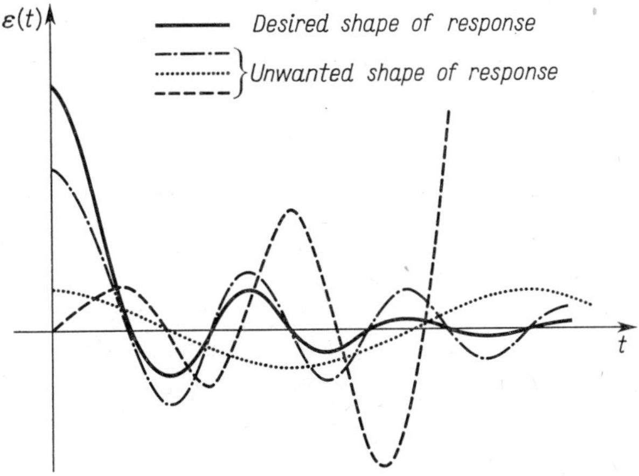

Fig. 8. Typical plant

Consider now the shift operator $(Sx)(t) = x(t-r)$. It is easy to verify that S is a linear operator transforming $E_\omega(\lambda_1, \ldots, \lambda_M)$ into itself. Moreover, S is an algebraic operator on this space. Indeed, let $x(t) = \sum_{l=1}^{M} e^{\lambda_l t} y_l(t)$, where $y_l \in E_\omega$, be an arbitrary function belonging to $E_\omega(\lambda_1, \ldots, \lambda_M)$. Since

$$S^N[e^{\lambda_l t} y_l(t)] = e^{\lambda_l(t-Nr)} y_l(t-Nr) = e^{-\lambda_l Nr} e^{\lambda_l t} y_l(t-\omega) = e^{-\lambda_l Nr} e^{\lambda_l t} y_l(t)$$

we find

$$\prod_{l=1}^{M} (S^N - e^{-\lambda_l Nr} I)x = 0.$$

But for any complex number z we have

$$S^N - z^N I = \prod_{\mu=1}^{N} (S-\varepsilon^\mu z I), \quad \text{where } \varepsilon = e^{2\pi i/N}$$

(compare Lemma II.1.3). This and the arbitrariness of x together imply that

$$\prod_{l=1}^{M}\prod_{\mu=1}^{N} (S-\varepsilon^\mu e^{-\lambda_l Nr} I) = 0 \quad \text{on } E_\omega(\lambda_1, \ldots, \lambda_M).$$

The characteristic polynomial

$$P(t) = \prod_{l=1}^{M} \prod_{\mu=1}^{N} (t - \varepsilon^{\mu} e^{-\lambda_l N r})$$

has simple roots only (provided that condition (1.3) is satisfied). The order of the operator S is $M \cdot N$. We write

(1.4) $e_{\nu} = \varepsilon^{\mu} e^{-\lambda_l r}$ for $\nu = M(\mu-1)+l$ ($\mu = 1, 2, \ldots, N$; $l = 1, 2, \ldots, M$).

This allows us to rewrite the polynomial $P(t)$ as

$$P(t) = \prod_{\nu=1}^{M \cdot N} (t - e_{\nu}), \tag{1.5}$$

where by assumption the roots e_{ν} are simple. Hence Theorem II.4.1 implies that the space $E_{\omega}(\lambda_1, \ldots, \lambda_M)$ is the direct sum of the spaces $X_{\nu} = P_{\nu} E_{\omega}(\lambda_1, \ldots, \lambda_M)$, where $P_{\nu} = \mathfrak{p}_{\nu}(S)$ (see Formula II.2.9) and the polynomials $\mathfrak{p}_{\nu}(S)$ are given by the formulae

$$\mathfrak{p}_{\nu}(t) = \prod_{\mu=1, \mu \neq \nu}^{M \cdot N} \frac{t - e_{\mu}}{e_{\nu} - e_{\mu}} \qquad (\nu = 1, 2, \ldots, M \cdot N).$$

According to Formula II.4.1, we write $x_{(\nu)} = P_{\nu} x$ and we have $S x_{(\nu)} = S P_{\nu} x = e_{\nu} x_{(\nu)}$ for $\nu = 1, 2, \ldots, M \cdot N$, and $x(t) = \sum_{\nu=1}^{M \cdot N} x_{(\nu)}(t)$.

Observe that the operator S commutes with derivation and with the operator of multiplication by an r-periodic function $a(t)$:

$$[(Sa - aS)x](t) = a(t-r) \sum_{l=1}^{M} e^{\lambda_l (t-r)} y_l(t-r) - a(t) \sum_{l=1}^{M} e^{\lambda_l (t-r)} y_l(t-r)$$

$$= [a(t-r) - a(t)] \sum_{l=1}^{M} e^{\lambda_l (t-r)} y_l(t-r) = [a(t) - a(t)] x(t-r) = 0.$$

We assume that the function $y(t)$ given in (1.1) belongs to the space $E_{\omega}(\lambda_1, \ldots, \lambda_M)$. Hence, just as in Section 1 of Chapter V, we infer that equation (1.1) is equivalent to the $N \cdot M$ differential equations without deviations:

$$A_{\nu} x_{\nu} = y_{(\nu)} \qquad (\nu = 1, 2, \ldots, M \cdot N), \tag{1.6}$$

where

$$y_{(\nu)} = P_\nu y \qquad (\nu = 1, 2, \ldots, M\cdot N),$$

(1.7) $$A_\nu u = \sum_{k=0}^{n} b_{k\nu}(t) u^{(k)}$$

$$b_{k\nu}(t) = \sum_{j=0}^{m} e_\nu^{\eta_j} a_{kj}(t) \qquad (\nu = 1, 2, \ldots, N\cdot M;\ k = 0, 1, \ldots, n).$$

This and Corollary II.5.1 together imply the following

THEOREM 1.1 (Włodarska-Dymitruk [1]). *Let $\omega_1, \ldots, \omega_m$ be commensurable and let $\omega_0 = 0$. Let $\omega = Nr$, where N is a common multiple of the positive integers $|\eta_1|, \ldots, |\eta_m|$, $\omega_j = \eta_j r$, $r \neq 0$. Let $y \in E_\omega(\lambda_1, \ldots, \lambda_M)$. Then equation* (1.1) *has a solution $x \in E_\omega(\lambda_1, \ldots, \lambda_M)$ if and only if all equations* (1.6) *have solutions belonging to $E_\omega(\lambda_1, \ldots, \lambda_M)$. The solution, if it exists, is of the form $x(t) = \sum_{\nu=1}^{M\cdot N} x_\nu(t)$, where x_ν is a solution belonging to $E_\omega(\lambda_1, \ldots, \lambda_M)$ of the νth equation* (1.4).

Just as before, this theorem is also valid for vector valued functions. The case, where the coefficients of the differential-difference operator A are ω-periodic will be studied in Section 2.

EXAMPLE 1.1. We are looking for solutions belonging to the class $E_{2\pi}(1, -1)$ of the equation:

(1.6) $$e^{\pi}x(t) - e^{\pi}x'(t) - x(t-\pi) - x'(t-\pi) = e^{-t}\sin t.$$

Here $r = \pi$, $\omega = 2\pi$, $N = 2$, $\varepsilon = -1$, $\lambda_1 = 1$, $\lambda_2 = -1$, $M = 2$ and $e_1 = -e^{-\pi}$, $e_2 = -e^{\pi}$, $e_3 = e^{-\pi}$, $e_4 = e^{\pi}$. Since

$$S(e^{-t}\sin t) = e^{-(t-\pi)}\sin(t-\pi) = -e^{\pi}e^{-t}\sin t = e_2 e^{-t}\sin t,$$

it follows that $e^{-t}\sin t \in X_2$. This implies that $y_{(1)} = y_{(3)} = y_{(4)} = 0$, and that $y_{(2)} = e^{-t}\sin t$. Hence we obtain the following system of differential equations:

$$(e^{\pi}+e^{-\pi})x_1 - (e^{\pi}-e^{-\pi})x_1' = 0, \quad 2e^{\pi}x_2' = e^{-t}\sin t,$$

$$(e^{\pi}-e^{-\pi})x_3 - (e^{\pi}+e^{-\pi})x_3' = 0, \quad -2e^{\pi}x_4' = 0.$$

It is easy to check that the first, the third and the fourth of these equations have only one solution belonging to the class $E_{2\pi}(1, -1)$, namely the solution identically equal to zero. The unique solution of the second equation is $x_2(t) = \frac{1}{4}e^{-\pi}e^{-t}(\sin t + \cos t) \in X_{(2)}$. Hence the solution we are looking for is $x(t) = \frac{1}{4}e^{-(t+\pi)}(\sin t + \cos t)$.

REMARK 1.1. Just as in Section 4 of Chapter V we can study the existence of solutions belonging to $E_\omega(\lambda_1, \ldots, \lambda_M)$ of a partial linear differential-difference equation with constant or periodic coefficients. We omit this problem here due to the rather complicated calculations involved.

2. NON-LINEAR DIFFERENTIAL-DIFFERENCE EQUATIONS

Consider the system of differential-difference equations

$$\frac{dx}{dt} = G(t, x(t), x(t-\omega_1), \ldots, x(t-\omega_m)), \tag{2.1}$$

where $x = (x_1, \ldots, x_q)$ is an unknown q-dimensional vector valued function and the given vector valued function $G(t, z_{0,1}, \ldots, z_{q,m})$ is a continuous vector valued function of its arguments. We assume as before that $\omega_1, \ldots, \omega_m$ are commensurable, i.e. that $\omega_j = \eta_j r$, $r \neq 0$, where the η_j are integers, and that N is a common multiple of the positive integers $|\eta_1|, \ldots, |\eta_m|$. Let $\omega = Nr$. We consider the space $E^q_\omega(\lambda_1, \ldots, \lambda_M)$ of all vector valued functions of the form

$$x(t) = \sum_{l=1}^{M} e^{\lambda_l t} y_l(t), \quad \text{where } y_l \in E^q_\omega, \tag{2.2}$$

E^q_ω is the space of all ω-periodic q-dimensional vector valued functions defined for $t \in R$, and the $\lambda_1, \ldots, \lambda_M$ are arbitrarily fixed complex numbers. Condition (1.3) is retained. As before, the shift operator $(Sx)(t) = x(t-r)$ is an algebraic operator on the space $E^q_\omega(\lambda_1, \ldots, \lambda_M)$ with characteristic polynomial $P(t) = \prod_{\nu=1}^{M\cdot N} (t-e_\nu)$, where $e_\nu = \varepsilon^\mu e^{-\lambda_l r}$ for $\nu = M(\mu-1)+l$ $(\mu = 1, 2, \ldots, N;$ $l = 1, 2, \ldots, M)$ and $\varepsilon = e^{2\pi i/N}$. We decompose the space $E^q_\omega(\lambda_1, \ldots, \lambda_M)$ into the direct sum of the spaces $X_\nu = P_\nu E^q_\omega(\lambda_1, \ldots, \lambda_M)$ having the same properties as these in Section 1. We now need a characterization of the spaces X_ν analogous to that given by Proposition 2.1 of Chapter V.

PROPOSITION 2.1 (Włodarska-Dymitruk [1]). *A vector valued function $u(t)$ belongs to X_ν if and only if*

$$u(t) = e^{\lambda_l t} e^{-2\pi i\mu t/\omega} v(t), \quad \textit{where } \nu = M(\mu-1)+l$$

($\mu = 1, 2, \ldots, N$; $l = 1, 2, \ldots, M$) *and the vector valued function $v(t)$ is r-periodic.*

Indeed, if $v(t)$ is an r-periodic vector valued function, then

$$u(t-r) = e^{\lambda_l(t-r)} e^{-2\pi i\mu(t-r)/\omega} v(t-r) = e^{\lambda_l t} e^{-\lambda_l r} e^{-2\pi i\mu t/\omega} e^{2\pi i\mu r/Nr} v(t)$$
$$= e_\nu e^{\lambda_l t} e^{-2\pi i\mu t/\omega} v(t) = e_\nu u(t).$$

Hence $u \in X_\nu$. Conversely, if $u \in X_\nu$, then $u(t-r) = e_\nu u(t) = e^{-\lambda_l r} \varepsilon^\mu u(t)$ and

$$v(t-r) = e^{-\lambda_l(t-r)} e^{2\pi i\mu(t-r)/\omega} u(t-r)$$
$$= e^{-\lambda_l t} e^{2\pi i\mu t/\omega} e^{\lambda_l r} e^{-2\pi i\mu r/Nr} e^{-\lambda_l r} \varepsilon^\mu u(t) = e^{-\lambda_l t} e^{2\pi i\mu t/\omega} u(t) = v(t),$$

which proves that $v(t)$ is r-periodic.

Now write

$$\tilde{x}_\nu(t) = e^{-\lambda_l t} e^{2\pi i\mu t/\omega} x_\nu(t) \qquad \text{for } \nu = M(\mu-1)+l$$

$$(2.3) \qquad (\mu = 1, 2, \dots, N;\ l = 1, 2, \dots, M),$$

$$\tilde{G}(t, \tilde{x}_1, \dots, \tilde{x}_N) = G\left(t, \sum_{k=1}^{M\cdot N} e^{\lambda_l t} e^{-2\pi i\mu t/\omega} \tilde{x}_k, \dots, \sum_{k=1}^{M\cdot N} e^{\lambda_l t} e^{-2\pi i\mu t/\omega} e^{2\pi i\mu\omega_m/\omega} \tilde{x}_k\right),$$

$$M(\mu-1)+l = k.$$

Proposition 2.1 implies that the $\tilde{x}_\nu$ are r-periodic vector valued functions. Acting on both sides of equation (2.1) with the projectors P_ν and using Proposition 2.1 and substitution (2.3) we can rewrite system (2.1) in the form of $q \cdot N \cdot M$ ordinary differential equations:

$$(2.4) \qquad \tilde{x}_\nu'(t) = \left(\frac{2\pi i\mu}{\omega} - \lambda_l\right) \tilde{x}_\nu(t) + \tilde{G}_\nu(t, \tilde{x}_1, \dots, \tilde{x}_{M\cdot N}),$$

$$\nu = M(\mu-1)+l,\ \mu = 1, 2, \dots, N;\ l = 1, 2, \dots, M,$$

where

$$(2.5) \qquad \tilde{G}_\nu(t, \tilde{x}_1, \dots, \tilde{x}_{M\cdot N}) = e^{-\lambda_l t} e^{2\pi i\mu t/\omega} P_\nu[\tilde{G}(t, \tilde{x}_1, \dots, \tilde{x}_{M\cdot N})]$$

and the operators P_ν acts on the variable t alone, since the vector-valued functions $\tilde{x}_\nu$ are r-periodic (compare the proof of Theorem 5.1 in Chapter V).

We finally obtain the following

THEOREM 2.1 (Włodarska-Dymitruk [1]). *The q-dimensional system* (2.1) *has a solution belonging to the space $E_\omega^q(\lambda_1, \dots, \lambda_M)$ if and only if the $q \cdot M \cdot N$-dimensional system* (2.4) *has an r-periodic solution. The solution, if it exists, is of the form*

$$x(t) = \sum_{\mu=1}^{N} \sum_{l=1}^{M} e^{\lambda_l t} e^{-2\pi i\mu t/\omega} x_{M(\mu-1)+l}(t),$$

where $(\tilde{x}_1, \dots, \tilde{x}_{M\cdot N})$ is an r-periodic solution of system (2.4).

If the vector valued function $G(t, z_{1,0}, \dots, z_{q,m})$ is ω-periodic with respect to the variable t, then the vector valued functions $\tilde{G}_\nu(t, \tilde{x}_1, \dots, \tilde{x}_{M\cdot N})$ given by Formulae (2.5) are r-periodic with respect to t.

In the same way we can study the linear differential-difference equation

$$(2.6) \qquad Ax \equiv \sum_{k=0}^{n} \sum_{j=0}^{m} a_{kj}(t) x^{(k)}(t-\omega_j) = y(t),$$

where the coefficients $a_{kj}(t)$ are ω-periodic and the given scalar valued function $y(t)$ belongs to the space $E_\omega(\lambda_1, \ldots, \lambda_M)$. (Here we have $q = 1$.) We can rewrite the operator A as follows:

$$Ax \equiv \sum_{k=0}^{n} \sum_{j\ 0}^{m} a_{kj}(t) x^{(k)}(t-\omega_j) = \sum_{k=0}^{n} \sum_{j=0}^{m} a_{kj}(t)\,(S^{\eta_j} x^{(k)})\,(t)$$

$$= \sum_{k=0}^{n} \sum_{j=0}^{m} a_{kj}(t) \Big[S^{\eta_j} \sum_{\nu=1}^{M\cdot N} x_{(\nu)}^{(k)} \Big](t) = \sum_{k=0}^{n} \sum_{j=0}^{m} a_{kj}(t) \sum_{\nu=1}^{N\cdot M} e_\nu^{\eta_j} x_{(\nu)}(t).$$

If we write

$$\gamma_\nu = \lambda_l - \frac{2\pi i \mu}{\omega} \quad \text{for } \nu = M(\mu-1)+l \ (\mu = 1, \ldots, N;\ l = 1, \ldots, M), \tag{2.7}$$

then Proposition 2.1 and the notation of (2.3) imply that $x_{(\nu)}(t) = e^{\lambda_l t} e^{-2\pi i \mu t/\omega} \tilde{x}_\nu(t) = e^{\gamma_\nu t} \tilde{x}_\nu(t)$. Hence, by the Leibniz formula,

$$x_{(\nu)}^{(k)}(t) = [e^{\gamma_\nu t} \tilde{x}_\nu(t)]^{(k)} = \sum_{\eta=0}^{k} \binom{k}{\eta} \gamma_\nu^{k-\eta} e^{\gamma_\nu t} \tilde{x}_\nu^{(\eta)}(t).$$

Therefore,

$$Ax \equiv \sum_{k=0}^{n} \sum_{j=0}^{m} \sum_{\nu=1}^{N\cdot M} \sum_{\eta=0}^{k} a_{kj}(t) e_\nu^{\eta_j} \binom{k}{\eta} \gamma_\nu^{k-\eta} e^{\gamma_\nu t} \tilde{x}_\nu^{(\eta)}(t)$$

$$= \sum_{\eta=0}^{n} \sum_{j=0}^{m} \sum_{\nu=0}^{M\cdot N} \sum_{k=\eta}^{n} \binom{k}{\eta} e_\nu^{\eta_j} \gamma_\nu^{k-\eta} e^{\gamma_\nu t} a_{kj}(t) \tilde{x}_\nu^{(\eta)}(t).$$

Acting on both sides of equation (2.6) with projectors P_σ ($\sigma = 1, 2, \ldots, M\cdot N$) we obtain

$$\sum_{\eta=0}^{n} \sum_{j=0}^{m} \sum_{\nu=1}^{M\cdot N} \sum_{k=\eta}^{n} \binom{k}{\eta} e_\nu^{\eta_j} \gamma_\nu^{k-\eta} \tilde{x}_\nu^{(\eta)}(t) P_\sigma[e^{\gamma_\nu t} a_{kj}]\,(t) = (P_\sigma y)\,(t) \tag{2.8}$$

since the functions $\tilde{x}_\nu^{(\eta)}$ being derivatives of r-periodic functions, are also r-periodic.

Applying Proposition 2.1 once again, we obtain

$$\begin{aligned} (P_\sigma y)\,(t) &= e^{\gamma_\sigma t} \tilde{y}_\sigma(t), \qquad \text{where } \tilde{y}_\sigma(t) \in E_r, \\ P_\sigma\big(e^{\gamma_\nu t} a_{kj}(t)\big) &= e^{\gamma_\sigma t} b_{kj\nu\sigma}(t), \qquad \text{where } b_{kj\nu\sigma} \in E_r. \end{aligned} \tag{2.9}$$

Multiplying both sides of (2.8) by $e^{-\gamma_\sigma t}$, we thus obtain

$$\sum_{\eta=0}^{n} \sum_{j=0}^{m} \sum_{\nu=1}^{M\cdot N} \sum_{k=\eta}^{n} \binom{k}{\eta} e_\nu^{\eta_j} \gamma_\nu^{k-\eta} b_{kj\nu\sigma}(t) \tilde{x}_\nu^{(\eta)}(t) = \tilde{y}_\sigma(t).$$

Now write

$$(2.10)\qquad \sum_{j=0}^{m}\sum_{k=\eta}^{n}\binom{k}{\eta}e_{\nu}^{\eta j}\gamma_{\nu}^{k-\eta}b_{kj\nu\sigma}(t) = d_{\nu\sigma}(t) \quad (\nu, \sigma = 1, 2, \ldots, N\cdot M),$$

where $b_{kj\nu\sigma}(t)$ is defined by the second formula of (2.9).

We finally obtain

THEOREM 2.2 (Włodarska-Dymitruk [1]). *Equation* (2.6) *with* $y \in E_{\omega}(\lambda_1, \ldots, \lambda_M)$ *has a solution belonging to* $E_{\omega}(\lambda_1, \ldots, \lambda_M)$ *if and only if the system*

$$(2.11)\qquad \sum_{\eta=0}^{n}\sum_{\nu=1}^{M\cdot N} d_{\nu\sigma}(t)\tilde{x}_{\nu}^{(\eta)}(t) = \tilde{y}_{\sigma}(t) \quad (\sigma = 1, 2, \ldots, N\cdot M)$$

has an r-periodic solution, where the $d_{\nu\sigma}(t)$ *are defined by Formula* (2.10). *The solution, if it exists, is of the form* $x(t) = \sum_{\nu=1}^{N\cdot M} e^{\gamma_{\nu}t}\tilde{x}_{\nu}(t)$, *where the* γ_{ν} *are defined by Formula* (2.7) *and* $(\tilde{x}_1, \ldots, \tilde{x}_{N\cdot M})$ *is an r-periodic solution of system* (2.11).

Note that even for the space $E_{\omega} = E_{\omega}(0, \ldots, 0)$ this last result is stronger than Theorem 2.2 of Chapter 2.2.

In the case, where the deviations are not commensurable, we have a result similar to Theorems 3.1 and 3.2 in Chapter V concerning perturbations of deviations.

Let C_{ω}^{n} be the space defined in Section 3 of Chapter V and let $C_{\omega}^{n}(\lambda_1, \ldots, \lambda_M)$ denote the space of all vector valued functions of the form

$$(2.12)\qquad x(t) = \sum_{l=1}^{M} e^{\lambda_l t}x_l(t), \quad \text{where } x_l(t) \in C_{\omega}^{n} \quad (l = 1, 2, \ldots, M).$$

We also write $C_{\omega}(\lambda_1, \ldots, \lambda_M) = C_{\omega}^{0}(\lambda_1, \ldots, \lambda_M)$. Consider the following linear differential-difference equation

$$(2.13)\qquad \frac{dx}{dt} + \sum_{j=0}^{m} a_j(t)x(t-h_j) = y(t) \quad (h_0 = 0),$$

where $a_1, \ldots, a_m$ are ω-periodic complex vector valued functions determined for $t \in \boldsymbol{R}$. We have the following

THEOREM 2.3 (Włodarska-Dymitruk[1]). *If the homogeneous equations* (2.13) (*i.e. this equation with* $y \equiv 0$) *has only zero as a solution belonging to* $C_{\omega}^{1}(\lambda_1, \ldots$ $\ldots, \lambda_M)$, *then equation* (2.13) *has a unique solution* $x_h \in C_{\omega}^{1}(\lambda_1, \ldots, \lambda_M)$ *for every*

[1] This result has been unpublished.

$y \in C_\omega(\lambda_1, \ldots, \lambda_M)$. *If equation* (2.13) *has a unique solution* $x_h \in C^1_\omega(\lambda_1, \ldots, \lambda_M)$, *then for arbitrary reals* $h'_1, \ldots, h'_m$ *such that the moduli* $|h'_j - h_j|$ *are sufficiently small* ($j = 1, 2, \ldots, m$), *the equation*

$$\frac{dx}{dt} + \sum_{j=0}^{m} a_j(t)x(t-h'_j) = y(t) \qquad (h'_0 = 0) \tag{2.14}$$

has a unique solution $x_{h'} \in C^1_\omega(\lambda_1, \ldots, \lambda_M)$. *Moreover, if* $h'_j \to h_j$ *for* $j = 1, 2, \ldots$ $\ldots, m$, *then* $x_{h'}$ *tends uniformly to* x_h.

PROOF. The space $C^1_\omega(\lambda_1, \ldots, \lambda_M)$ can be decomposed into a direct sum of the spaces X_l ($l = 1, \ldots, M$) such that $x_l(t) \in X_l$ if and only if $x_l(t) = e^{\lambda_l t}\hat{x}_l(t)$, where $\hat{x}_l(t) \in C^1_\omega$. Let $y(t) = \sum_{l=1}^{M} e^{\lambda_l t}\hat{y}_l(t)$. By this decomposition we obtain the following system of the M independent equations:

$$\frac{d\hat{x}(t)}{dt} + \sum_{j=0}^{M} \hat{a}_{jl}(t)\hat{x}_l(t-h_j) = \hat{y}_l(t) \qquad (l = 1, \ldots, M), \tag{2.15}$$

where

$$\hat{y}_l(t) = e^{-\lambda_l t}y_l(t),$$

$$\hat{a}_{jl}(t) = \begin{cases} a_0 + \lambda_l & \text{for } j = 0 \ (h_0 = 0), \\ e^{-\lambda_l h_j}a_j(t) & \text{for } j = 1, \ldots, m, \end{cases}$$

and $\hat{x}_l \in C^1_\omega$, $\hat{y}_l$, $\hat{a}_{jl} \in C_\omega$.

Since all the given functions are ω-periodic, applying Theorems 3.1 and 3.2 of Chapter V to each equation of (2.15) and writing $x(t) = \sum_{l=1}^{M} e^{\lambda_l t}\hat{x}_l(t)$, we obtain our theorem.

In the same way we prove this theorem for the equation

$$\frac{d^n x}{dt^n} = \sum_{k=0}^{n-1}\sum_{j=0}^{m} a_{kj}(t)x^{(k)}(t-h_j) = y(t) \qquad (n > 1, h_0 = 0), \tag{2.16}$$

where we are looking for solutions belonging to $C^n_\omega(\lambda_1, \ldots, \lambda_M)$.

Consider now, as in Section 5 of Chapter V, a system dependent on a parameter μ (scalar or vector):

$$\frac{dx}{dt} = \boldsymbol{G}(t, \boldsymbol{x}(t), \boldsymbol{x}(t-\omega_1), \ldots, \boldsymbol{x}(t-\omega_m), \mu), \tag{2.17}$$

where ω_i and $\boldsymbol{G}$ satisfy all the same conditions as in system (2.1). Moreover, we assume that $\boldsymbol{G}(t, z_{1,0}, \ldots, z_{q,m}, \mu)$ and $\dfrac{\partial \boldsymbol{G}}{\partial z_{ij}}$ $(i = 1, \ldots, q;\ j = 0, 1, \ldots, m)$ are vector valued functions belonging to $E^q_\omega(\lambda_1, \ldots, \lambda_M)$, continuous with respect to the variable t.

Suppose that for $\mu = 0$ there is a solution $\boldsymbol{p}(t) \in E^q_\omega(\lambda_1, \ldots, \lambda_M)$ of system (2.17) and consider the linearized system

$$\frac{dy_i}{dt} = \sum_{j=0}^{m} \sum_{k=0}^{n} \left.\frac{\partial G_i}{\partial z_{k,j}}\right|_{(t,p(t),0)} y_i(t-\omega_j) \quad (i = 1, 2, \ldots, q) \tag{2.18}$$

corresponding to system (2.17), where G_i is the ith coordinate of the vector valued function $\boldsymbol{G}$.

THEOREM 2.3 (Włodarska-Dymitruk [1]). *If the linearized system* (2.18) *only has one solution belonging to* $E^q_\omega(\lambda_1, \ldots, \lambda_M)$ *namely that which is identically zero, then for* $|\mu|$ *sufficiently small system* (2.17) *has the unique solution* $\boldsymbol{q}(t, \mu) \in E^q_\omega(\lambda_1, \ldots, \lambda_M)$ *dependent continuously on* (t, μ) *and such that* $\boldsymbol{q}(t, 0) = \boldsymbol{p}(t)$.

Indeed, Theorems 2.1 and 2.2 imply that the existence of solutions belonging to $E^q_\omega(\lambda_1, \ldots, \lambda_M)$ of systems (2.17) and (2.18) is equivalent to the existence of r-periodic solutions of the corresponding systems of ordinary differential equations. Theorem 2.3 thus follows immediately from the classical theorem on perturbations of periodic ordinary differential equations (cf. Coddington and Levinson [1], Chapter XIV, § 1).

CHAPTER VII

Linear Differential Equations with Reflection and Rotation of Argument

The assumptions made at the beginning of Chapter V are retained. An equation in which, as well as the unknown function $x(t)$ and its derivatives, appear the value $x(-t)$ and, perhaps, the derivatives at the point $-t$, is called a *differential equation with reflection.* An equation in which as well as the unknown function $x(t)$ and its derivatives, the values $x(\varepsilon t-\alpha_1), \ldots, x(\varepsilon t-\alpha_N)$ and the corresponding values of the derivatives, where $\varepsilon_1, \ldots, \varepsilon_N$ are Nth roots of unity and $\alpha_1, \ldots, \alpha_N$ are complex numbers, is called a *differential equation with rotation.* For $N = 2$ this last definition includes the previous one.

Some results for non-linear differential equations with reflection are given in Section 4 of Chapter VIII.

1. ORDINARY LINEAR DIFFERENTIAL EQUATIONS WITH REFLECTION

Suppose we are given a differential equation with reflection of order one,

$$a_0 x(t)+b_0 x(-t)+a_1 x'(t)+b_1 x'(-t) = y(t), \tag{1.1}$$

where the given function $y(t)$ is defined on the real line (or perhaps on a subset of the real line symmetric with respect to zero) and a_0, a_1, b_0, b_1 are real or complex scalars.

Denote the reflection operator by S: $(Sx)(t) = x(-t)$ (Fig. 9). Then on any space X of functions defined on the real line[1] S is an involution; i.e. $S^2 = I$.

[1] We do not impose any restriction on this space, because it does not play any role in what follows. Any linear space of functions defined on the real line will do here, even a linear space of vector valued functions, or functions defined on the real line with values in any Banach space.

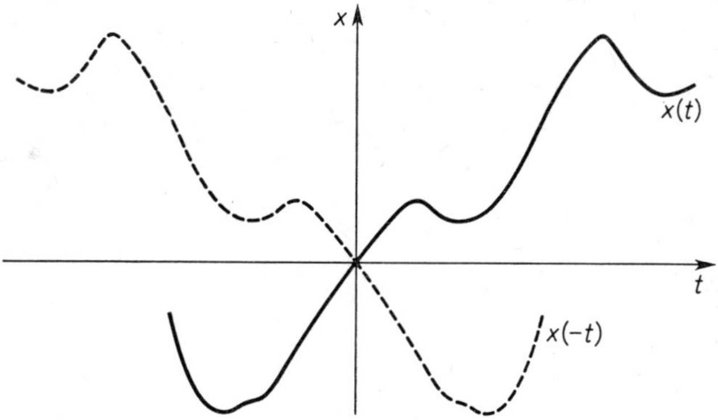

Fig. 9. Reflection

Note that in this situation we have for all $x \in X$:

$$x^+(t) = (P^+x)(t) = \frac{x(t)+x(-t)}{2} \quad \text{is an even function,}$$

$$x^-(t) = (P^-x)(t) = \frac{x(t)-x(-t)}{2} \quad \text{is an odd function.}$$

Hence the reflection S decomposes all functions into a direct sum of even and odd functions.

The differentiation operator $Dx = x'$ anticommutes with S:

$$(SDx)(t) = -x'(-t), \qquad (DS)(t) = -x'(-t) = -(SD)(t),$$

which implies that $SD+DS = 0$ on X. This enables to us to use the theory developed in Section 1 of Chapter III. Suppose now, that $a_1^2-b_1^2 \neq 0 \neq a_0^2-b_0^2$, and observe that the general form of the solution of the equation $(D-\sqrt{\lambda t})z_1 = 0$, where $\lambda = (a_0^2-b_0^2)/(a_1^2-b_1^2)^{-1} \neq 0$, is $z_1(t) = Ce^{\sqrt{\lambda}t}$, where C is an arbitrary constant. We also easily find a solution $\tilde{x}(t)$ of the equation

$$(D^2-\lambda I)\tilde{x} = y. \tag{1.2}$$

Namely, since $e^{\sqrt{\lambda}t}$ and $e^{-\sqrt{\lambda}t}$ are two linearly independent solutions of (1.2) and the Wronskian $W(t)$ of these function is $W(t) = -2\sqrt{\lambda} \neq 0$ by our assumptions, we obtain

$$\tilde{x}(t) = -\frac{1}{2\sqrt{\lambda}}\left\{e^{\sqrt{\lambda}t}\int e^{-\sqrt{\lambda}t}[-y(t)]\,dt + e^{-\sqrt{\lambda}t}\int e^{\sqrt{\lambda}t}y(t)\,dt\right\}$$

$$= \frac{1}{2\sqrt{\lambda}}\left[e^{\sqrt{\lambda}t}\int e^{-\sqrt{\lambda}t}y(t)\,dt - e^{-\sqrt{\lambda}t}\int e^{\sqrt{\lambda}t}y(t)\,dt\right].$$

By simple calculations we find

$$\tilde{x}'(t) = \frac{1}{2\sqrt{\lambda}}\left[e^{\sqrt{\lambda}t}\int e^{-\sqrt{\lambda}t}\left[-y(t)\right]dt + e^{-\sqrt{\lambda}t}\int e^{\sqrt{\lambda}t}y(t)\,dt\right].$$

Finally, applying Theorem III.1.3, we can formulate the following

THEOREM 1.1. *If $a_0^2-b_0^2 \neq 0 \neq a_1^2-b_1^2$, then any solution of equation* (1.1) *is of the form*

$$x(t) = (R_A\tilde{x})(t) + C[(a_0-a_1\sqrt{\lambda})e^{\sqrt{\lambda}t} - (b_0+b_1\sqrt{\lambda})e^{-\sqrt{\lambda}t}],$$

where C is an arbitrary constant (real or complex, depending on the coefficients), $\lambda = (a_0^2-b_0^2)/(a_1^2-b_1^2) \neq 0$ and

$$\begin{aligned}(R_A\tilde{x})(t) &= -(a_1^2-b_1^2)^{-1}[a_0\tilde{x}(t)-b_0\tilde{x}(-t)-a_1\tilde{x}'(t)-b_1\tilde{x}'(-t)]\\ &= -(a_1^2-b_1^2)^{-1}\frac{1}{2\sqrt{\lambda}}\Big[(a_0-a_1\sqrt{\lambda})e^{\sqrt{\lambda}t}\int y(t)e^{-\sqrt{\lambda}t}dt-\\ &\quad -(a_0+a_1\sqrt{\lambda})e^{-\sqrt{\lambda}t}\int y(t)e^{\sqrt{\lambda}t}dt-\\ &\quad -(b_0-b_1\sqrt{\lambda})e^{-\sqrt{\lambda}t}\Big(\int y(s)e^{\sqrt{\lambda}s}ds\Big)_{s=-t}+\\ &\quad +(b_0+b_1\sqrt{\lambda})e^{\sqrt{\lambda}t}\Big(\int y(s)e^{-\sqrt{\lambda}s}ds\Big)_{s=-t}\Big].\end{aligned}$$

We now consider equation (1.1) when the condition $a_0^2-b_0^2 \neq 0 \neq a_1^2-b_1^2$ is not satisfied. We suppose, of course, that a_0, b_0, a_1, b_1 do not vanish simultaneously. If $a_1+b_1 = a_1-b_1 = 0$, equation (1.1) is a *functional equation with reflection* and we obtain by Theorem III.5.2) the following

THEOREM 1.2. *The functional equation* (1.1) *in the case $a_1+b_1 = a_1-b_1 = 0$ has the following solutions*:

(a) *If $a_0^2-b_0^2 \neq 0$, then* (1.1) *has the unique solution*

$$x(t) = (a_0^2-b_0^2)^{-1}[a_0y(t)-b_0y(-t)].$$

(b) *If $a_0+b_0 = 0$ but $a_0-b_0 \neq 0$ a necessary and sufficient condition for the solvability of equation* (1.1) *is $y(t)+y(-t) = 0$, which means that $y(t)$ is an odd function and under this condition the solution is of the form*

$$x(t) = \tfrac{1}{2}a_0^{-1}y(t)+x^+(t), \quad \textit{where } x^+(t) \textit{ is an arbitrary even function.}$$

(c) *If $a_0-b_0 = 0$, but $a_0+b_0 \neq 0$ we obtain similarly a solution*

$$x(t) = \tfrac{1}{2}a_0^{-1}y(t)+x^-(t), \quad \textit{where } x^-(t) \textit{ is an arbitrary odd function,}$$

under the necessary and sufficient condition that $y(t)$ be an even function.

If $a_0+b_0 = a_0-b_0 = 0$ but $a_1^2-b_1^2 \neq 0$, then the left-hand side of equation (1.1) can be written in the form

$$a_1 x'(t)+b_1 x'(-t) = [(a_1 I+b_1 S)Dx](t).$$

Hence we can solve equation (1.1) with respect to the new unknown function $u(t) = x'(t)$ by Theorem 1.2 and thus obtain a new equation

$$x'(t) = (a_1^2-b_1^2)^{-1}[a_1 y(t)-b_1 y(-t)].$$

This implies

THEOREM 1.3. *The differential equation with reflection* (1.1) *in the case where* $a_0+b_0 = a_0-b_0 = 0$ *and* $a_1^2-b_1^2 \neq 0$ *has the solution*

$$x(t) = (a_1^2-b_1^2)^{-1}\int [a_1 y(t)-b_1 y(-t)dt]+C,$$

where C *is an arbitrary constant.*

In a similar way, using points B.IV and B.V of Section 1 in Chapter III we obtain further

THEOREM 1.4. *Equation* (1.1) *in the case where* $a_1+b_1 = 0$ *but* $a_1-b_1 \neq 0$ *has the following solutions*:

(a) *If* $a_0^2-b_0^2 \neq 0$, *then the solution is unique and*

$$x(t) = (a_0^2-b_0^2)^{-1}\left[a_0 y(t)-b_0 y(-t)-(a_1-b_1)\frac{y'(t)-y'(-t)}{2}\right].$$

(b) *If* $a_0+b_0 = 0$ *but* $a_0-b_0 \neq 0$, *then a solution* $x(t)$ *exists if and only if* $y(t)$ *is an odd function and then*

$$x(t) = x_1(t)+(a_0-b_0)^{-1}y(t)-(a_1-b_1)x_1'(t),$$

where $x_1(t)$ *is an arbitrary even function.*

(c) *If* $a_0-b_0 = 0$ *but* $a_0+b_0 \neq 0$, *then a solution exists if and only if*

$$\frac{y(t)-y(-t)}{2}-\frac{a_1-b_1}{a_0+b_0}\cdot\frac{y'(t)-y'(-t)}{2} = 0^{1}.$$

Under this condition the solution is

$$x(t) = (a_0+b_0)^{-1}\frac{y(t)+y(-t)}{2}+x_1(t),$$

where $x_1(t)$ *is an arbitrary odd function.*

---

[1] Recall here that $(Dy^+)(t) = \left(\frac{y(t)+y(-t)}{2}\right)' = \frac{y'(t)-y'(-t)}{2}$.

(d) *If* $a_0-b_0 = a_0+b_0 = 0$, *then a necessary and sufficient condition for the existence of a solution of equation* (1.1) *is that* $y(t) \equiv 0$. *In this case* $x(t) = \int z(t)\,dt$, *where* $z(t)$ *is an arbitrary even function.*

If $a_1-b_1 = 0$ but $a_1+b_1 \neq 0$, then interchanging the roles of a_1+b_1 and a_1-b_1 and of even and odd functions, we obtain a theorem analogous to Theorem 1.4.

Note that in the case where $a_1-b_1 = a_1+b_1 = 0$, i.e. in the case of a functional equation with reflection (Theorem 1.2), if the given function is a scalar valued function, the assumption that all the coefficients are constant is not essential. It is enough to assume that a_0 and b_0 commute with S. For example, if these coefficients are even functions, this condition will be satisfied. Indeed, if $a(t)$ is an even function, then $(Sa)(t) = a(-t) = a(t)$, whence $[(Sa-aS)x](t) = a(-t)x(-t)-a(t)x(-t) = 0$ for all x. We therefore can apply Theorem II.5.2. This is also true for Theorem 1.3.

EXAMPLE 1.1. Consider equation (1.1) with the right-hand side function $y(t) \equiv 1$. To begin with, suppose that $a_0^2-b_0^2 \neq 0 \neq a_1^2-b_1^2$. According to Theorem 1.1 we find

$$\tilde{x}(t) = \frac{1}{2\sqrt{\lambda}}\left[e^{\sqrt{\lambda}t}\int e^{-\sqrt{\lambda}t}dt - e^{-\sqrt{\lambda}t}\int e^{\sqrt{\lambda}t}dt\right]$$

$$= \frac{1}{2\sqrt{\lambda}}\left[e^{\sqrt{\lambda}t}\left(-\frac{e^{-\sqrt{\lambda}t}}{\sqrt{\lambda}}\right) - e^{-\sqrt{\lambda}t}\frac{e^{\sqrt{\lambda}t}}{\sqrt{\lambda}}\right] = -\frac{1}{\lambda},$$

where $\lambda = (a_0^2-b_0^2)/(a_1^2-b_1^2)^{-1} \neq 0$ and the solution is

$$x(t) = \frac{1}{a_0+b_0} + C[(a_0-a_1\sqrt{\lambda})e^{\sqrt{\lambda}t} - (b_0+b_1\sqrt{\lambda})e^{-\sqrt{\lambda}t}],$$

where C is an arbitrary constant.

If $a_1+b_1 = a_1-b_1 = 0$, we obtain by Theorem 1.2:

(a) if $a_0^2-b_0^2 \neq 0$, then $x(t) = \dfrac{1}{a_0+b_0}$;

(b) if $a_0+b_0 = 0$ but $a_0-b_0 \neq 0$, there is no solutions since the condition for solvability is not satisfied. In fact $y(t) \equiv 1$ is an even function.

(c) if $a_0-b_0 = 0$ but $a_0+b_0 \neq 0$, the condition for solvability is satisfied and $x(t) = \dfrac{1}{2a_0} + x_1(t)$, where $x_1(t)$ is an arbitrary odd function.

Theorem 1.3 implies that for $a_0+b_0 = a_0-b_0 = 0$ and $a_1^2-b_1^2 \neq 0$ we obtain $x(t) = \dfrac{t}{a_1+b_1} + C$, where C is an arbitrary constant, and so on.

EXAMPLE 1.2. We settle the following problem, called "the problem of reciprocal trajectories". Namely, we have to find a plane curve $L\colon y = y(x)$, such that the reflection L' of L with respect to the axis Oy shifted parallelwise in the direction of the axis Oy intersects L at a given angle. The case where this angle is constant was solved by J. Bernoulli and L. Euler (see Lacroix [1]). The problem reduces to that of solving a differential equation with reflection:

$$\arctan\frac{1}{y'(x)}+\arctan\frac{1}{y'(-x)} = 2a.$$

If $z(x) = \arctan\dfrac{1}{y'(x)}$, we obtain a functional equation with reflection: $z(x)+z(-x) = 2a$. It follows from Theorem 1.2 that $z(x) = a+u(x)$, where $u(x)$ is an arbitrary odd function (the condition for solvability is automatically satisfied, for $y(t) = 2a$ is an even function). Hence

$$y'(x) = \cot(a+u(x)) \quad \text{and} \quad y(x) = C+x\cot(a+u(x))-2\int\frac{xu'(x)dx}{1-\cos(u(x)-a)}.$$

So far we have considered only differential equations with reflection of order one. The problem is much more difficult in the case of a differential equation with reflection of order greater than one. All we do here is to give without proof a simple consequence of Theorem III.2.5.

THEOREM 1.5 (Mażbic-Kulma [1]). *Suppose we are given a differential equation with reflection of order n with constant coefficients*:

$$\sum_{k=0}^{n}[a_k x^{(k)}(t)+b_k x^{(k)}(-t)] = y(t). \tag{1.3}$$

We suppose that

(a) $a_n^2-b_n^2 \neq 0$,

(b) $a_{j-k}a_k-b_{j-k}b_k \neq 0$ *for* $k = 0, 1, \ldots, n$ *and* $j = k+1, \ldots, k+n$,

(c) *the polynomial* $\sum_{j=0}^{n}\lambda_{2j}t^j$ *has simple roots* u_q *only, where*

$$\lambda_j = \begin{cases}\sum_{k=0}^{j} c_{jk} & \text{for } 0 \leqslant j \leqslant n,\\ \sum_{k=j-n}^{n} c_{jk} & \text{for } n \leqslant j \leqslant 2n,\end{cases} \qquad c_{jk} = (-1)^{n+j-k}(a_n^2-b_n^2)(a_{j-k}a_k-b_{j-k}b_k).$$

Then every solution of equation (1.3) *is of the form*

$$x(t) = (-1)^n(a_n^2-b_n^2)^{-1}\sum_{m=0}^{n}[(-1)^m a_m\tilde{x}(t)-b_m\tilde{x}(-t)]+$$
$$+\sum_{q=1}^{n}\sum_{k=0}^{n}(C_k u_q^{k/2}(a_k e^{\sqrt{u_q}t}-b_k e^{-\sqrt{u_q}t}),$$

where the C_k *are arbitrary constants and* $\tilde{x}(t)$ *is a solution of the equation*

$$\prod_{q=1}^{n}\left(\frac{d^2}{dt^2}-u_q\right)\tilde{x}(t) = y(t).$$

When we consider a system of functional equations with reflection which have constant coefficients:

$$\sum_{j=1}^{n}[a_{jk}x(t)+b_{jk}x(-t)] = y(t) \qquad (k = 1, 2, \ldots, n), \tag{1.4}$$

we have to use a different method. Namely, consider the space X of all vector valued functions $\boldsymbol{x}(t) = \big(x_1(t), \ldots, x_n(t)\big)$ defined for $t \in \boldsymbol{R}$, and let us denote the matrices (a_{jk}) and (b_{jk}) by $\boldsymbol{a}$ and $\boldsymbol{b}$, respectively, and let $(S\boldsymbol{x})(t) = \boldsymbol{x}(-t)$. Then system (1.4) can be written as a functional equation with reflection:

$$(\boldsymbol{a}+\boldsymbol{b}S)\boldsymbol{x} = \boldsymbol{y}, \quad \boldsymbol{y}(t) = \big(y_1(t), \ldots, y_n(t)\big) \tag{1.5}$$

with coefficients $\boldsymbol{a}$ and $\boldsymbol{b}$ which do not commute in general, but which both commute with the reflection operator. Using Proposition II.6.1 we obtain

THEOREM 1.6. *If the given matrices* $\boldsymbol{a}$ *and* $\boldsymbol{b}$ *are invertible, then system* (1.4) (*i.e. equation* (1.5)) *has the unique solution*

$$\boldsymbol{x}(t) = [\boldsymbol{a}^2-\boldsymbol{aba}^{-1}\boldsymbol{b}]^{-1}[\boldsymbol{ay}(t)-\boldsymbol{aba}^{-1}\boldsymbol{y}(-t)].$$

This theorem also holds in the case where all the coefficients are even functions for the same reasons as in Theorems 1.2 and 1.3.

A different method for solving an equation with reflection in some particular cases is given in Example 4.1 of Chapter XV. For non-linear equations with reflection we can apply results of Section 4 of the next Chapter.

2. ORDINARY LINEAR DIFFERENTIAL EQUATIONS WITH ROTATION

Consider in the complex plane the differential equation

$$\sum_{k=0}^{N-1} a_k x(\varepsilon^k t) - \sum_{k=0}^{N-1} b_k x'(\varepsilon^k t) = y(t), \tag{2.1}$$

where the a_k, b_k are complex constant, $\varepsilon = e^{2\pi i/N}$ and $N > 2$ (the case $N = 2$ was completely solved in the last section). Consider the operator given by

$$(Sx)(t) = x(\varepsilon t).$$

This is an involution of order N on the space X of all complex-valued differentiable functions of a complex variable. Indeed, $(S^m x)(t) = x(\varepsilon^m t)$ for $m = 1, 2, \ldots$ Hence

$$(S^N x)(t) = x(\varepsilon^N t) = x(t) \quad \text{for all } x \in X.$$

The differentiation operator $(Dx)(t) = x'(t)$ permutes S. Indeed,

$$(DSx)(t) = [x(\varepsilon t)]' = \varepsilon x'(\varepsilon t) = \varepsilon(SDx)(t) \quad \text{for all } x \in X$$

(cf. Section 3 of Chapter III).

It follows from Theorem II.4.1 that the space X is the direct sum of the spaces $X_\nu = P_\nu X$, where

$$(2.2) \qquad (P_\nu x)(t) = \frac{1}{N} \sum_{k=0}^{N-1} \varepsilon^{-k\nu} x(\varepsilon^k t) \quad (\nu = 1, 2, \ldots, N).$$

As a simple consequence of Theorem III.4.1 we obtain the following

THEOREM 2.1. *Suppose that* $a(\varepsilon^j) \neq 0 \neq b(\varepsilon^j)$ *for* $j = 1, 2, \ldots, N$[1]. *Then every solution of equation* (2.1) *is of the form*

$$x(t) = -\tilde{\tilde{x}}^{(N-1)}(t) - \frac{1}{N} \sum_{j=1}^{N} \sum_{k=0}^{N-1} \varepsilon^{-kj} [c_{N-1+j} \tilde{\tilde{x}}^{(N-2)}(\varepsilon^k t) +$$

$$+ c_{N-1+j} c_{N-2+j} \tilde{\tilde{x}}^{(N-3)}(\varepsilon^k t) + \ldots + c_{N-1+j} \ldots c_{1+j} \tilde{\tilde{x}}(\varepsilon^k t)] + c \sum_{k=0}^{N-1} d_k e^{\varepsilon^k \lambda_0 t},$$

where

$\tilde{x}(t)$ *is a solution of the equation* $\tilde{x}^{(N)} - \lambda \tilde{x} = y$,

$$\tilde{\tilde{x}}(t) = \{[b(S)]^{-1} \tilde{x}\}(t) = \frac{1}{N} \sum_{n=0}^{N-1} \sum_{m=0}^{N-1} \frac{\varepsilon^{-mj}}{b(\varepsilon^j)} \tilde{x}(\varepsilon^m t),$$

$$\lambda = \prod_{j=1}^{N} c_j, \quad c_{N+j} = c_j = \frac{a(\varepsilon^j)}{b(\varepsilon^j)} \quad (j = 0, 1, 2, \ldots, N-1),$$

C *is an arbitrary complex number,*

$$d_k = \sum_{m=1}^{N} \lambda_0^{-m} c_1 \ldots c_m V_{k,m} \quad (k = 0, 1, \ldots, N-1),$$

$$\lambda_0 = \sqrt[N]{|\lambda|} e^{\varphi i/N}, \quad \varphi = \operatorname{Arg} \lambda \quad (0 \leqslant \operatorname{Arg} \lambda < 2\pi),$$

---

[1] As before, we write: $a(t) = \sum_{k=0}^{N-1} a_k t^k$, $b(t) = \sum_{k=0}^{N-1} b_k t^k$.

and $V_{k,m}$ *is the subdeterminant obtained by cancelling the* $(k+1)$*th column and the* m*th row of the Vandermonde determinant* V *of the numbers* $\varepsilon^2, \ldots, \varepsilon^N, \varepsilon^1$ $(m = 1, 2, \ldots, N;\ k = 0, 1, \ldots, N-1)$.

We now drop the assumption that $a(\varepsilon^j) \neq 0 \neq b(\varepsilon^j)$ for $j = 1, 2, \ldots, N$ and we discuss some of the most typical cases. As in Formula III. (4.5), equation (2.1) can be written as an equivalent system of equations:

$$\begin{aligned} a(\varepsilon^m)x_{(m)} - b(\varepsilon^m)x'_{(m+1)} &= y_{(m)} \quad \text{for } m = 1, 2, \ldots, N-1, \\ a(\varepsilon^N)x_{(N)} - b(\varepsilon^N)x'_1 &= y_{(1)}, \end{aligned} \tag{2.3}$$

where $x_{(m)} = P_m x$, $y_{(m)} = P_m y$ and the P_m are defined by Formula (2.2). We, of course, assume that $a(\varepsilon^m), b(\varepsilon^m)$, $m = 1, 2, \ldots, N$, do not all vanish simultaneously.

If $b(\varepsilon^m) = 0$ for $m = 1, 2, \ldots, N$, equation (1.1) is a *functional equation with rotation* and Theorem II.5.2 implies

THEOREM 2.2. *The functional equation with rotation*

$$\sum_{k=0}^{N-1} a_k x(\varepsilon^k t) = y(t) \tag{2.4}$$

has the solution

$$x(t) = \frac{1}{N} \sum_{m:a(\varepsilon^m)\neq 0} \frac{1}{a(\varepsilon^m)} \sum_{k=0}^{N-1} \varepsilon^{-km} y(\varepsilon^k t) + \frac{1}{N} \sum_{m:a(\varepsilon^m)=0} \sum_{k=0}^{N-1} \varepsilon^{-km} z(\varepsilon^k t)$$

(the first sum runs over all m *such that* $1 \leqslant m \leqslant N$ *and* $a(\varepsilon^m) \neq 0$ *and the second sum over all* m *such that* $1 \leqslant m \leqslant N$ *and* $a(\varepsilon^m) = 0$*) under the necessary and sufficient condition that*

$$\sum_{k=0}^{N-1} \varepsilon^{-km} y(\varepsilon^k t) = 0 \quad \textit{for all } m \textit{ such that } a(\varepsilon^m) = 0 \ (m = 1, 2, \ldots, N),$$

where $z(t)$ *is an arbitrary function of a complex variable.*

REMARK 2.1. Theorem 2.2 remains true if the coefficients of equation (2.4) are variable, i.e. $a_k = a_k(t)$ but invariant under rotation: $a_k(\varepsilon t) = a_k(t)$. It is easy to check that the operator of multiplication by such a function commute with rotation (compare also Remark 5.1 in Chapter IV). It is therefore enough to assume that the $a(\varepsilon^m)$ are either invertible or identically equal to zero for $m = 1, 2, \ldots, N$.

If $a(\varepsilon^m) = 0$ for $m = 1, 2, \ldots, N$, then we solve equation (2.1) with respect to the new unknown function $u(t) = x'(t)$ and, as in Theorem 1.3, we obtain from Theorem 2.2

THEOREM 2.3. *If $a(\varepsilon_m) = 0$ for $m = 1, 2, \ldots, N$, then the differential equation with rotation* (2.1) *has the solution*

$$x(t) = -\frac{1}{N}\sum_{m:b(\varepsilon^m)\neq 0}\frac{1}{b(\varepsilon^m)}\sum_{k=0}^{N-1}\varepsilon^{-km}\int y(\varepsilon^k t)\,dt - \frac{1}{N}\sum_{m:b(\varepsilon^m)=0}\sum_{k=0}^{N-1}\varepsilon^{-km}\int z(\varepsilon^k t)\,dt$$

under the necessary and sufficient condition that

$$\sum_{k=0}^{N-1}\varepsilon^{-km}y(\varepsilon^k t) = 0 \quad \text{for all } m \text{ such that } b(\varepsilon^m) = 0,$$

where $z(t)$ is an arbitrary (locally integrable) function of a complex variable.

Suppose that $a(\varepsilon^m) \neq 0$ for all m and that $b(\varepsilon^m) = 0$ for at least one m. For instance, let $b(\varepsilon^N) = 0$. From the last equation of (2.3) we obtain $x_{(N)} = [a(\varepsilon^N)]^{-1}y_{(N)}$ and, solving this system successively, as in Section 3 of Chapter III, we obtain

THEOREM 2.4. *If $a(\varepsilon^m) \neq 0$ for $m = 1, 2, \ldots, N$ and $b(\varepsilon^N) = 0$, then the differential equation with rotation* (2.1) *has the solution*

$$x(t) = \frac{1}{N}\sum_{m=1}^{N}\frac{1}{a(\varepsilon^m)}\sum_{k=0}^{N-1}\varepsilon^{-km}y(\varepsilon^k t) + \frac{b(\varepsilon^m)}{a(\varepsilon^{m+1})}y'(\varepsilon^k t) + \ldots +$$
$$+ \frac{b(\varepsilon^m)\ldots b(\varepsilon^{N-1})}{a(\varepsilon^{m+1})\ldots a(\varepsilon^N)}y^{(N-m)}(\varepsilon^k t).$$

We can similarly find the solution of equation (2.1) if $b(\varepsilon^m) = 0$ for some $m \neq N$.

Suppose that $b(\varepsilon^m) \neq 0$ for all m and that $a(\varepsilon^m) = 0$ for at least one m. As before, we consider the case $a(\varepsilon^N) = 0$. We then find $x_{(1)}$ from the Nth equation (2.3): $x'_{(1)} = -\dfrac{1}{b(\varepsilon^N)}y_{(N)}$.

Having found $x_{(1)}$, we solve the first equation (2.3), from which we find

$$x'_{(2)} = \frac{a(\varepsilon)}{b(\varepsilon)}x_{(1)} - \frac{1}{b(\varepsilon)}y_{(1)},$$

and so on.

A similar method can be used for solving equation (2.1) when $a(\varepsilon^m) \neq 0$ for some $m \neq N$.

REMARK 2.2. All the considerations of this Section are also valid for an equation with rotation of the type

$$\sum_{k=0}^{N-1} a_k x(\varepsilon^k t+\alpha_k) - \sum_{k=0}^{N-1} b_k x'(\varepsilon^k t+\alpha_k) = y(t), \tag{2.5}$$

where, as before, a_k, b_k are complex constants, $\varepsilon = e^{2\pi i/N}$, $N > 2$, and

$$\alpha_k = \alpha_0(\varepsilon^{k-1}+ \ldots +\varepsilon+1) \quad \text{for } k = 1, 2, \ldots, N. \tag{2.6}$$

Indeed, if we write $(Sx)(t) = x(\varepsilon t+\alpha_0)$, we find for $m = 1, 2, \ldots, N$

$$(S^m x)(t) = x\big(\varepsilon^m t+\alpha_0(\varepsilon^{m-1}+ \ldots +1)\big) = x(\varepsilon^m t+\alpha_m).$$

Since $\varepsilon^N = 1$ and $\varepsilon^{N-1}+ \ldots +\varepsilon+1 = 0$ (Lemma II.1.3), then

$$(S^N x)(t) = x\big(\varepsilon^N t+\alpha_0(\varepsilon^{N-1}+ \ldots +\varepsilon+1)\big) = x(t).$$

Hence S is an involution of order N. The differentiation operator also permutes this involution:

$$(DSx)(t) = [x(\varepsilon t+\alpha_0)]' = \varepsilon x'(\varepsilon t+\alpha_0) = \varepsilon(SDx)(t).$$

EXAMPLE 2.1. We examine the equation

$$\mu x(t)-x'(-it) = e^{\mu t}, \quad \text{where } \mu \text{ is a real parameter.}$$

Here $(Sx)(t) = x(it)$, $N = 4$, $\varepsilon = i$, and so $\varepsilon^2 = -1$, $\varepsilon^3 = -i$, $\varepsilon^4 = 1$ and we can rewrite our equation as

$$(\mu I-S^3 D)x = e^{\mu t}. \tag{2.7}$$

1° If $\mu = 0$, we obtain the equation $-S^3Dx = 1$, and, since $S^4 = I$, we have $Dx = -S(1)$, or $x'(t) = -1$, which implies $x(t) = -t+C$, where C is an arbitrary complex constant.

2° Now suppose that $\mu \neq 0$. We find $a(t) = \mu$, $b(t) = t^3$, and $b(i) = -i$, $b(-1) = -1$, $b(-i) = i$, $b(1) = 1$. Hence we can apply Theorem 2.1. We have $c_1 = \mu i$, $c_2 = -\mu$, $c_3 = -\mu i$, $c_4 = \mu$, and $\lambda = c_1c_2c_3c_4 = -\mu^4 \neq 0$ by assumption. Solving the differential equation $\tilde{x}^{(4)}+$ $+\mu^4\tilde{x} = e^{\mu t}$, we obtain

$$\tilde{x}(t) = \frac{1}{2\mu^4}e^{\mu t}, \quad \tilde{\tilde{x}}(t) = \{[b(S)]^{-1}\tilde{x}\}(t) = \sum_{j=1}^{4} [b(\varepsilon^j)]^{-1}(P_j x)(t)$$

$$= \frac{1}{4}\sum_{j=1}^{4}\sum_{k=0}^{3} i^j i^{-kj}\frac{1}{2\mu^4}e^{i^k \mu t} = \frac{1}{2\mu^4}e^{i\mu t}.$$

Further, we have $\operatorname{Arg}\lambda = \operatorname{Arg}(-\mu^4) = \pi$ and

$$\lambda_0 = \sqrt[4]{\mu^4 e^{\pi i}}{}^{/4} = \mu\left(\cos\frac{\pi}{4} + i\sin\frac{\pi}{4}\right) = \mu\frac{1+i}{\sqrt{2}},$$

$$V = \begin{vmatrix} 1 & 1 & 1 & 1 \\ -1 & -i & 1 & i \\ 1 & -1 & 1 & -1 \\ -1 & i & 1 & -i \end{vmatrix},$$

$$V_{01} = V_{02} = V_{03} = V_{04} = -4i, \quad V_{11} = 4i, \quad V_{13} = -4i, \quad V_{12} = V_{10} = -4,$$

$$V_{21} = V_{23} = -4i, \quad V_{22} = V_{24} = 4i, \quad V_{31} = 4i, \quad V_{32} = -4, \quad V_{33} = -4i, \quad V_{34} = 4,$$

$$d_0 = 8i, \quad d_1 = 8, \quad d_2 = 8, \quad d_3 = 8\frac{1+i}{\sqrt{2}}.$$

Since in our case the general form of the solution is

$$x(t) = -\frac{1}{2\mu^4}\Big\{\mu^3 e^{\mu t} + \tfrac{1}{4}\sum_{j=1}^{4}\sum_{k=0}^{3} i^{-kj-3j}[c_{3+j}\mu^2 + c_{3+j}\,c_{2+j}\mu +$$

$$+ c_{3+j}\,c_{2+j}c_{1+j}]e^{i^{k+1}\mu t} + C\sum_{k=0}^{3} d_k e^{i^k\lambda_0 t}\Big\},$$

we obtain after simplification

$$x(t) = \frac{i}{\mu}(e^{-\mu t} - e^{-i\mu t}) + 8C\Big\{i\exp\left(\frac{1+i}{\sqrt{2}}\mu t\right) + \exp\left(-\frac{1-i}{\sqrt{2}}\mu t\right) +$$

$$+\exp\left(-\frac{1+i}{\sqrt{2}}\mu t\right) + \frac{i-1}{\sqrt{2}}\exp\left(\frac{1-i}{\sqrt{2}}\mu t\right)\Big\}e^{i\mu t},$$

where C is an arbitrary complex constant.

Using the notion of the algebraic derivative and Theorem IV.3.5, we can formulate the following

THEOREM 2.5. *Under the assumptions of Theorem* 2.1 *every solution of equation* (2.1) *is of the form*

$$x = R_A\Big[(I-\lambda R^N)^{-1}R^N y + \sum_{k=1}^{N} C_k y_k\Big] + \sum_{k=0}^{N-1} d_k S^k y_0,$$

where

$$(Rx)(t) = \int_0^t x(t)\,dt,$$

$$R_A = \frac{d^{N-1}}{dt^{N-1}} - \frac{1}{N}\sum_{k=0}^{N-1}\sum_{j=0}^{N} \varepsilon^{-kj}S^k\left(c_{N+j-1}\frac{d^{N-2}}{dt^{N-2}} + \ldots + c_{N+j-1}\ldots c_{j+1}\right)[b(S)]^{-1},$$

$y_k = 1 - e^{\lambda_k t}$, C_k *are arbitrary complex constants*, $\lambda_k = \varepsilon^k \lambda_0$ $(k = 0, 1, \ldots, N-1, N)$, $\varepsilon = e^{2\pi i/N}$ *and* $c_1, \ldots, c_{2N}, \lambda, d_k, V_{k,m}, V, \lambda_0$ *are as defined in Theorem* 2.1.

Indeed, in our case $z_k = \tilde{C}_k$, where the $\tilde{C}_k$ are arbitrary complex constants and (compare Example IV.2.1)

$$[(I - \lambda_k R)^{-1} z_k](t) = \int_0^t \tilde{C}_k e^{-\lambda_k(t-s)} ds = \frac{\tilde{C}_k}{\lambda_k}(1 - e^{\lambda_k t}) = C_k y_k,$$

where the C_k are arbitrary constants. The required formula now follows.

CHAPTER VIII

Functional-Differential Equations of Carleman Type

A function $g(t) \not\equiv t$ defined on a set Ω with values in Ω satisfies the *Carleman condition* if

$$(*) \qquad g(g(t)) \equiv t \quad \text{on } \Omega.$$

For an arbitrary function $g(t)$ defined on Ω and taking values in Ω we write:

$$g_{n+1}(t) = g(g_n(t)) \quad \text{for } n = 1, 2, \dots \quad \text{and} \quad g_1(t) = g(t).$$

A function $g(t)$ defined on Ω and taking values in Ω satisfies the *Carleman condition of order m* if there exists an integer $m \geqslant 2$ such that

$$(**) \qquad g_m(t) \equiv t \quad \text{on } \Omega\,[1], \quad g_k(t) \not\equiv t \quad \text{for } k = 1, \dots, m-1.$$

In other words, the function $g(t)$ generates a *cyclic group of automorphisms of the set Ω of order m.*

It is easy to check that the following functions satisfy the Carleman condition:

$$(1) \qquad g(t) = -t \quad \text{on } \boldsymbol{R}.$$

$$(2) \qquad g(t) = \frac{1}{t} \quad \text{on } (0, \infty).$$

$$(3) \qquad g(t) = \begin{cases} -at & \text{for } t > 0, \\ 0 & \text{for } t = 0, \\ -\dfrac{1}{a}t & \text{for } t < 0, \end{cases}$$ where $a \neq 0$ is an arbitrary positive real.

[1] The Hilbert boundary problem for analytic functions with transformation of argument satisfying condition (*) was first considered by Carleman [2]. Thus in all papers concerning the corresponding singular integrals it is called the Carleman condition and the respective equations are said to be *of Carleman type* (compare Chapter XIII, Section 3). Identities (*) and (**) considered as functional equations are called the *Babbage equation* (Babbage [1], [2], [3], for further references see Kuczma [1]). For the properties of continuous mappings of this type on the real line see McShane [1], in $\boldsymbol{R}^n$ — Montgomery and Zippin [1], and on differentiable manifolds — Conner and Floyd [1].

(4) $$g(t) = \begin{cases} t^{-1/k} & \text{for } t \geqslant 1, \\ t^{-k} & \text{for } 0 < t \leqslant 1, \end{cases} \quad \text{where } k \text{ is an arbitrary positive integer},$$

(5) $g(t) = c-t$ on $\boldsymbol{R}$, where c is an arbitrary real.

It should be pointed out here that in all the preceding examples there is only one *fixed-point* a of the function g, i.e. a point a such that $g(a) = a$. Namely $a = 0$ in Examples (1), (3), $a = 1$ in Examples (2), (4) and $a = c/2$ in Example (5).

The following function satisfies the Carleman condition of order m on the complex plane:

(6) $$g(t) = \varepsilon^k t, \quad \text{where } \varepsilon = e^{2\pi i/m}.$$

A functional-differential equation is said to be *of Carleman type* if it is of the form

$$x^{(n)}(t) = F\big[t, x(t), x\big(g(t)\big), \ldots, x\big(g_{m-1}(t)\big), \ldots, x^{(n-1)}(t), \ldots, x^{(n-1)}\big(g_{m-1}(t)\big)\big],$$

where the function $g(t)$ satisfies the Carleman condition of order $m \geqslant 2$.

Equations with functions (1) and (6) were studied in Chapter VII on differential equations with reflection and rotation. The convention regarding "what is meant by a solution" made at the beginning of Chapter V is retained, although for simplicity, we restrict ourselves here to continuously differentiable solutions, i.e. we say that x is a solution if it has a continuous nth derivative and if it satisfies the equation at each point $t \in \Omega$.

1. PROPERTIES OF FUNCTIONS SATISFYING CARLEMAN CONDITION OF ORDER $m \geqslant 2$

In this section we study properties of functions satisfying the Carleman condition of order $m \geqslant 2$.

PROPOSITION 1.1. *Every function $g(t) \not\equiv t$ satisfying the Carleman condition of order $m \geqslant 2$ on a set Ω is one-to-one.*

Indeed, let $g_m(t) \equiv t$ on Ω and let $t_1, t_2 \in \Omega$, $t_1 \neq t_2$. Suppose that $g(t_1) = g(t_2)$. Hence $g_2(t_1) = g\big(g(t_1)\big) = g\big(g(t_2)\big) = g_2(t_2)$. After $m-1$ steps we obtain

$$t_1 = g_m(t_1) = g\big(g_{m-1}(t_1)\big) = g\big(g_{m-1}(t_2)\big) = g_m(t_2) = t_2,$$

which is a contradiction. Then $g(t_1) \neq g(t_2)$.

PROPOSITION 1.2. *If $g(t) \not\equiv t$ is a Carleman function on $\boldsymbol{R}$ (i.e. $g\big(g(t)\big) \equiv t$) with a fixed-point a, then the function $g_0(t) = g(t+a)-a$ is a Carleman function with the fixed-point 0 (conversely, $g(t) = g_0(t-a)+a$ is a Carleman function with the fixed-point a, provided that 0 is a fixed-point of g_0).*

Indeed, let $g(g(t)) \equiv t$ on $\boldsymbol{R}$ and $g(a) = a$. Then for every $t \in \boldsymbol{R}$:

$$g_0(g_0(t)) = g_0(g(t+a)-a) = g[g(t+a)-a+a]-a = g(g(t+a))-a$$
$$= t+a-a = t.$$

Moreover, $g_0(0) = g(a)-a = 0$.

PROPOSITION 1.3. *Let $g(t) \not\equiv t$ be a continuous Carleman function on $\boldsymbol{R}$. Then*

1° $g(t)$ is decreasing;

2° the derivative $g'(t)$ exists almost everywhere and $g(t) = \int\limits_a^t g'(s)ds$;

3° $g(t)$ has a unique fixed point a_g;

4° if $g(t)$ is odd, then $g(t) = -t$.

PROOF. Since $g(t)$ is one-to-one, $g(t)$ is strictly monotone. Suppose, that g is increasing. Since $g(t) \not\equiv t$, there is a t_0 such that $g(t_0) \neq t_0$. Then either $g(t_0) > t_0$ or $g(t_0) < t_0$. The first inequality implies that $t_0 = g(g(t_0)) > g(t_0) > t_0$, which is a contradiction. A similar proof holds for the inequality $g(t_0) < t_0$. Hence g is decreasing. Since $g(t)$ is continuous and monotone, its derivative exists almost everywhere and $g(t) = \int\limits_a^t g'(s)ds$, as is well known from the calculus of one real variable.

It follows immediately from our assumptions and Property 1° that

$$\lim_{t\to+\infty} g(t) = -\infty, \qquad \lim_{t\to-\infty} g(t) = +\infty. \tag{1.1}$$

Since the function $g(t)$ is a continuous decreasing function on $\boldsymbol{R}$ and Formulae (1.1) hold, there exists a unique point a_g such that $g(a_g)-a_g = 0$, which was to be proved.

Suppose now that $g(t)$ is an odd function. Since $g(-t) = -g(t)$, we have $g(0) = -g(0)$. Hence $g(0) = 0$ and 0 is the unique fixed-point of g. Suppose that $g(t) \neq -t$. If $g(t) > -t$, we have $t = g(g(t)) < g(-t) = -g(t)$, as the function $g(t)$ is decreasing. This implies that $-t > g(t) > -t$, which is a contradiction. If $g(t) < -t$, then $t = g(g(t)) > g(-t) = -g(t)$, which implies that $-t < g(t) < -t$, which is a contradiction. Hence $g(t) = -t$.

PROPOSITION 1.4. *Every continuous Carleman function $g(t) \not\equiv t$ on $\boldsymbol{R}$ with a fixed-point a is of the form:*

$$g(t) = g_0(t-a)+a, \tag{1.2}$$

where

$$g_0(t) = \begin{cases} \tilde{g}(t) & \textit{for } t \geqslant 0, \\ \tilde{g}^{-1}(t) & \textit{for } t < 0, \end{cases}$$

$\tilde{g}(t)$ is a continuous function on $\boldsymbol{R}$ such that $\tilde{g}(0) = 0$, $\tilde{g}(t) < 0$ for $t > 0$ and $\tilde{g}(t)$ is decreasing for $t > 0$. We denote by $\tilde{g}^{-1}$ the inverse function, which exists by our assumptions. Conversely, every function of the form (1.2) *is a continuous Carleman function on the real line with the fixed point a such that $g(t) \not\equiv t$.*

PROOF. Let $g(t)$ be a continuous Carleman function with the fixed-point a. Then from Proposition 1.2 $g_0(0) = 0$ and $g_0(g_0(t)) \equiv t$, which implies that $g_0^{-1}(t) = g_0(t)$ and the function $g(t)$ can be written in the form (1.2). Conversely, suppose that the function $\tilde{g}(t)$ is continuous on $\boldsymbol{R}$, that $\tilde{g}(0) = 0$, that $\tilde{g}(t) < 0$ for $t > 0$ and that $\tilde{g}(t)$ is decreasing for $t > 0$. Since $\tilde{g}(t) \leqslant 0$ for $t \geqslant 0$, we have $g_0(g_0(t)) = \tilde{g}^{-1}(\tilde{g}(t)) \equiv t$. If $t < 0$, we write $u = \tilde{g}^{-1}(t)$. Then $\tilde{g}(u) = t < 0$, which implies $\tilde{g}^{-1}(t) = u > 0$. Hence $g_0(g_0(t)) = \tilde{g}(\tilde{g}^{-1}(t)) \equiv t$ for $t < 0$.

Since $g(t) = g_0(t-a)+a$ is continuous at each point and is decreasing, $g(t)$ has a unique fixed-point, namely $t = a$.

The construction of a Carleman function on the real line is shown on Fig. 10.

Propositions 1.2 and 1.4 immediately imply

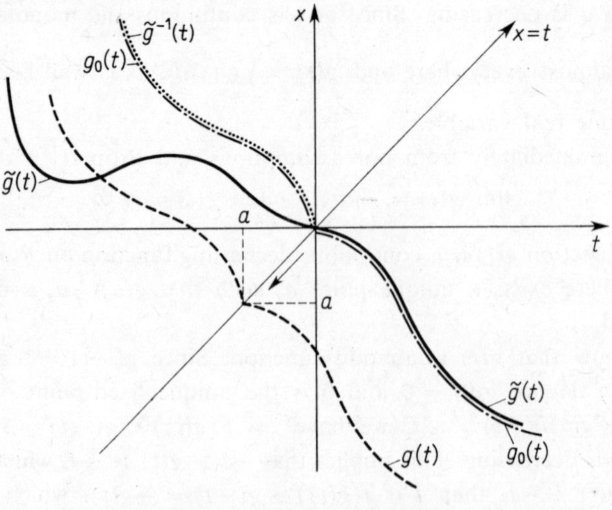

Fig. 10. Construction of a Carleman function on the real line

COROLLARY 1.1. *Every continuous Carleman function $g(t) \not\equiv t$ on $\boldsymbol{R}$ has a graph symmetric with respect to the line $x = t$.*

A *homeomorphism* of a set Ω_1 into a set Ω_2 is a continuous one-to-one function h, defined on Ω_1 and with values in Ω_2, such that the inverse function h^{-1} (which exists by assumption) is also continuous. Two sets Ω_1 and Ω_2 are said to be *homeomorphic* if there is a homeomorphism h such that $h(\Omega_1) = \Omega_2$.

The superposition of two homeomorphism (if it exists) is also a homeomorphism. For example, it is easy to check that the function

$$y = \tan\frac{\pi}{b-a}\left(t-\frac{b+a}{2}\right)$$

is a homeomorphism of the open interval (a, b) onto $\boldsymbol{R}$. The function $y = \ln t$ is a homeomorphism of the positive open half-axis onto $\boldsymbol{R}$. The function $y = -t$ is a homeomorphism of $(-\infty, 0)$ onto $(0, +\infty)$.

A set L in the real or complex plane is said to be an *open arc* if it is homeomorphic with an open interval. By definition each open arc is homeomorphic with $\boldsymbol{R}$.

In the sequel we will denote the superposition of two functions x and y by $x \circ y$, i.e., by definition,

$$(x \circ y)(t) = x\big(y(t)\big), \tag{1.3}$$

if this superposition exists.

THEOREM 1.1. *Every continuous Carleman function $g(t) \neq t$ on $\boldsymbol{R}$ with a fixed point a is the superposition of a homomorphism h of $\boldsymbol{R}$ onto itself, a reflection, and of the inverse homemorphism h^{-1},*

$$g = h^{-1} \circ (-h), \quad \text{where } h = h_1 \circ h_2, \tag{1.4}$$

h_1 is a homeomorphism of $\boldsymbol{R}$ onto itself such that $h_1(0) = 0$ and $h_2(t) = t-a$, is a shift. Conversely, every function of the form (1.4) *is a continuous Carleman function on $\boldsymbol{R}$ with the fixed-point a.*

PROOF. Let g be of the form (1.4). Since $h_2^{-1}(t) = t+a$, we can write $g(t) = h_1^{-1}\big(-h_1(t-a)\big)+a$. For an arbitrary homeomorphism h_1 having the required properties, the function $g_0(t) = h_1^{-1}\big(-h_1(t)\big)$ is a continuous Carleman function with the fixed-point 0. Indeed,

$$g_0\big(g_0(t)\big) = h_1^{-1} \circ \big[-h_1 \circ h_1^{-1} \circ \big(-h_1(t)\big)\big] = h_1^{-1} \circ \big[-\big(-h_1(t)\big)\big]$$
$$= h_1^{-1} \circ h_1(t) \equiv t$$

and $g_0(0) = h_1^{-1}\big(-h_1(0)\big) = h_1^{-1}(0) = 0$.

From Proposition 1.2 we therefore infer that $g(t)$ is a continuous Carleman function with the fixed-point a. Conversely, suppose that $g(t)$ is a continuous Carleman function on $\boldsymbol{R}$ with the fixed-point a. Let

$$h_1(t) = t-g_0(t), \quad \text{where } g_0(t) = g(t+a)-a. \tag{1.5}$$

Since $g_0(t)$, as being a continuous Carleman function, is decreasing, the function $h_1(t)$ is increasing. Since h_1 is also continuous, it is a homeomorphism of $\boldsymbol{R}$

onto itself and $h_1(0) = -g_0(0) = 0$. Moreover, $h_1(g_0(t)) = g_0(t) - g_0(g_0(t)) = g_0(t) - t = -h_1(t)$. Hence $g_0(t) = h_1^{-1}(-h_1(t))$ and $g(t) = h_1^{-1}(-h_1(t-a)) + a = h_2^{-1} \circ h_1^{-1} \circ (-h_1 \circ h_2(t))$, where $h_2 = t-a$. Writing $h = h_1 \circ h_2$ we obtain the required formula for $g(t)$.

COROLLARY 1.2. *Every continuous Carleman function $g(t)$ on an open arc L is of the form: $g(t) = h^{-1}(-h(t))$, where $h = h_1 \circ h_2 \circ h_3$, h_3 is a homeomorphism of L onto $\mathbf{R}$, $h_2(s) = s-a$, a is the fixed-point of the function $h_3 \circ g \circ h_3^{-1}$, and h_1 is a homeomorphism of $\mathbf{R}$ onto itself such that $h_1(0) = 0$. Conversely, every function of this form is a Carleman function on L with the fixed-point $a_0 = h_3^{-1}(a)$*

PROOF. Let h_3 be a homeomorphism of the arc L onto $\mathbf{R}$ and let $s = h_3(t)$. Then the function

$$g_1(s) = (h_3 \circ g \circ h_3^{-1})(s) = h_3 \circ g(t), \qquad s \in \mathbf{R}, \tag{1.6}$$

is a continuous Carleman function on $\mathbf{R}$, and is therefore of the form

$$g_1(s) = h_2^{-1} \circ h_1^{-1}(-h_1 \circ h_2(s)), \qquad \text{where } h_1(0) = 0,\ h_2(s) = s-a, \tag{1.7}$$

a is the fixed-point of $g_1(s)$, and h_1 is a homeomorphism of $\mathbf{R}$ onto itself. Formulae (1.6) and (1.7) imply that

$g(t) = h^{-1}(-h(t))$, where $h = h_1 \circ h_2 \circ h_3$. Since $h_2(a) = a-a = 0$, we have $h_2^{-1}(0) = a$ and writing $a_0 = h_3^{-1}(a)$ we find

$$\begin{aligned} g(a_0) &= h_3^{-1} \circ h_2^{-1} \circ h_1^{-1}[-h_1 \circ h_2 \circ h_3(h_3^{-1}(a))] = h_1^{-1} \circ h_2^{-1} \circ h_3^{-1}[-h_1 \circ h_2(a)] \\ &= h_3^{-1} \circ h_2^{-1} \circ h_1^{-1}(-h_1(0)) = h_3^{-1} \circ h_2^{-1} \circ h_1^{-1}(0) \\ &= h_3^{-1} \circ h_2^{-1}(0) = h_3^{-1}(a) = a_0. \end{aligned}$$

Hence a_0 is the fixed-point of g.

THEOREM 1.2. *Continuous functions other than the functions identically equal to t satisfying the Carleman condition of order $m > 2$ on $\mathbf{R}$ do not exist.*

PROOF. Suppose that there exists a function $g(t) \not\equiv t$ such that for some $m > 2$

$$g_m(t) \equiv t \quad \text{on } \mathbf{R}. \tag{1.8}$$

Let m be an odd number. Then, if $g(t)$ is decreasing, we have

$$\begin{aligned} t_1 < t_2 \Leftrightarrow g(t_1) > g(t_2) \Leftrightarrow \ldots \Leftrightarrow g_{m-1}(t_1) \\ < g_{m-1}(t_2) \Leftrightarrow t_1 = g_m(t_1) > g_m(t_2) = t_2, \end{aligned}$$

which is a contradiction. If $g(t)$ is increasing, then there is a t_0 such that $g(t_0) > t_0$ and from identity (1.8) we find

$$t_0 < g(t_0) < g_2(t_0) < \ldots < g_{m-1}(t_0) < g_m(t_0) = t_0.$$

which is a contradiction. Thus m cannot be odd. Let $m = 2k$, $k > 1$. Since (1.8) holds, the function $g_k(t)$ is a continuous Carleman function and is therefore decreasing. Hence we conclude by a similar argument that k cannot be an even number. Hence k is odd and we can write $k = 2p+1$, which implies $m = 2k = 2(2p+1)$. Writing $\breve{g}(t) = g_2(t)$, we obtain $\breve{g}_{2p+1}(t) = g_m(t) \equiv t$ on $\boldsymbol{R}$, which means that $g(t)$ is a continuous Carleman function of the odd order $2p+1$. But this is impossible. Hence $g_m(t) \not\equiv t$ for every $m > 2$, which was to be proved.

COROLLARY 1.3. *Continuous functions satisfying the Carleman condition of order $m > 2$ on an open arc L do not exist, apart from the function identically equal to t.*

PROOF. Suppose that $g_m(t) \equiv t$ on L for some $m > 2$. Let h be a homeomorphism of L onto $\boldsymbol{R}$ of the type described in Corollary 1.2. We denote $s = h(t) \in \boldsymbol{R}$. We define on $\boldsymbol{R}$ the function $g^*(s) = h \circ g \circ h^{-1}(s)$, for $s = h(t)$, $t \in L$. Then

$$g_2^* = g^* \circ g^* = (h \circ g \circ h^{-1}) \circ (h \circ g \circ h^{-1}) = h \circ g \circ g \circ h^{-1} = h \circ g_2 \circ h^{-1},$$

$$\cdots\cdots\cdots\cdots\cdots\cdots\cdots\cdots\cdots\cdots\cdots\cdots\cdots\cdots\cdots\cdots$$

$$g_m^* = g^* \circ g_{m-1}^* = (h \circ g \circ h^{-1})(h \circ g_{m-1} \circ h^{-1}) = h \circ g \circ g_{m-1} \circ h^{-1} = h \circ g_m \circ h^{-1}$$

and

$$g_m^*(s) = h \circ g_m \circ h^{-1}(s) = h \circ g_m \circ (t) = h(t) = s \quad \text{for } s \in \boldsymbol{R},$$

which contradicts Theorem 1.2.

We turn next to the question of differentiability of a Carleman function. The definition immediately implies the following

PROPOSITION 1.5. *If $g(t) \not\equiv t$ is a continuous Carleman function on L, where L is either $\boldsymbol{R}$ or an open interval, or the positive or negative half-axis, then the derivative of the function $g \circ g$ exists at each point and $(g \circ g)' = 1$.*

PROPOSITION 1.6. *If $g(t) \not\equiv t$ is a differentiable Carleman function on $\boldsymbol{R}$, then*

$$g'(t)g'(g(t)) \equiv 1, \tag{1.9}$$

$$g'(t) \neq 0 \quad \textit{for all } t \in \boldsymbol{R}. \tag{1.10}$$

Indeed, from Proposition 1.5 we find $1 = [g(g(t))]' = g'(t)g'(g(t))$ for all $t \in \boldsymbol{R}$. This implies that $g'(t) \neq 0$ for $t \in \boldsymbol{R}$.

A homeomorphism h is said to be a *diffeomorphism* if both h and h^{-1} are continuously differentiable. Two sets Ω_1 and Ω_2 are *diffeomorphic* if there is a diffeomorphism h such that $h(\Omega_1) = \Omega_2$.

It follows from what has been said above that an open interval and the positive and negative half-axes are diffeomorphic with $\boldsymbol{R}$. This enables us to consider, instead of a general subset $L \subset \boldsymbol{R}$ diffeomorphic with $\boldsymbol{R}$, just an open interval (a, b), where we allow the values $a = -\infty$ and $b = +\infty$.

PROPOSITION 1.7. *If $g(t) \neq t$ is a Carleman function on an interval (a, b), such that $g(t) = h^{-1}(-h(t))$ and h is a diffeomorphism of (a, b) onto $\boldsymbol{R}$, then*

(i) *the derivative g' exists at each point $t \in (a, b)$;*

(ii) $g'(t)g'(g(t)) \equiv 1$;

(iii) $g'(t) \neq 0$ *for all* $t \in (a, b)$.

PROOF. Let h be a diffeomorphism of (a, b) onto $\boldsymbol{R}$ and let $g(t) = h^{-1}(-h(t))$. Since the derivatives h' and $(h^{-1})'$ both exist by assumption, it follows that g' also exists. Since $h' \neq 0$, we have

$$g'(t) = [h^{-1}(-h(t))]' = \frac{1}{h'(h^{-1}(s))}\bigg|_{s=-h(t)} [-h'(t)] = \frac{1}{h'[h^{-1}(-h(t))]} [-h'(t)]$$
$$= -\frac{h'(t)}{h'(g(t))}.$$

Therefore

$$g'(t)g'(g(t)) = -\frac{h'(t)}{h'(g(t))} \cdot \left[-\frac{h'(g(t))}{h'[g(g(t))]}\right] = \frac{h'(t)}{h'(g(t))} \cdot \frac{h'(g(t))}{h'(t)} = 1$$

at each point $t \in (a, b)$. Moreover, we conclude from (ii) that $g'(t) \neq 0$ for all $t \in (a, b)$.

2. FIRST ORDER LINEAR FUNCTIONAL-DIFFERENTIAL EQUATIONS OF CARLEMAN TYPE

In this section we will consider the following functional-differential equation of Carleman type [1]

$$x'(t) = x(g(t)) + y(t), \tag{2.1}$$

where the differentiable function $g(t) \not\equiv t$ satisfies the Carleman condition on an open interval (a, b), where we possibly have $a = -\infty$, $b = +\infty$, and the given continuous function is also defined on (a, b). Let X be the linear space of all real-valued continuous functions defined on (a, b). We define a linear operator on X by means of the equality

$$(Sx)(t) = x(g(t)) \quad \text{on } (a, b).$$

[1] Cf.: Silberstein [1], Kuller [1], Viner [3].

Since $g(g(t)) \equiv t$ on (a, b), the operator S is an involution: $S^2 = I$ on X. This involution is *multiplicative*, i.e. for all $x, y \in X$ we have

$$S(xy) = (Sx)(Sy). \tag{2.2}$$

Indeed, $[S(x, y)](t) = x(g(t))y(g(t)) = (Sx)(t)(Sy)(t)$ for all $t \in (a, b)$. Unfortunately S does not commute with derivation:

$$(Sx)' = g'Sx' \qquad (x' = \frac{dx}{dt}). \tag{2.3}$$

Indeed,

$$(Sx)'(t) = [x(g(t))]' = g'(t)x'(g(t)) = g'(t)(Sx')(t).$$

From Propositions 1.5, 1.6, 1.7 we immediately obtain

PROPOSITION 2.1. *If $g(t) \not\equiv t$ is a differentiable Carleman function on (a, b), then*

$$\left.\begin{array}{l}\text{(i) } (Sg)' = 1\\ \text{(ii) } g'Sg' = 1\\ \text{(iii) } g' \neq 0\end{array}\right\} \text{on } (a, b).$$

REMARK 2.1. If we assume only that $g(t)$ is absolutely continuous on (a, b), then g' exists almost everywhere and conditions (i)–(iii) are satisfied a.e.

THEOREM 1.1. *If $g(t) \not\equiv t$ is a differentiable Carleman function on an interval (a, b) and y is a continuously differentiable function on (a, b), then equation (2.1) is equivalent to the ordinary differential equation*

$$x'' - g'x = y' + g'Sy \tag{2.4}$$

with the condition

$$x'(a_g) = x(a_g) + y(a_g), \quad \textit{where } a \textit{ is the fixed-point of } g. \tag{2.5}$$

PROOF. *Necessity*. Suppose that x is a twice differentiable solution of equation (2.1). Then $x'(a_g) = x(g(a_g)) + y(a_g) = x(a_g) + y(a_g)$ and condition (1.5) is satisfied. Furthermore writing equation (2.1) in the form

$$x' = Sx + y \tag{2.6}$$

we find differentiating both sides and using Formula (2.3):

$$\begin{aligned} x'' &= (Sx+y)' = (Sx)'+y' = g'Sx'+y' = g'S(Sx+y)+y' \\ &= g'S^2x+g'Sy+y' = g'x+g'Sy+y', \end{aligned} \tag{2.7}$$

sine x is a solution of equation (2.6). Hence equation (2.4) is also satisfied.

Sufficiency. Suppose now that x is a solution of equation (2.4) satisfying condition (2.5). We write $u = x' - Sx - y$ and we show that $u \equiv 0$, i.e. that equation (2.1) is satisfied. Note that u satisfies the differential equation

$$u' + g'Su = 0 \tag{2.8}$$

with the initial condition

$$u(a) = 0. \tag{2.9}$$

Indeed,

$$u' + g'Su = (x' - Sx - y)' + g'S(x' - Sx - y) = x'' - (Sx)' - y' + g'Sx' - \\ - g'S^2x - g'Sy = x'' - g'Sx' - y' + g'Sx' - g'x - g'Sy = 0,$$

for x is a solution of equation (2.4). Moreover,

$$u(a_g) = x'(a_g) - x(g(a_g)) - y(a_g) = x'(a_g) - x(a_g) - y(a_g) = 0$$

as condition (2.5) is satisfied. Acting on both sides of equation (2.8) with the operator S we find from Proposition 2.1

$$0 = S(u' + g'Su) = Su' + (Sg')S^2u = Su' + (Sg')u$$

$$= \frac{1}{g'}(g'Su' + g(Sg')u) = \frac{1}{g'}[(Su)' + u].$$

Differentiating the equation $(Su)' + u = 0$ we obtain

$$0 = (Su)'' + u' = (Su)'' + (-g'Su) = (Su)'' - g'Su = 0.$$

Putting $v = Su$ we conclude that v satisfies the ordinary differential equation $v'' - g'v = 0$ with the initial condition $v(a) = 0$, $v'(a) = 0$, because $v(a) = u(g(a)) = u(a) = 0$, $v'(a) = (Su)'(a) = g'(a)u'(g(a)) = g'(a)u'(a) = -g'(a)^2u(g(a)) = -g'(a)^2u(a) = 0$. Hence $v \equiv 0$. Therefore $u = Sv \equiv 0$, which was to be proved.

COROLLARY 2.1. *Under the assumptions of Theorem* 2.1, *if* $[t_1, t_2] \subset (a, b)$ *is an arbitrary closed interval such that* $a_g \in [t_1, t_2]$, *then the general form of the solutions of equation* (2.1) *is*

$$x(t) = [x(a_g) + y(a_g)]t + (1 - a_g)x(a_g) - a_g y(a_g) + \\ + \int_{a_g}^{t} (t-s)[y'(s) + g'(s)y(g(s))]ds + \\ + \int_{a_g}^{t} \mathfrak{R}(t,s)\{[x(a_g) + y(a_g)]s + (1 - a_g)x(a_g) - a_g y(a_g)\}ds + \\ + \int_{a_g}^{t} \mathfrak{R}(t,s)\int_{a_g}^{t} (s-u)[y'(u) + g'(u)y(g(u))]du\,ds, \tag{2.10}$$

where $x(a_g)$ is an arbitrary constant, $\mathfrak{N}(t,s) = \sum_{n=0}^{\infty} N_n(t,s)$ is absolutely and uniformly convergent, $N_0(t,s) = (t-s)g'(s)$, $N_n(t,s) = \int_s^t N_0(t,u)N_{n-1}(u,s)\,du$ ($n = 1, 2, \ldots$).

PROOF. In order to prove Formula (2.11) we follow the same procedure as in Example 4.1 of Chapter IV, where we put $p_0 = 0$, $q = g'$. This gives us the general form of the solutions of equation (2.4):

$$x(t) = Ct + C_1 + \int_{a_g}^{t} (t-s)[y'(s)+g'(s)y(g(s))]\,ds + \int_{a_g}^{t} \mathfrak{N}(t,s)(Cs+C_1)\,ds +$$

$$+ \int_{a_g}^{t} \mathfrak{N}(t,s) \int_{a_g}^{s} (s-u)[y'(u)+g'(u)y(g(u))]\,du\,ds.$$

We are looking for solutions such that $x'(a_g) = x(a_g)+y(a_g)$. To begin with, we remark that $N_n(t,t) = 0$ for $n = 0, 1, 2, \ldots$, whence $\mathfrak{N}(t,t) = 0$ and that

$$\frac{\partial}{\partial t} N_0(t,s) = g'(s), \qquad \frac{\partial}{\partial t} N_n(t,s) = g'(s)\int_s^t N_{n-1}(u,s)\,du.$$

Therefore

$$\frac{\partial}{\partial t}\mathfrak{N}(t,s) = g'(s)\left[1+\sum_{n=1}^{\infty}\int_s^t N_{n-1}(u,s)\,du\right] = g'(s)\left[1+\int_s^t \mathfrak{N}(u,s)\,du\right].$$

Hence

$$x'(t) = C + \int_{a_g}^{t} [y'(s)+g'(s)y(g(s))]\,ds + \mathfrak{N}(t,t)(Ct+C_1) +$$

$$+ \int_{a_g}^{t} \frac{\partial}{\partial t}\mathfrak{N}(t,s)\,(Cs+C_1)\,ds + \mathfrak{N}(t,t)\int_{a_g}^{t} (t-u)[y'(u)+g'(u)y(g(u))]\,du +$$

$$+ \int_{a_g}^{t} \frac{\partial}{\partial t}\mathfrak{N}(t,s) \int_{a_g}^{t} (s-u)[y'(u)+g'(u)y(g(u))]\,du\,ds.$$

Thus $x'(a) = C$. But $x(a) = Ca+C_1$. Since we required $x'(a_g) = x(a_g)+y(a_g)$, we obtain

$$\begin{aligned} C_1 &= -Ca_g + x(a_g) = -x'(a_g)a_g + x(a_g) = -[x(a_g)+y(a_g)]a_g + x(a_g) \\ &= (1-a_g)x(a_g) - a_g\,y(a_g), \end{aligned}$$

which gives the required Formula (2.10).

All the results of this section remain true if y is a vector valued function. Other forms of equation of Carleman type of first order will be studied in the next section.

EXAMPLE 2.1. Consider the following equation of Carleman type:

$$x'(t) = x\left(\frac{1}{t}\right), \quad 0 < t < \infty. \tag{2.11}$$

The function $g(t) = 1/t$ is a Carleman function with the fixed-point $a_g = 1$. We have $g'(t) = -1/t^2$. Hence equation (2.11) is equivalent to the equation

$$x'' + \frac{1}{t^2}x = 0 \quad (\text{or: } t^2x''+x = 0) \tag{2.12}$$

with the condition $x'(1) = x(1)$. All the solutions of equation (2.13) are of the form

$$x(t) = \sqrt{t}\left[C\cos\left(\frac{\sqrt{3}}{2}\ln t\right)+C_1\sin\left(\frac{\sqrt{3}}{2}\ln t\right),\right.$$

where C, C_1 are arbitrary constants [1]. Since

$$x'(t) = \frac{1}{2\sqrt{t}}\left[C\cos\left(\frac{\sqrt{3}}{2}\ln t\right)+C_1\sin\left(\frac{\sqrt{3}}{2}\ln t\right)\right]+$$

$$+\sqrt{t}\left[-C\sin\left(\frac{\sqrt{3}}{2}\ln t\right)+C_1\cos\left(\frac{\sqrt{3}}{2}\ln t\right)\right]\frac{1}{t}\frac{\sqrt{3}}{2},$$

we find: $x(1) = C$, $x'(1) = \frac{C}{2}+\frac{\sqrt{3}}{2}C_1$ and $C_1 = \frac{1}{\sqrt{3}}C = \frac{1}{\sqrt{3}}x(1)$. Thus all solutions of equation (2.11) are of the form

$$x(t) = x(1)\sqrt{t}\left[\cos t\left(\frac{\sqrt{3}}{2}\ln t\right)+\frac{1}{\sqrt{3}}\sin\left(\frac{\sqrt{3}}{2}\ln t\right)\right],$$

where $x(1)$ is an arbitrary constant.

3. HIGHER ORDER LINEAR FUNCTIONAL-DIFFERENTIAL EQUATIONS OF CARLEMAN TYPE

We retain the notation of the previous section. Consider the following functional-differential equation of Carleman type:

$$Ax = Sx+y, \tag{3.1}$$

[1] It is well known that all the solutions of equation (2.12), called the *Euler equation* are determined if we substitute $x = t^r$.

where

$$A = \sum_{k=0}^{n} a_k(t) \frac{d^k}{dt^k}, \quad n \geqslant 1, \tag{3.2}$$

$g(t)$ is an n-times continuously differentiable function on an open interval (a, b) and $a_0, \ldots, a_n, y$ are given functions defined on (a, b). Denote by Γ the operator given by

$$(\Gamma x)(t) = \frac{1}{g'(t)} x'(t) \quad \text{on } (a, b). \tag{3.3}$$

We have the following (cf. Viner [3]):

THEOREM 3.1. *If the given functions $a_0, \ldots, a_n, y, g$ belong to the class $C^n(a, b)$* [1] *and if $g(t) \not\equiv t$ is a Carleman function on (a, b), then every $2n$-times differentiable solution of equation* (3.1) *satisfies the ordinary differential equation of order $2n$:*

$$\sum_{k=0}^{n} (Sa_k)\Gamma^k Ax - x = \sum_{k=0}^{n} (Sa_k)\Gamma^k y + Sy, \tag{3.4}$$

with the condition

$$(\Gamma^k Ax)(a_g) = x^{(k)}(a_g) + (\Gamma^k y)(a_g) \quad (k = 0, 1, \ldots, n-1), \tag{3.5}$$

where a_g is the unique fixed-point of the function g.

PROOF. Let x be a $2n$-times differentiable solution of equation (3.1). By differentiation we obtain from (3.1) and from Formulae (2.3), (3.3) the n equalities:

$$\begin{aligned} Sx &= Ax - y = \Gamma^0 Ax - \Gamma^0 y, \\ Sx' &= \frac{1}{g'}(Ax)' - \frac{1}{g'} y' = \Gamma Ax - \Gamma y, \\ Sx'' &= \frac{1}{g'}(\Gamma Ax)' - \frac{1}{g'}(\Gamma y)' = \Gamma^2 Ax - \Gamma^2 y, \\ &\cdots\cdots\cdots\cdots\cdots\cdots \\ Sx^{(n)} &= \frac{1}{g'}(\Gamma^{n-1} Ax)' - \frac{1}{g'}(\Gamma^{n-1} y) = \Gamma^n Ax - \Gamma^n y. \end{aligned} \tag{3.6}$$

Multiplication of $Sx^{(k)}$ by Sa_k followed by the summation of all the resulting equations yields the following equality:

$$SAx = \sum_{k=0}^{n} (Sa_k)(Sx^{(k)}) = \sum_{k=0}^{n} (Sa_k)(\Gamma^k Ax - \Gamma^k y).$$

[1] I.e. are n-times continuously differentiable on (a, b).

Since x is a solution of equation (3.1), we have $SAx = S^2x+Sy = x+Sy$. Hence

$$\sum_{k=0}^{n} (Sa_k)\,[\Gamma^k Ax - \Gamma^k y] = x+Sy$$

and

$$\sum_{k=0}^{n} (Sa_k)\Gamma^k Ax - x = \sum_{k=0}^{n} (Sa_k)\Gamma^k + Sy$$

which is the required form of equation (3.4). From equalities (3.6) we obtain

$$\Gamma^k Ax = \Gamma^k y + Sx^{(k)} \qquad \text{for } k = 0, 1, 2, \ldots, 2n-1.$$

Since $(Sx^{(k)})(a_g) = x^{(k)}(g(a_g)) = x^{(k)}(a_g)$, we find

$$(\Gamma^k Ax)(a_g) = x^{(k)}(a_g) + (\Gamma^k y)(a_g) \qquad (k = 0, 1, \ldots, n-1).$$

Hence conditions (3.5) are satisfied.

An example shows that the converse statement is not true, i.e. not each solution of equation (3.4) satisfying condition (3.5) is a solution of equation (3.1).

Theorem 3.1 remains true if y is a vector valued function and the coefficients are scalar valued functions. Multiplication of a vector valued function by a scalar function is understood to be multiplication of each coordinate function by the scalar function.

4. NON-LINEAR FUNCTIONAL-DIFFERENTIAL EQUATIONS OF CARLEMAN TYPE

Consider now a non-linear functional-differential equation of Carleman type

$$x' = F(t, x, Sx), \tag{4.1}$$

where

$$(Sx)(t) = x(g(t)), \quad g(t) \neq t$$

is a Carleman function on an open interval (a, b).

We obtain the following (cf. Viner [3])

THEOREM 4.1. *Let $g(t) \not\equiv t$ be a differentiable Carleman function on the interval (a, b). Let the given function $F(t, x, y)$ be continuously differentiable with respect to all its arguments and such that the equation $z = F(t, x, y)$ is solvable with respect to y, with the unique solution*

$$y = G(t, x, z). \tag{4.2}$$

Then every twice-differentiable solution of equation (4.1) *satisfies the ordinary differential equation*

$$x'' = H(t, x, x') \tag{4.3}$$

with the condition

$$x'(a_g) = F\big(a_g, x(a_g), x(a_g)\big), \tag{4.4}$$

where a_g is the fixed-point of g and

$$\begin{aligned} H(t, x, x') = {} & F_t'\big(t, x, G(t, x, x')\big) + x'F_x'\big(t, x, G(t, x, x')\big) + \\ & + g'F\big[g(t), G(t, x, x'), x\big]F_y'\big(t, x, G(t, x, x')\big). \end{aligned} \tag{4.5}$$

PROOF. Differentiating both sides of equation (4.1) we obtain from Formula (2.3)

$$\begin{aligned} x'' &= F_t'(t, x, Sx) + F_x'(t, x, Sx)x' + F_y'(t, x, Sx)(Sx)' \\ &= F_t'(t, x, Sx) + x'F_x'(t, x, Sx) + g'Sx'F_y'(t, x, Sx). \end{aligned}$$

But from (4.1) $Sx' = F[g(t), Sx, x]$, and from (4.2) $Sx = G(t, x, x')$. Therefore

$$\begin{aligned} x'' = {} & F_t'\big(t, x, G(t, x, x')\big) + x'F_x'\big(t, x, G(t, x, x')\big) + \\ & + g'F[g(t), G(t, x, x'), x]F_y'\big(t, x, G(t, x, x')\big), \end{aligned}$$

which gives the required equation (4.3). From equality (4.1) we find

$$x'(a_g) = F\big[a_g, x(a_g), x\big(g(a_g)\big)\big] = F\big(a_g, x(a_g), x(a_g)\big),$$

and so condition (4.4) is also satisfied.

Theorem 4.1 remains valid if F is a vector valued function and equation (4.1) is a system of functional-differential equations of Carleman type.

The following example shows that the converse theorem is not true, i.e. that not every solution of equation (4.3) satisfying condition (4.4) is a solution of equation (4.1).

EXAMPLE 4.1. We discuss the equation (with reflection):

$$x'(t) = \frac{1}{x(-t)} \quad \text{for } t \in \boldsymbol{R}. \tag{4.6}$$

Here we have $F(t, x, y) = 1/y$, $F_t' = F_x' = 0$, $F_y' = -1/y^2$, $G(t, x, x') = 1/x'$, $g(t) = -t$, $g'(t) = -1$, $a = 0$. Condition (4.4) is thus of the form

$$x'(0) = \frac{1}{x(0)}. \tag{4.7}$$

Equation (4.3) is in our case

$$x'' = -1\frac{1}{x}\left(-\frac{1}{(1/x')^2}\right).$$

After simplification we find

$$x'' = \frac{1}{x}x'^2 \quad \text{or} \quad xx''-x'^2 = 0. \tag{4.8}$$

Since $x \neq 0$ (equation (4.6)), we can rewrite this equation as follows:

$$\frac{xx''-x'^2}{x^2} = 0 \quad \text{or} \quad \left(\frac{x'}{x}\right)' = 0.$$

Hence $x'/x = A$, where A is an arbitrary constant. Therefore $x = Be^{At}$, where B is an arbitrary constant. From condition (4.7) we have $AB = 1/A$, which implies $B = 1/A^2$ and $x(t) = A^{-2}e^{At}$. We already have solutions of equation (4.8) satisfying condition (4.7). Now we have to find those solutions which are solutions of the given equation (4.6), i.e. for which $x'(t)x(-t) = 1$. Since $x'(t) = \frac{1}{A}e^{At}$, $x(-t) = \frac{1}{A^2}e^{-At}$, we have the following equality $1 = A^{-3}$, which implies $A = 1$.

The unique solution of equation (4.6) is thus $x(t) = e^t$. We finally conclude that equation (4.8) with condition (4.7) is not equivalent to equation (4.6) because the two equations have different number of linearly independent solutions.

REMARK. All the results of the present chapter can be extended to a much larger class of functional-differential equations using the notion of the algebraic derivative introduced in Chapter IV (cf. the author [16], [17]).

CHAPTER IX

Optimal Control Problems with Time Delays

In this Chapter we consider *time lag systems* with constant *delays*. Such a system can be described by the differential-difference equation:

$$(0.1) \quad \frac{d^p}{dt^p}x(t) = F\big(t, x(t), x(t-\omega_1), \ldots, x(t-\omega_q), \ldots, x^{(p-1)}(t), x^{(p-1)}(t-\omega_1),$$

$$\ldots, x^{(p-1)}(t-\omega_q), u(t), u(t), u(t-\omega_1), \ldots, u(t-\omega_q)\big),$$

where $x(t)$ is an n-dimensional vector valued function, called the *state*, $u(t)$ is an m-dimensional vector-valued function called the *control*, and $F(t, x_{10}, \ldots$ $\ldots, x_{nqp}, u_{10}, \ldots, u_{mq})$ is a given n-dimensional vector valued function of $(np+m)(q+1)+1$ variables which is measurable and locally integrable in $\boldsymbol{R}^{(np+m)(q+1)+1}$. The control $u(t)$ is supposed to be measurable and locally integrable for $t \in \boldsymbol{R}$ (Fig. 11).

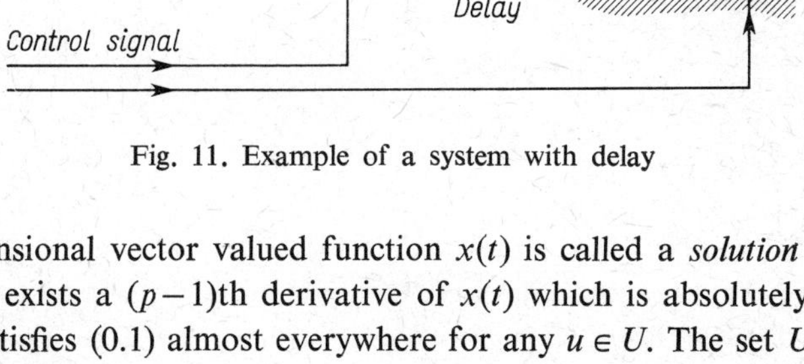

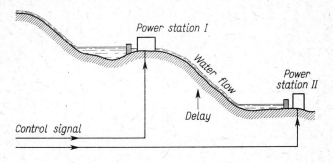

Fig. 11. Example of a system with delay

An n-dimensional vector valued function $x(t)$ is called a *solution* of equation (0.1) if there exists a $(p-1)$th derivative of $x(t)$ which is absolutely continuous and if $x(t)$ satisfies (0.1) almost everywhere for any $u \in U$. The set U of controls is determined by engineering constraints imposed on u. If delays appear only in the control $u(t)$, these delays are not essential and equation (0.1) becomes an ordinary differential equation. However, from the engineering point of view, if $u(t)$ is unknown, it is important to define it in such a way that the required conditions on $x(t)$ are satisfied (Fig. 12).

Control of interconnected systems with delay

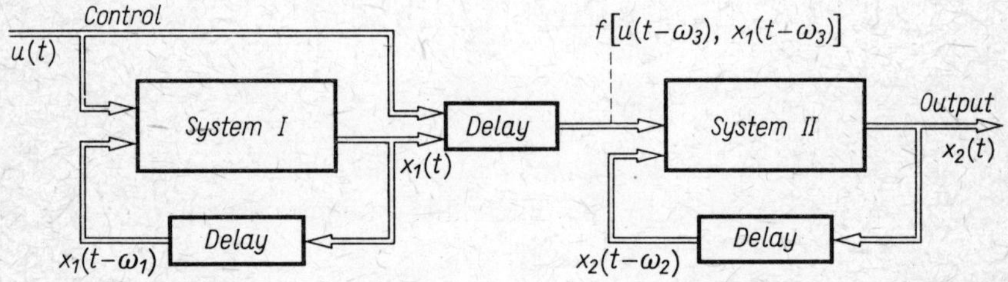

System I and System II are described by ordinary differential equations

$$\frac{d}{dt}\begin{bmatrix} x_1(t) \\ x_2(t) \end{bmatrix} = F(x_1(t),\ x_2(t),\ x_1(t-\omega_1),\ x_1(t-\omega_3),\ x_2(t-\omega_2),\ u(t),\ u(t-\omega_3)).$$

Delay follows from transport phenomena, etc.

Fig. 12. Typical configuration of chemical industrial plants

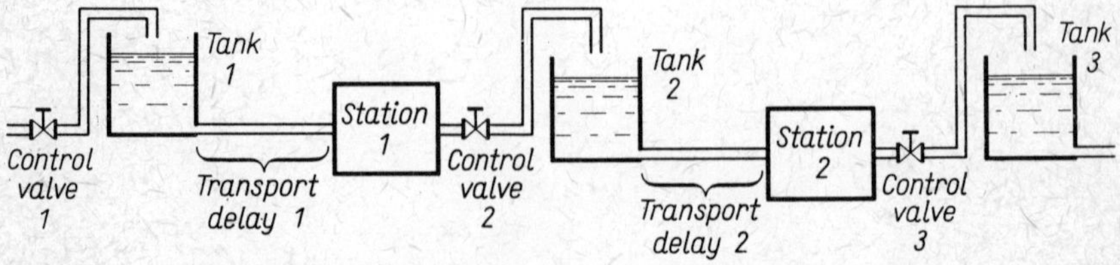

The model is described by linear differential-difference equations with delayed argument in state and control variables. The performance functional is assumed to be quadratic (see Findeisen, Pułaczewski, Manitius [1])

Fig. 13. Dynamic model of the sugar beet plant

The problem under consideration is that of finding an *optimal control*, i.e. to find a control $u_0(t)$ such that the pair (x, u_0) minimizes some *performance functional* $\boldsymbol{F}(x, u)$ defined for all $u \in U$ (Fig. 13).

For details see: Lee and Markus [1], Oğuztörelli [1], Gnoenskiĭ, Kamenskiĭ and Elsgolc [1].

1. PERIODIC LINEAR PROBLEMS

Consider a linear time lag system

$$\sum_{k=0}^{p}\sum_{j=0}^{q} F_{kj}(t)x^{(k)}(t-\omega_j) = \sum_{j=0}^{q} G_j(t)u(t-\omega_j), \tag{1.1}$$

where the state $x(t)$ is an n-dimensional vector valued function, the control $u(t)$ is an m-dimensional vector valued function, $F_{kj}(t)$ are $n\times n$ matrix-functions, and $G_j(t)$ are $n\times m$ matrix-functions, all of the real variable t. We assume that all the functions $u(t)$, $F_{kj}(t)$, $G_j(t)$ are measurable and locally integrable.

Let $0 = \omega_0 < \omega_1 < \ldots < \omega_q$ and let $\omega_1, \ldots, \omega_q$ be commensurable. We thus have an $r \neq 0$ and positive integers n_j such that $\omega_j = n_j r$ for $j = 1, 2, \ldots, q$. Assume that the functions $F_{kj}(t)$ and $G_j(t)$ are r-periodic, let N be a common multiple of the numbers $a_1, \ldots, a_q$ (not necessarily the least one) and let $\omega = Nr$. We shall study system (1.1) in the class E^n_ω of all ω-periodic vector valued functions.

Suppose we are given the following performance functional:

$$\mathbf{K}(u, x) = \int_0^\omega K\big(t, [x(t)-x^0(t), u(t)-u^0(t)]\big)dt, \tag{1.2}$$

where $[x, u]$ is an $(n+m)$-dimensional vector $(x_1, \ldots, x_n, u_1, \ldots, u_m)$ for each fixed t and $K(t, [x, u])$ is a non-negative quadratic form defined on an $(n+m)$-dimensional space and $x^0(t)$, $u^0(t)$ are given functions. We suppose that $K(t, [x, u])$ is an r-periodic square integrable function with respect to t.

We study the following problem:

To minimize the performance functional (1.2) *under the assumption that* $x(t)$ *and* $u(t)$ *satisfy equation* (1.1) *and are* ω*-periodic.*

Let X be a space of n-dimensional measurable ω-periodic square integrable vector valued functions $x(t) = \big(x_1(t), \ldots, x_n(t)\big)$ of the real variable t. Let U be a space of m-dimensional measurable ω-periodic square integrable vector valued functions $u(t) = \big(u_1(t), \ldots, u_m(t)\big)$, and consider the space $Y = X\times U$. The performance functional (1.2) induces an inner product in the space $Y = X\times U$. Namely

$$([x, u], [\xi, \eta]) = \int_0^\omega \tilde{K}\big(t, [x(t), u(t)], [\overline{\xi(t)}, \overline{\eta(t)}]\big)dt, \tag{1.3}$$

where $\tilde{K}(t, [x, u], [\xi, \eta])$ is the bilinear form induced by the quadratic form $K(t, [x, u])$.

We now define an operator S acting in X (respectively in U and in Y)

$$(Sx)(t) = x(t-r) \quad \text{for } x \in X, \qquad (Su)(t) = u(t-r) \quad \text{for } u \in U,$$
$$(S[x, u])(t) = [x(t-r), u(t-r)] \quad \text{for } [x, u] \in Y.$$

Since X and U are spaces of ω-periodic functions, the operator S is an involution of order N in X and in U and also in Y: $S^N = I$ (cf. Section 1 of Chapter V). By Theorem 4.1 of Chapter II we can decompose the spaces X and U into direct sums

$$X = X_1 \oplus \dots \oplus X_N, \qquad U = U_1 \oplus \dots \oplus U_N$$

such that

$$(1.4) \qquad \left.\begin{aligned} (Sx)(t) &= \varepsilon^j x \quad \text{for } x \in X_j,\ X_j = P_j X \\ (Su)(t) &= \varepsilon^j u \quad \text{for } u \in U_j,\ U_j = P_j U \end{aligned}\right\}$$

$$\varepsilon = e^{2\pi i/N}, \qquad P_j = \frac{1}{N}\sum_{k=0}^{N-1} \varepsilon^{-kj} S^k,$$

$(j = 1, 2, \dots, N)$. Formulae (1.4) imply that for $[x, u] \in Y_j = X_j \times U_j$ equation (1.1) is transformed into a differential equation without time lags (obtained in the same way as in Section 1 of Chapter V):

$$(1.5) \qquad \sum_{k=0}^{p} \tilde{F}_{kj}(t) x_{(j)}^{(k)}(t) = \tilde{G}_j(t), \quad x_{(j)} \in X_j,\ u_{(j)} \in U_j \ (j = 1, 2, \dots, N),$$

where

$$\tilde{F}_{kj}(t) = \sum_{\nu=0}^{q} \varepsilon^{\nu n_j} F_{k\nu}(t), \quad \tilde{G}_j(t) = \sum_{\nu=0}^{q} \varepsilon^{\nu n_j} G_\nu(t).$$

Formulae (1.4) also imply that the spaces $Y_j = X_j \times U_j$ are mutually orthogonal with respect to the inner product (1.3). Indeed, since $K(t, [x, u])$ is r-periodic with respect to t, S is a unitary transformation for

$$\begin{aligned}(S[x, u], S[\xi, \eta]) &= \int_0^\omega \tilde{K}(t, [x(t-r), u(t-r)], [\overline{\xi(t-r)}, \overline{\eta(t-r)}])dt \\ &= \int_{-r}^{\omega-r} \tilde{K}(t-r, [x(t), u(t)], [\overline{\xi(t)}, \overline{\eta(t)}])dt \\ &= \int_0^\omega \tilde{K}(t, [x(t), u(t)], [\overline{\xi(t)}, \overline{\eta(t)}])dt = ([x, u], [\xi, \eta]).\end{aligned}$$

Hence Theorem 6.1 of Chapter II implies that

$$(y_{(j)}, y_{(k)}) = ([x_{(j)}, u_{(j)}], [x_{(k)}, u_{(k)}]) = 0 \quad \text{if } j \neq k.$$

The method of solving our problem is as follows. We decompose the space Y into an orthogonal direct sum $Y = Y_1 \oplus \dots \oplus Y_N$. This decomposition also induces a decomposition of the element $y^0(t) = [x^0(t), u^0(t)]$:

$$\begin{aligned} x^0(t) &= x^0_{(1)}(t) + \dots + x^0_{(N)}(t), \quad \text{where } x^0_{(j)} \in X_j \\ u^0(t) &= u^0_{(1)}(t) + \dots + u^0_{(N)}(t), \quad \text{where } u_{(j)} \in U_j \end{aligned} \qquad (j = 1, 2, \dots, N).$$

We consider the jth differential equation (1.5) in the space Y_j and minimize the functional

$$\int_0^{\omega} K(t, [x_{(j)}(t) - x^0_{(j)}(t), u_{(j)}(t) - u^0_{(j)}(t)])dt. \tag{1.6}$$

It is easy to check that for any $[x_{(j)}, u_{(j)}] \in Y_{(j)}$

$$\int_0^{\omega} K(t, [x_{(j)}(t) - x^0_{(j)}(t), u_{(j)}(t) - u^0_{(j)}(t)])dt \tag{1.7}$$

$$= N \int_0^{r} K(t, [x_{(j)}(t) - x^0_{(j)}(t), u_{(j)}(t) - u^0_{(j)}(t)])dt.$$

The solution of the initial problem is of the form

$$x(t) = x_{(1)}(t) + \dots + x_{(N)}(t), \quad u(t) = u_{(1)}(t) + \dots + u_{(N)}(t),$$

where $[x_{(j)}, u_{(j)}] \in Y_j$ is a pair minimizing functional (1.7) under condition (1.5).

The method described above can also be used in cases where we have some additional constraints. We here give two examples.

Consider system (1.1) with the additional condition that $N = 2$. We minimize functional (1.2). Assume that the control $u(t) = (u_1(t), \dots, u_m(t))$ is constrained in each coordinate, i.e. that $|u_j(t)| \leqslant M_j$ $(j = 1, 2, \dots, m)$.

It is not difficult to verify that $|u_j(t)| \leqslant M_j$ if and only if $|u_{(1)i}(t)| + |u_{(2)i}(t)| \leqslant M_i$, where $u_{(1)}(t) = \frac{1}{2}[u(t) - u(t-r)]$, $u_{(2)}(t) = \frac{1}{2}[u(t) + u(t-r)]$. We can thus regard $u(t)$ as a $2m$-dimensional vector $[u_{(1)1}(t), u_{(2)1}(t), \dots, u_{(1)m}(t), u_{(2)m}(t)]$. Similarly, $x(t)$ can be regarded as a $2n$-dimensional vector $[x_{(1)1}(t), x_{(2)1}(t), \dots$ $\dots, x_{(1)n}(t), x_{(2)n}(t)]$. System (1.5) can thus be regarded as a single $2n$-dimensional system.

Formula (1.7) implies that we minimize the performance functional

$$2\Big[\int_0^r K\big(t, [x_{(1)}(t)-x^0_{(1)}(t), u_{(1)}(t)-u^0_{(1)}(t)]\big)dt+$$

$$+\int_0^r K\big(t, [x_{(2)}(t)-x^0_{(2)}(t), u_{(2)}(t)-u^0_{(2)}(t)]\big)dt\Big]$$

with the constraints

$$|u_{(1)j}(t)|+|u_{(2)j}(t)| \leqslant M_j \quad (j=1,2,\dots,m).$$

The second example is the following. We consider a non-negative quadratic form $C(t, [x, u])$ defined for each fixed t on an $(n+m)$-dimensional space. We assume that $C(t, [x, u])$ is r-periodic and square integrable with respect to t, and that the constraint is the following:

$$C(x, u) = \int_0^\omega C\big(t, [x(t), u(t)]\big)dt \leqslant M^2. \tag{1.8}$$

Clearly the spaces Y_j are also orthogonal with respect to the inner product induced by $C(x, u)$, and so the method for minimizing functional (1.2) under condition (1.1) with constraint (1.8) is the following: We assume that

$$|y_{(j)}(t)| \leqslant M_j \quad (j=1,2,\dots,M), \tag{1.9}$$

and we minimize functional (1.7) under condition (1.5) with constraint (1.9). This minimum depends on the constants $M_1, \dots, M_N$. We obtain in this way the minimum of functional (1.2) under condition (1.1) with constraints (1.9). This minimum also depends on the constants $M_1, \dots, M_N$. We then minimize this minimum with respect to $M_1, \dots, M_N$ under the condition

$$M_1^2+\dots+M_N^2 \leqslant M^2.$$

We thus obtain a minimum of functional (1.2) under condition (1.1) with constraint (1.8).

The case in which the data are ω-periodic will be included in the result of the next section.

2. PERIODIC NON-LINEAR PROBLEMS

In this section we consider the non-linear time lag system

$$\frac{dx(t)}{dt} = G\big(t, x(t), x(t-\omega_1), \dots, x(t-\omega_q), u(t), u(t-\omega_1), \dots, u(t-\omega_q)\big), \tag{2.1}$$

where the state x is an absolutely continuous n-dimensional vector-valued function, and the control $u(t)$ is a measurable and locally integrable m-dimensional vector-valued function. The given n-dimensional vector-valued function $G(t, x_{10}, \ldots, x_{nq}, u_{10}, \ldots, u_{1q})$ of $(n+m)(q+1)+1$ variables is continuous.

Let $\omega_0, \ldots, \omega_q$, ω be defined as in Section 1. We consider system (2.1) in the class E^n_ω of ω-periodic n-dimensional vector-valued functions.

Suppose we are given the performance functional

$$K(x, u) = \int_0^\omega K(t, [x(t)-x^0(t), u(t)-u^0(t)])dt, \tag{2.2}$$

where $[x, u]$ is an $(n+m)$-dimensional vector $(x_1, \ldots, x_n, u_1, \ldots, u_m)$ and, for each fixed t, $K(t, [x, u])$ is a non-negative quadratic form defined on an $(n+m)$-dimensional space and $x^0(t), u^0(t)$ are given vector valued functions. We assume that $K(t, [x, u])$ is an ω-periodic square integrable function with respect to t.

Our problem is to minimize the performance functional (2.2) under the condition that the pair $[x(t), u(t)]$ satisfies equation (1.1) and that $x(t), u(t)$ are ω-periodic.

Let X be a space of n-dimensional measurable ω-periodic square integrable vector valued functions $x(t) = (x_1(t), \ldots, x_n(t))$ of the real variable t. Let U be a space of m-dimensional measurable ω-periodic square integrable vector valued functions $u(t) = (u_1(t), \ldots, u_m(t))$. Consider the space $X \times U = Y$. The performance functional (2.2) induces in the space Y the inner product given by

$$([x, u], [\xi, \eta]) = \int_0^\omega \tilde{K}(t, [x(t), u(t)], [\overline{\xi(t)}, \overline{\eta(t)}])dt, \tag{2.3}$$

where $\tilde{K}(t, [x, u], [\xi, \eta])$ is the bilinear form induced by the quadratic form $K(t, [x, u])$.

We define the operator S in the space X, U and Y just as in Section 1 and we obtain the same decomposition of these spaces. We recall that (Proposition 2.1 of Chapter V)

$$x \in X_j \text{ if and only if } x(t) = e^{-2\pi ijt/\omega}\tilde{x}(t),$$

where $\tilde{x}(t)$ is an r-periodic vector valued function $(j = 1, 2, \ldots, N)$.

Similarly,

$$u \in U_j \text{ if and only if } u(t) = e^{-2\pi ijt/\omega}\tilde{u}(t),$$

where $\tilde{u}(t)$ is an r-periodic vector valued function $(j = 1, 2, \ldots, N)$.

We write:

$$\tilde{x}_j(t) = e^{2\pi ijt/\omega} x_j(t), \quad \tilde{u}_j(t) = e^{2\pi ijt/\omega} \tilde{u}_{(j)}(t) \quad (j = 1, 2, \ldots, N).$$

The vector valued functions $\tilde{x}_j(t)$ and $\tilde{u}_j(t)$ are r-periodic by definition.

This implies that system (2.1) can be rewritten in the form of a system of $n \cdot N$ ordinary differential equations (without delays)

$$(2.4) \qquad \dot{\tilde{x}}_j(t) = \frac{2\pi ij}{\omega} \tilde{x}_j(t) + \tilde{G}_j\big(t, \tilde{x}_1(t), \ldots, \tilde{x}_N(t), \tilde{u}_1(t), \ldots, \tilde{u}_N(t)\big) \qquad (j = 1, 2, \ldots, N),$$

where $\tilde{G}_j(t, \tilde{x}_1, \ldots, \tilde{x}_N, \tilde{u}_1, \ldots, \tilde{u}_N) = e^{2\pi ijt/\omega} P_j \tilde{G}(t, \tilde{x}_1, \ldots, \tilde{x}_N, \tilde{u}_1, \ldots, \tilde{u}_N)$ and the operators P_j act on the variable t alone (compare the proof of Theorem 4.1 of Chapter V),

$$\tilde{G}(t, \tilde{x}_1, \ldots, \tilde{x}_N, \tilde{u}_1, \ldots, \tilde{u}_N)$$
$$= G\Big(t, \sum_{k=1}^{N} e^{-2\pi ikt/\omega} \tilde{x}_k, \ldots, \sum_{k=1}^{N} e^{-2\pi ikt/\omega} e^{2\pi ik\omega_q/\omega} \tilde{x}_k,$$
$$\sum_{k=1}^{N} e^{-2\pi ikt/\omega} \tilde{u}_k, \ldots, \sum_{k=0}^{N} e^{-2\pi ikt/\omega} e^{2\pi ik\omega_q/\omega} \tilde{u}_k\Big).$$

Formulae (1.4) and Corollary 1.6 of Chapter II imply that the spaces $Y_j = X_j \times U_j$ are mutually orthogonal with respect to the following inner product induced by the inner product (2.3):

$$([x, u], [\xi, \eta])_S = \sum_{k=0}^{N-1} (S^k[x, u], S^k[\xi, \eta]).$$

Hence the space Y is an orthogonal direct sum: $Y = Y_1 \oplus \ldots \oplus Y_N$. This decomposition also induces a decomposition of the element $y^0(t) = [x^0(t), u^0(t)]$

$$x^0(t) = \sum_{j=1}^{N} x^0_{(j)}(t) = \sum_{j=1}^{N} e^{-2\pi ijt/\omega} x^0_{(j)}(t),$$
$$u^0(t) = \sum_{j=1}^{N} u^0_{(j)}(t) = \sum_{j=1}^{N} e^{-2\pi ijt/\omega} u^0_{(j)}(t),$$

where $\tilde{x}^0_{(j)}(t)$, $\tilde{u}^0_{(j)}(t)$ are r-periodic vector valued functions.

We consider the differential system (2.4) in the space E^{n+m} of r-periodic vector valued functions $[\tilde{x}_j, \tilde{u}_j]$ and we minimize the functional

$$(2.5) \quad \int_0^{\omega} K_1\big(t, \tilde{x}_1(t) - \tilde{x}^0_1(t), \ldots, \tilde{x}_N(t) - \tilde{x}^0_N(t), \tilde{u}_1(t) - \tilde{u}^0_1(t), \ldots, \tilde{u}_N(t) - \tilde{u}^0_N(t)\big) dt,$$

where

$$K_1(t, \tilde{x}_1, \dots, \tilde{x}_N, \tilde{u}_1, \dots, \tilde{u}_N) = K\Big(t, \Big[\sum_{k=1}^{N} e^{-2\pi ikt/\omega}\tilde{x}_k, \sum_{k=1}^{N} e^{-2\pi ikt/\omega}\tilde{u}_k\Big]\Big).$$

The solution of the initial problem is

$$x(t) = \sum_{j=1}^{N} e^{-2\pi ijt/\omega}\tilde{x}_j(t), \quad u(t) = \sum_{j=1}^{N} e^{-2\pi ijt/\omega}\tilde{u}_j(t),$$

where $(\tilde{x}_1, \dots, \tilde{x}_N, \tilde{u}_1, \dots, \tilde{u}_N)$ is the system minimizing the functional (2.5) under condition (2.4).

If the given function G is ω-periodic with respect to t, then $\tilde{G}_j$ is r-periodic with respect to t.

3. PROBLEMS WITH DAMPING

Consider now a linear time lag system

$$\sum_{k=0}^{p}\sum_{j=0}^{q} F_{kj}(t)x^{(k)}(t-\omega_j) = \sum_{j=0}^{q} G_j(t)u(t-\omega_j), \tag{3.1}$$

where the state $x(t)$ is an n-dimensional vector valued function, the control $u(t)$ is an m-dimensional vector valued function, F_{kj} are $n \times m$ matrix-functions, and G_j are $n \times m$ matrix functions, all of the real variable t. We assume that all the matrix-functions F_{kj} and G_j are measurable and locally integrable on the real line.

Let $\omega_0 = 0 < \omega_1 < \dots < \omega_q$, and let the numbers $\omega_1, \dots, \omega_q$ be commensurable, so that we have a real $r \neq 0$ and positive integers n_j such that $\omega_j = n_j r$ for $j = 1, 2, \dots, q$. Let N be a common multiple of the numbers $n_1, \dots, n_q$ (not necessarily the least one) and let $\omega = Nr$. In addition we put $n_0 = 0$.

As before (Chapter VI), we denote by E^n_ω the space of all ω-periodic n-dimensional vector valued functions and write

$$E^n_\omega(\lambda) = \{x(t) = e^{\lambda t}y(t) \colon y(t) \in E^n_\omega\}.$$

We assume that $\lambda \neq 0$ is a real.

Observe that for a $\lambda = -\mu$, where $\mu > 0$, the functions of the form $e^{-\mu t}y(t)$, where $y(t)$ is a periodic function, represent damped oscillations (Fig. 7 in Section 1 of Chapter VI).

Assume that the matrix-functions F_{kj} and G_j are r-periodic. We discuss system (3.1) in the space $E^n_\omega(\lambda)$. Consider the performance functional

$$\boldsymbol{K}(x, u) = \int_0^\omega K\big(t, [x(t)-x^0(t), u(t)-u^0(t)]\big)dt, \tag{3.2}$$

where $[x, u]$ is an $(n+m)$-dimensional vector $(x_1, \ldots, x_n, u_1, \ldots, u_m)$, $K(t, [x, u])$ for each fixed t is a non-negative quadratic form defined on an $(n+m)$-dimensional space and $x^0(t), u^0(t)$ are given functions. We also assume that $K(t, [x, u])$ is an r-periodic square integrable function with respect to t.

The problem[1] we are going to solve is the following: To minimize the performance functional (3.2) under the condition that the pair $[x(t), u(t)]$ satisfies equation (3.1) and that $x(t) \in E_\omega^n(\lambda)$, $u(t) \in E_\omega^m(\lambda)$.

Let $X \subset E_\omega^n(\lambda)$ be a subspace of n-dimensional square integrable measurable vector valued functions $x(t) = (x_1(t), \ldots, x_n(t))$. Let $U \subset E_\omega^m(\lambda)$ be a subspace of m-dimensional square integrable measurable vector valued functions $u(t) = (u_1(t), \ldots, u_m(t))$. Consider the space $Y = X \times U$ and define in the space Y a *weighted inner product*

$$(3.3) \qquad ([x, u], [\xi, \eta]) = \int_0^\omega p(t)\tilde{K}(t, [x(t), u(t)], [\overline{\xi(t)}, \overline{\eta(t)}])dt,$$

where $\tilde{K}(t, [x, u], [\xi, \eta])$ is the bilinear form induced by the quadratic form $K(t, [x, u])$ and $p(t) = e^{2\lambda t}$.

We now shall define an operator S acting in X (resp. in U and Y) by means of the equalities:

$$(Sx)(t) = x(t-r) \quad \text{for } x \in X, \qquad (Su)(t) = u(t-r) \quad \text{for } u \in U,$$
$$(Sy)(t) = (S[x, u])(t) = [x(t-r), u(t-r)] \quad \text{for } [x, u] \in Y.$$

The operator S is an algebraic operator of order N in X, in U and also in Y (cf. Section 1 of Chapter VI). We can thus decompose the spaces X and Y into direct sums

$$X = X_1 \oplus \ldots \oplus X_N, \qquad U = U_1 \oplus \ldots \oplus U_N,$$

such that

$$(3.4) \qquad \begin{array}{ll} (Sx)(t) = e^{-\lambda r}\varepsilon^j x(t) & \text{for } x \in X_j, \\ (Su)(t) = e^{-\lambda r}\varepsilon^j u(t) & \text{for } u \in U_j, \end{array} \qquad \varepsilon = e^{2\pi i/N} \quad (j = 1, 2, \ldots, N).$$

Formulae (3.4) also imply that for $[x, u] \in Y_j = X_j \times U_j$ equation (3.1) is transformed into a system of N differential equations without time lags

$$(3.5) \qquad \sum_{k=0}^{p} \tilde{F}_{kj}(t)x_{(j)}^{(k)}(t) = \tilde{G}_j(t)u_{(j)}(t) \quad (j = 1, 2, \ldots, N),$$

[1] This problem was solved by Włodarska-Dymitruk [4].

where

$$x_{(j)} = P_j x, \quad u_{(j)} = P_j u, \quad P_j = \mathfrak{p}_j(S), \; \mathfrak{p}_j(t) = \prod_{\substack{k=1 \\ k \neq j}}^{N} \frac{t - e^{-\lambda r}\varepsilon^k}{e^{-\lambda r}\varepsilon^j - e^{-\lambda r}\varepsilon^k},$$

$$\left.\begin{aligned} \tilde{F}_{kj}(t) &= \sum_{\nu=0}^{q} e^{-\lambda\omega_\nu \varepsilon^{jn_\nu}} F_{k\nu}(t) \quad (k = 0, 1, \dots, p) \\ \tilde{G}_j(t) &= \sum_{\nu=0}^{q} e^{-\lambda\omega_\nu \varepsilon^{jn_\nu}} G_\nu(t) \end{aligned}\right\} \quad (j = 1, 2, \dots, N).$$

Formulae (3.4) also imply that the spaces Y_j are mutually orthogonal with respect to the inner product (3.3). Indeed, since $K(\bar{t}, [x, u])$ is r-periodic with respect to t, we have

$$(y_{(j)}, y_{(k)}) = e^{2\lambda r}(Sy_{(j)}, Sy_{(k)}) \quad \text{for } y_{(j)} \in Y_j, \; y_{(k)} \in Y_k,$$

and from (3.4)

$$(Sy_{(j)}, Sy_{(k)}) = e^{-\lambda r}\varepsilon^j e^{-\lambda r}\varepsilon^{-k}(y_{(j)}, y_{(k)}) = e^{-2\lambda r}\varepsilon^{j-k}(y_{(j)}, y_{(k)}).$$

Then

$$(y_{(j)}, y_{(k)}) = e^{2\lambda r} e^{-2\lambda r}\varepsilon^{j-k}(y_{(j)}, y_{(k)}) = e^{2\pi i(j-k)/N}(y_{(j)}, y_{(k)}).$$

Hence $(y_{(j)}, y_{(k)}) = 0$ if $j \neq k$.

We now decompose the space Y into an orthogonal direct sum

$$Y = Y_1 \oplus \dots \oplus Y_N.$$

This decomposition also induced a decomposition of the element $y^0(t) = [x^0(t), u^0(t)]$:

$$x^0(t) = \sum_{j=1}^{N} x^0_{(j)}(t), \quad \text{where } x^0_{(j)} \in X_j,$$

$$u^0(t) = \sum_{j=1}^{N} u_{(j)}(t), \quad \text{where } u^0_{(j)} \in U_j.$$

We minimize the functional

$$\int_0^\omega K(t, [x_{(j)}(t) - x^0_{(j)}(t), u_{(j)}(t) - u^0_{(j)}(t)])\,dt \quad (j = 1, 2, \dots, N) \tag{3.6}$$

under the condition that $[x_{(j)}, u_{(j)}] \in X_j \times U_j$ is a solution of the jth equation (3.5).

The solution of the initial problem is then of the form

$$x(t) = \sum_{j=1}^{N} x_{(j)}(t), \quad u(t) = \sum_{j=1}^{N} u_{(j)}(t),$$

where $[x_{(j)}(t), u_{(j)}(t)] \in Y_j = X_j \times U_j$ is a pair minimizing functional (3.6) under condition (3.5).

Suppose that $K(t, [x, u])$ is an ω-periodic function with respect to t, and define in the space Y the following inner product

$$([x, u], [\xi, \eta])_S = \sum_{j=1}^{N} (S^j[x, u], S^j[\xi, \eta]). \tag{3.7}$$

It is easy to check that the spaces Y_j are mutually orthogonal with respect to the inner product (3.7). Indeed, $(y_{(j)}, y_{(k)})_S = e^{2\lambda r}(Sy_{(j)}, Sy_{(k)})_S$ and from Formulae (3.4) we find $(y_{(j)}, y_{(k)})_S = 0$ if $j \neq k$.

As in Section 2, using the method described in Section 2 of Chapter VI, we can solve the corresponding non-linear problem with the performance functional (3.2) (see: Włodarska-Dymitruk [4]).

4. ALMOST PERIODIC LINEAR PROBLEMS

Consider the linear time lag system

$$\sum_{k=0}^{p} \sum_{j=0}^{q} F_{kj}(t) x^{(k)}(t-\omega_j) = \sum_{j=0}^{q} G_j(t) u(t-\omega_j), \tag{4.1}$$

where the state $x(t)$ is an n-dimensional vector valued function, the control $u(t)$ is an m-dimensional vector valued function, $F_{kj}(t)$ are $n \times n$ matrix-functions, and $G_j(t)$ are $n \times m$ matrix-functions, all of the real variable t. We assume that all the matrix-functions $F_{kj}(t)$, $G_j(t)$ are measurable and locally integrable on the real line.

Let $0 = \omega_0 < \omega_1 < \ldots < \omega_q$ and let $\omega_1, \ldots, \omega_q$ be commensurable, i.e. we have a real $r \neq 0$ and positive integers n_j such that $\omega_j = n_j r$ for $j = 1, 2, \ldots, q$.

We put $n_0 = 0$. Let N be a common multiple of numbers $n_1, \ldots, n_q$ (not necessarily the least one) and let $\omega = Nr$.

Let E^n_ω be the space of all ω-periodic vector valued functions n-dimensional defined on the real line. Let $\lambda_1, \ldots, \lambda_M$ be real numbers. By $E^n_\omega(i\lambda_1, \ldots, i\lambda_M)$ we mean the space of all n-dimensional vector valued functions of the form

$$x(t) = \sum_{j=1}^{M} e^{i\lambda_j t} y_j(t), \quad \text{where } y \in E^n_\omega. \tag{4.2}$$

$E^n_\omega(i\lambda_1, \ldots, i\lambda_M)$ is a subspace of the space of almost periodic functions (cf. Section 1 of Chapter VI).

Without loss of generality we can assume that

$$\lambda_j \neq \lambda_k + \frac{2\pi}{\omega} n \quad \text{for } j, k = 1, 2, \ldots, M \text{ and } n = 0, \pm 1, \pm 2, \ldots$$

(compare Section 1 of Chapter VI).

Assume that the functions $F_{kj}(t)$ and $G_j(t)$ are r-periodic. We consider system (4.1) in the space $E^n_\omega(i\lambda_1, \ldots, i\lambda_M)$. Suppose the following performance functional is defined for any $T > 0$:

$$\boldsymbol{K}_T(x, u) = \int_0^T K\big(t, [x(t) - x^0(t), u(t) - u^0(t)]\big)dt,$$

where $[x, u]$ is an $(n+m)$-dimensional vector $(x_1, \ldots, x_n, u_1, \ldots, u_m)$, $K(t, [x, u])$ for each fixed t is a non-negative quadratic form defined on an $(n+m)$-dimensional space and $x^0(t)$, $u^0(t)$ are given functions of form (3.2). We assume that $K(t, [x, u])$ is an r-periodic square integrable function with respect to t.

It is easy to see that the limit

$$\lim_{T\to\infty} \frac{1}{T} \int_0^T K\big(t, [x(t) - x^0(t), u(t) - u^0(t)]\big)dt \tag{4.3}$$

exists by virtue of the theorem concerning the existence of a mean value (see Levitan [1], p. 32).

The problem which we want to solve is to minimize the performance functional (4.3) under the condition that the pair $[x(t), u(t)]$ satisfies equation (4.1) and that $x \in E^n_\omega(i\lambda_1, \ldots, i\lambda_M)$, $u \in E^m_\omega(i\lambda_1, \ldots, i\lambda_M)$[1].

Let $X \subset E^n_\omega(i\lambda_1, \ldots, i\lambda_M)$ be a space of n-dimensional square integrable vector valued functions $x(t)$ and let $U \subset E^m_\omega(i\lambda_1, \ldots, i\lambda_M)$ be a space of m-dimensional vector value functions $u(t)$. Consider the space $Y = X \times U$. The performance functional induces in the space Y an inner product of the form

$$([x, u], [\xi, \eta]) = \lim_{T\to\infty} \frac{1}{T} \int_0^T \tilde{K}\big(t, [x(t), u(t)], [\overline{\xi(t)}, \overline{\eta(t)}]\big)dt, \tag{4.4}$$

where $\tilde{K}(t, [x, u], [\xi, \eta])$ is the bilinear form induced by the quadratic form $K(t, [x, u])$.

We now define the operator S acting in the spaces X, U and Y by means of the equalities:

[1] This problem has been solved by Włodarska-Dymitruk [2].

$$(Sx)(t) = x(t-r) \quad \text{for } x \in X, \qquad (Su)(t) = u(t-r) \quad \text{for } u \in U,$$

$$(S[x, u])(t) = [x(t-r), u(t-r)] \quad \text{for } [x, u] \in Y.$$

Just as in Section 1 of Chapter VI, we conclude that S is an algebraic operator of order $N \cdot M$ in X, in U and also in Y. We can therefore decompose the spaces X and U into direct sums

$$\begin{aligned} X &= X_1 \oplus \ldots \oplus X_{N\cdot M}, \quad \text{where } X_j = P_j X, \\ U &= U_1 \oplus \ldots \oplus U_{N\cdot M}, \quad \text{where } U_j = P_j U, \end{aligned} \qquad P_j = \prod_{\substack{\nu=1 \\ \nu \neq j}}^{N\cdot M} \frac{S-e_\nu}{e_j - e_\nu},$$

$$e_j = e^{-i\lambda_\mu r}\varepsilon^\nu \quad \text{for } j = M(\nu-1)+\mu \quad (\nu = 1, 2, \ldots, N;\ \mu = 1, 2, \ldots, M), \qquad \varepsilon = e^{2\pi i/N}, \tag{4.5}$$

such that

$$\begin{aligned} (Sx)(t) &= e_j x(t) \quad \text{for } x \in X_j \\ (Su)(t) &= e_j u(t) \quad \text{for } u \in U_j \end{aligned} \qquad (j = 1, \ldots, N \cdot M). \tag{4.6}$$

Formulae (4.5) imply that for $[x, u] \in Y_j = X_j \times U_j$ equation (4.1) is transformed into the system of $N \cdot M$ differential equations without time lags (obtained in the same way as in Section 1 of Chapter VI)

$$\sum_{k=0}^{p} \tilde{F}_{kj}(t) x_{(j)}^{(k)}(t) = \tilde{G}_j(t) u_j(t) \quad (j = 1, 2, \ldots, N \cdot M), \tag{4.7}$$

where

$$x_{(j)} = P_j x, \qquad u_{(j)} = P_j u,$$

$$\tilde{F}_{kj}(t) = \sum_{\nu=0}^{q} e_j^{n_\nu} F_{k\nu}(t) \quad (k = 0, 1, \ldots, p;\ j = 1, 2, \ldots, N \cdot M),$$

$$\tilde{G}_j(t) = \sum_{\nu=0}^{q} e_j^{n_\nu} G_j(t).$$

Formulae (4.5) also imply that the spaces Y_j are mutually orthogonal with respect to the inner product (4.4). Indeed, since $K(t, [x, u])$ is r-periodic with respect to t, S is a unitary transformation (cf. Section 1).

Theorem 6.1 of Chapter II thus implies that

$$(y_{(j)}, y_{(k)}) = 0 \quad \text{if } j \neq k \ (j, k = 1, 2, \ldots, N \cdot M).$$

Consequently we can decompose the space Y into the orthogonal direct sum $Y = Y_1 \oplus \ldots \oplus Y_{N\cdot M}$. This decomposition also induces a decomposition of the element $y^0(t) = [x^0(t), u^0(t)]$:

$$x^0(t) = \sum_{j=0}^{N \cdot M} x^0_{(j)}(t), \qquad \text{where } x^0_{(j)} \in X_j,$$

$$u^0(t) = \sum_{j=0}^{N \cdot M} u^0_{(j)}(t), \qquad \text{where } u^0_{(j)} \in U_j.$$

We study the differential equation (4.7) in the space Y_j and we minimize the functional

$$\text{(4.8)} \quad \lim_{T\to\infty} \frac{1}{T} \int_0^T K(t, [x^0_{(j)}(t) - x^0_{(j)}(t), u_{(j)}(t) - u^0_{(j)}])dt \quad (j = 1, 2, \ldots, N \cdot M).$$

The solution of the initial problem is then of the form

$$x(t) = \sum_{j=0}^{N \cdot M} x_{(j)}(t), \qquad u(t) = \sum_{j=0}^{N \cdot M} u_{(j)}(t),$$

where $[x_{(j)}, u_{(j)}]$ is the pair minimizing functional (4.8) under the jth condition (4.7).

The solution of the same problem with ω-periodic data will be included in the results of the next Section.

5. ALMOST PERIODIC NON-LINEAR PROBLEMS

Consider a system of non-linear differential-difference equations

$$\text{(5.1)} \quad \frac{dx(t)}{dt} = G(t, x(t)x(t-\omega_1), \ldots, x(t-\omega_q), u(t), u(t-\omega_1), \ldots, u(t-\omega_q)),$$

where the state $x(t)$ is an absolutely continuous n-dimensional vector valued function, the control $u(t)$ is an absolutely continuous m-dimensional vector valued function and the given n-dimensional vector valued function

$$G(t, x_{10}, \ldots, x_{nq}, u_{10}, \ldots, u_{nq})$$

of $(n+m)(q+1)+1$ variables is continuous, all with respect to the real variable t.

Let $0 = \omega_0 < \omega_1 < \ldots < \omega_q$ and let the numbers $\omega_1, \ldots, \omega_q$ be commensurable, i.e. let $\omega_j = n_j r$, where $r \neq 0$ is real and the n_j are positive integers. Let N be a common multiple of numbers $n_1, \ldots, n_q$ and let $\omega = Nr$.

We study system (5.1) in the space $E^n_\omega(i\lambda_1, \ldots, i\lambda_M)$ (cf. the preceding section). Let the performance functional be defined by Formula (4.3).

Let $X \subset E^n_\omega(i\lambda_1, \ldots, i\lambda_M)$ and $U \subset E^m_\omega(i\lambda_1, \ldots, i\lambda_M)$ be spaces of square integrable n-dimensional and m-dimensional vector valued functions respectively.

We wish to minimize the performance functional (4.3) under condition (5.1) in the space $X \times U$[1].

The performance functional (4.3) induces an inner product of the form (4.4) in the space $Y = X \times U$.

The shift operator S defined in Section 1 induces a decomposition of the spaces X, U and Y into the corresponding direct sums. Just as in Section 2 of Chapter VI, we obtain an equivalent system of $n \cdot M \cdot N$ equations without shifts:

$$\frac{d\tilde{x}_j}{dt} = \left(\frac{2\pi i\nu}{\omega} - i\lambda_\mu\right)\tilde{x}_j + \tilde{G}_j(t, \tilde{x}_1, \ldots, \tilde{x}_{N\cdot M}, \tilde{u}_1, \ldots, \tilde{u}_{N\cdot M}), \tag{5.2}$$

$$j = M(\nu-1)+\mu \;(\mu = 1, 2, \ldots, M;\; \nu = 1, 2, \ldots, N),$$

where

$$\tilde{x}_j(t) = e^{-i\lambda_\mu t} e^{2\pi i\nu t/\omega}(P_j x)(t), \qquad \tilde{u}_j(t) = e^{-i\lambda_\mu t} e^{2\pi i\nu t/\omega}(P_j u)(t),$$

are r-periodic vector valued functions and

$$\tilde{G}_j(t, \tilde{x}_1, \ldots, \tilde{x}_{N\cdot M}, \tilde{u}_1, \ldots, \tilde{u}_{N\cdot M}) = e^{-i\lambda_\mu t} e^{2\pi i\nu t/\omega} P_j[\tilde{G}(t, \tilde{x}_1, \ldots, \tilde{x}_{N\cdot M}, \tilde{u}_1, \ldots, \tilde{u}_{N\cdot M})],$$

$$\tilde{G}(t, \tilde{x}_1, \ldots, \tilde{x}_{N\cdot M}, \tilde{u}_1, \ldots, \tilde{u}_{N\cdot M})$$

$$= G\Big(t, \sum_{\mu=1}^{M}\sum_{\nu=1}^{N} e^{i\lambda_{\bar{\mu}} t} e^{-2\pi i\nu t/\omega}\tilde{x}_j, \ldots, \sum_{\mu=1}^{M}\sum_{\nu=1}^{N} e^{i\lambda_\mu t} e^{-2\pi i\nu t/\omega} e^{-i\lambda_\mu \omega_q} e^{2\pi i\nu n_q/N}\tilde{x}_j,$$

$$\ldots, \sum_{\mu=1}^{M}\sum_{\nu=1}^{N} e^{i\lambda_\mu t} e^{-2\pi i\nu t/\omega}\tilde{u}_j, \ldots, \sum_{\mu=1}^{M}\sum_{\nu-1}^{N} e^{i\lambda_\mu t} e^{-2\pi i\nu t/\omega} e^{-i\lambda_\mu \omega_q} e^{2\pi i\nu n_q/N}\tilde{u}_j\Big).$$

Using Formulae (4.6) we can easily check (as in Section 2) that the spaces Y_j are mutually orthogonal with respect to the following inner product

$$([x, u], [\xi, \eta])_S = \sum_{k=0}^{N\cdot M-1} (S^k[x, u], S^k[\xi, \eta])$$

induced by the inner product (4.3). The space Y is therefore the orthogonal direct sum: $Y = Y_1 \oplus \ldots \oplus Y_{N\cdot M}$. This decomposition also implies the corresponding decomposition of the element $y^0(t) = [x^0(t), u^0(t)]$. Now write

$$\begin{aligned} \tilde{x}_j^0(t) &= e^{-i\lambda_\mu t} e^{2\pi i\nu t/\omega}(P_j x^0)(t) \\ \tilde{u}_j^0(t) &= e^{-i\lambda_\mu t} e^{2\pi i\nu t/\omega}(P_j u^0)(t) \end{aligned} \qquad (j = M(\nu-1)+\mu).$$

[1] The solution has been given by Włodarska-Dymitruk [3].

These functions are r-periodic by definition.

We minimize the functional

$$\lim_{T\to\infty}\frac{1}{T}\int_0^T K_1\big(t,\,[\tilde{x}_1(t)-\tilde{x}_1^0(t),\,\tilde{u}_1(t)-\tilde{u}_1^0(t)],\, \ldots,\,[\tilde{x}_{N\cdot M}(t)-\tilde{x}_{N\cdot M}^0(t),\,\tilde{u}_{N\cdot M}(t)-\tilde{u}_{N\cdot M}^0(t)]\big)dt \tag{5.3}$$

(where K_1 is obtained from K in the same way as in Formula (2.5)), under the condition that $(\tilde{x}_1, \ldots, \tilde{x}_{N\cdot M}, \tilde{u}_1, \ldots, \tilde{u}_{N\cdot M})$ satisfies system (5.4) and is r-periodic.

The solution of the initial problem is then

$$x(t) = \sum_{\nu=1}^{N}\sum_{\mu=1}^{M} e^{-i\lambda_\mu t}e^{-2\pi i\nu t/\omega}\tilde{x}_j(t)$$

$$(j = M(\nu-1)+\mu),$$

$$u(t) = \sum_{\nu=1}^{N}\sum_{\mu=1}^{M} e^{-i\lambda_\mu t}e^{-2\pi i\nu t/\omega}\tilde{u}_j(t)$$

where $(\tilde{x}_1, \ldots, \tilde{x}_{N\cdot M}, \tilde{u}_1, \ldots, \tilde{u}_{N\cdot M})$ is an r-periodic system minimizing functional (5.3) under condition (5.2).

REMARK. Note the assumption made in this chapter regarding the numbers $\omega_1, \ldots, \omega_q$: $0 < \omega_1 < \ldots < \omega_q$, does not play any part in our studies. We make this assumption because of the fact that time lag systems appear much more frequently than other kinds of systems.

CHAPTER X

Boundary Problems with Transformed Argument

This chapter includes the simplest boundary problems for partial differential equations which can be solved by our method. These examples point the way to further, more complicated, applications.

1. BOUNDARY PROBLEMS FOR THE LAPLACE EQUATION

Let S_α be a rotation through the angle α of the plane $\boldsymbol{R}^2$, i.e. a linear operator pefined by means of the matrix

$$S_\alpha = \begin{pmatrix} \cos\alpha & -\sin\alpha \\ \sin\alpha & \cos\alpha \end{pmatrix}$$

(Fig. 14). In the sequel we will only consider rotations through an angle commensurable with 2π. Without loss of generality we can assume that

$$\alpha = 2\pi/N. \tag{1.1}$$

Under this hypothesis, it is easy to verify that

$$S_{2\pi/N}^N = I. \tag{1.2}$$

Indeed, for an arbitrary α we have

$$S_\alpha^2 = \begin{pmatrix} \cos\alpha & -\sin\alpha \\ \sin\alpha & \cos\alpha \end{pmatrix} \cdot \begin{pmatrix} \cos\alpha & -\sin\alpha \\ \sin\alpha & \cos\alpha \end{pmatrix} = \begin{pmatrix} \cos^2\alpha - \sin^2\alpha & -2\sin\alpha\cos\alpha \\ 2\sin\alpha\cos\alpha & \cos^2\alpha - \sin^2\alpha \end{pmatrix}$$

$$= \begin{pmatrix} \cos 2\alpha & -\sin 2\alpha \\ \sin 2\alpha & \cos 2\alpha \end{pmatrix} = S_2,$$

and by simple induction

$$S_\alpha^k = S_{k\alpha} \quad (k = 1, 2, \ldots).$$

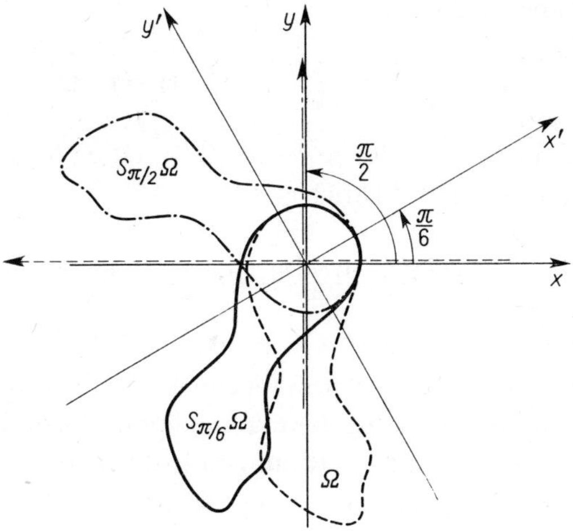

Fig. 14. Rotated domain

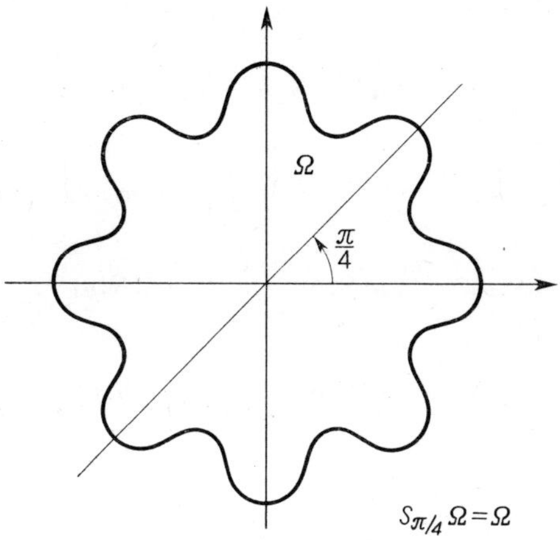

Fig. 15. Domain invariant with respect to rotation

This last formula justifies assumption (1.1). Consequently, if $\alpha = 2\pi/N$, then

$$S_{2\pi/N}^{N} = S_{N \cdot 2\pi/N} = S_2 = \begin{pmatrix} 1 & 0 \\ 0 & 1 \end{pmatrix} = I.$$

A domain $\Omega \subset \boldsymbol{R}^2$ is said to be *invariant with respect to the rotation* S_α if $S_\alpha \Omega = \Omega$ (Fig. 15).

Consider the Laplace equation

$$\Delta u \equiv \frac{\partial^2 u}{\partial x^2} + \frac{\partial^2 u}{\partial y^2} = 0, \quad (x, y) \in \Omega, \tag{1.3}$$

where Ω is a domain invariant with respect to the rotation $S_{2\pi/N}$, with a boundary condition with rotation of the Dirichlet type:

$$\lim_{p \to p_0} \sum_{k=0}^{N-1} a_k u(S_{2\pi/N}^k p) = b(p_0), \quad p = (x, y) \in \Omega, \tag{1.4}$$

where p_0 belongs to the boundary $\partial\Omega$ of the domain Ω.

We are looking for complex-valued functions, continuous on $\bar{\Omega} = \Omega \cup \partial\Omega$, which satisfy equation (1.3) inside Ω and condition (1.4) on the boundary.

Denote by X the space of all twice continuously differentiable functions defined in Ω. We introduce a linear operator $\tilde{S}_\alpha$ in the space X induced by the operator S_α in the following way

$$(\tilde{S}_\alpha u)(p) = u(S_\alpha p), \quad \text{where } u \in X \text{ and } p = (x, y) \in \Omega. \tag{1.5}$$

PROPOSITION 1.1. *For all α the operators $\tilde{S}_\alpha$ commute with the Laplacian on the space X.*

PROOF. Write $p = (x, y)$ and $p' = S_\alpha p = (x', y')$. Then

$$\begin{pmatrix} x' \\ y' \end{pmatrix} = p' = S_\alpha p = \begin{pmatrix} \cos\alpha & -\sin\alpha \\ \sin\alpha & \cos\alpha \end{pmatrix} \begin{pmatrix} x \\ y \end{pmatrix}$$

and

$$\frac{\partial u(S_\alpha p)}{\partial x'} = \frac{\partial u(S_\alpha p)}{\partial x} \cos\alpha - \frac{\partial u(S_\alpha p)}{\partial y} \sin\alpha,$$

$$\frac{\partial u(S_\alpha p)}{\partial y'} = \frac{\partial u(S_\alpha p)}{\partial x} \sin\alpha + \frac{\partial u(S_\alpha p)}{\partial y} \cos\alpha.$$

Therefore

$$\frac{\partial^2 u(S_\alpha p)}{\partial x^2} = \frac{\partial^2 u(S_\alpha p)}{\partial x^2} \cos^2\alpha - \frac{\partial^2 u(S_\alpha p)}{\partial x\, \partial y} \cos\alpha \sin\alpha - \frac{\partial^2 u(S_\alpha p)}{\partial x\, \partial y} \cos\alpha \sin\alpha + {}$$
$$+ \frac{\partial^2 u(S_\alpha p)}{\partial y^2} \sin^2\alpha,$$

$$\frac{\partial^2 u(S_\alpha p)}{\partial y'^2} = \frac{\partial^2 u(S_\alpha p)}{\partial x^2} \sin^2\alpha + \frac{\partial^2 u(S_\alpha p)}{\partial x\, \partial y} \cos\alpha \sin\alpha + \frac{\partial^2 u(S_\alpha p)}{\partial x\, \partial y} \cos\alpha \sin\alpha + {}$$
$$+ \frac{\partial^2 u(S_\alpha p)}{\partial y^2} \cos^2\alpha.$$

Hence

$$(\Delta\tilde{S}_\alpha u)(p) = \frac{\partial^2 u(S_\alpha p)}{x'^2} + \frac{\partial^2 u(S_\alpha p)}{\partial y'^2} = \frac{\partial^2 (S_\alpha p)}{x^2} + \frac{\partial^2 (S_\alpha p)}{y^2} = (\tilde{S}_\alpha \Delta u)(p)$$

for all $u \in X$ and $p \in \Omega$. This is which was to be proved.

The operator $\tilde{S}_{2\pi/N}$ is an involution of order N on the space X. Indeed, since (1.2) holds, we have

$$(\tilde{S}^N_{2\pi/N} u)(p) = u(S^N_{2\pi/N} p) = u(p) \quad \text{for all } x \in X \text{ and } p \in \Omega.$$

Thus

$$\tilde{S}^N_{2\pi/N} = I \quad \text{on } X.$$

Observe that the operator $\tilde{S}_{2\pi/N}$ also induces an involution of order N defined in the same way on the space $\tilde{X}$ of all continuous functions on the boundary $\partial\Omega$. We can denote this involution by the same symbol $\tilde{S}_{2\pi/N}$, without risk of misunderstanding. We thus have

$$(\tilde{S}_{2\pi/N} u)(q) = u(S_{2\pi/N} q) \quad \text{for all } u \in \tilde{X} \text{ and } q \in \partial\Omega.$$

Since $\tilde{S}_{2\pi/N}$ is an involution of order N on the two spaces X and $\tilde{X}$, we have N projectors $\tilde{P}_j$ (see Section 4 of Chapter II):

$$\tilde{P}_j = \frac{1}{N}\sum_{k=0}^{N-1} \varepsilon^{-kj} \tilde{S}^k_{2\pi k/N} = \frac{1}{N}\sum_{k=0}^{N-1} \varepsilon^{-kj} \tilde{S}_{2\pi k/N}, \quad \text{where } \varepsilon = e^{2\pi i/N} \ (j = 1, 2, \ldots, N).$$

Then, if $p = (x, y)$ belongs either to Ω or to $\partial\Omega$, we have

(1.6)
$$(\tilde{P}_j u)(x, y) = \frac{1}{N}\sum_{k=0}^{N-1} \varepsilon^{-kj} (\tilde{S}_{2\pi k/N} u)(p) = \frac{1}{N}\sum_{k=0}^{N-1} \varepsilon^{-kj} u(S_{2\pi k/N} p)$$

$$= \frac{1}{N}\sum_{k=0}^{N-1} \varepsilon^{-kj} u\left(x\cos\frac{2\pi k}{N} - y\sin\frac{2\pi k}{N},\ x\sin\frac{2\pi k}{N} + y\cos\frac{2\pi k}{N}\right)$$

$$(j = 1, 2, \ldots, N).$$

The spaces X and $\tilde{X}$ are both direct sums

$$X = X_1 \oplus \ldots \oplus X_N, \quad \tilde{X} = \tilde{X}_1 \oplus \ldots \oplus \tilde{X}_N, \quad \text{where } X_j = P_j X,\ \tilde{X}_j = P_j \tilde{X}$$

such that

$$\tilde{S}_{2\pi/N} u = \varepsilon^j u \quad \text{for } u \in X_j \text{ (and for } u \in \tilde{X}_j \text{ respectively) } (j = 1, 2, \ldots, N)$$

(Section 4 of Chapter II). Proposition 1.1 therefore implies that equation (1.3) with the boundary condition (1.4), where the function $b(q)$ is continuous on $\partial\Omega$, is equivalent to the system of N independent Laplace equations

(1.7)
$$\Delta u_{(j)} = 0 \quad \text{on } \Omega \ (j = 1, 2, \ldots, N)$$

with the boundary condition for the jth equation of Dirichlet type:

$$\lim_{p \to p_0} u_{(j)}(p) = \tilde{b}_j(p_0), \quad p_0 \in \partial\Omega \ (j = 1, 2, \dots, N), \tag{1.8}$$

where

$$\tilde{b}_j = \frac{1}{b_j} P_j b, \quad \text{provided that } b_j = \sum_{k=0}^{N-1} a_k \varepsilon^{kj} \neq 0 \ (j = 1, 2, \dots, N). \tag{1.9}$$

An n-dimensional surface $S \subset \boldsymbol{R}^{n+1}$ satisfies the *Liapounov condition* [1], if

(i) S has a tangent plane at each point $p \in S$ and the angle $\nu_{t,s}$ between two normals to S at arbitrary points $t, s \in S$ satisfies the inequality:

$$|\nu_{t,s}| < C|t-s|^{\chi}, \quad 0 < \chi \leqslant 1, \ C > 0.$$

(ii) There exists a number $\delta > 0$ so small that a sphere of radius δ and centre $t \in S$ cuts out from S a part $S_\delta \subset S$ such that an arbitrary line parallel to the normal at t intersects S_δ at most one point (Fig. 16).

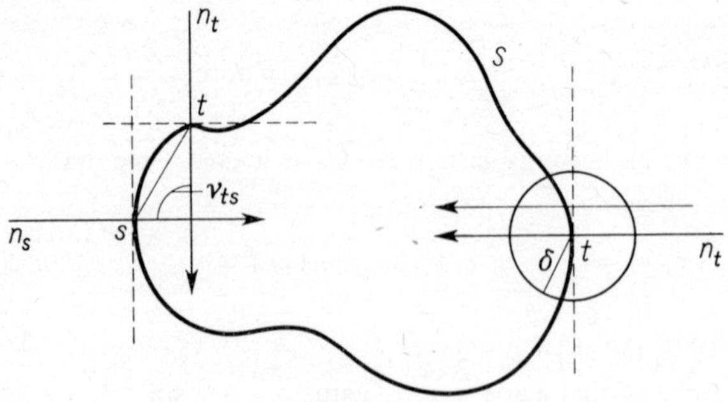

Fig. 16. One-dimensional Liapounov surface

We obtain the following

THEOREM 1.1. *Let $\Omega \subset \boldsymbol{R}^2$ be a domain invariant with respect to the rotation $S^{2\pi/N}$ with boundary $\partial\Omega$ satisfying the Liapounov condition. If $b_j = \sum_{k=0}^{N-1} a_k \varepsilon^{kj} \neq 0$ for $j = 1, 2, \dots, N$, where $\varepsilon = e^{2\pi i/N}$ and if the function $b(p)$ is continuous on $\partial\Omega$, then there exists a solution of equation* (1.3) *which is twice continuously*

[1] See, for instance, Pogorzelski [1].

differentiable in Ω *and continuous in* $\bar{\Omega}$ *which satisfies the boundary condition* (1.4). *This solution is of the form*

$$u(t) = \sum_{j=1}^{N} u_j,$$

where u *is a solution of the jth equation* (1.7) *with the jth condition* (1.8).

The proof follows immediately from the fact that the jth equation (1.7) with the jth boundary condition (1.8) is the classical Dirichlet boundary problem, which has a unique solution having the required properties.

Without any change in the proof we can study the boundary condition (1.4) with variable coefficients a_k such that

$$a_k(S_{2\pi/N}p) = a_k(p) \quad \text{for } p \in \Omega, \tag{1.10}$$

because the operator of multiplication by such a function commutes with the involution $\tilde{S}_{2\pi/N}$ (cf. Section 3 of Chapter VII: equations with rotation). It is enough to assume that $b_j(p) = \sum_{k=0}^{N-1} a_k(p)\varepsilon^{kj} \neq 0$ on $\partial\Omega$.

Now consider the boundary problem of Neumann type for equation (1.3). This means that we are looking for complex-valued functions continuous in $\bar{\Omega}$, which satisfy equation (1.3) in Ω and the condition

$$\lim_{p=p_0} \sum_{k=0}^{N-1} \frac{\partial u(q)}{\partial n}\bigg|_{q=S^k_{2\pi/N}p} = b(p_0), \qquad p_0 \in \partial\Omega,\ p \in \Omega, \tag{1.11}$$

where by $\partial/\partial n$ we mean the derivative in the direction of the interior normal to the boundary $\partial\Omega$. We assume that $b(p)$ is a continuous function on $\partial\Omega$. Just as for the Dirichlet problem, we conclude that equation (1.3) with the boundary condition (1.11) is equivalent to N independent equations (1.7) with the boundary condition for the jth equation:

$$\lim_{p\to p_0} \sum_{k=0}^{N-1} \frac{\partial u_{(j)}(q)}{\partial n}\bigg|_{q=S^k_{2\pi/N}p} = \tilde{b}_j(p_0), \qquad p_0 \in \partial\Omega,\ p \in \Omega\ (j = 1, 2, \ldots, N), \tag{1.12}$$

where

$$\tilde{b}_j = \frac{1}{b_j} P_j b \quad \text{and } b_j(p_0) = \sum_{k=0}^{N-1} a_k(p_0)\varepsilon^{kj} \neq 0 \qquad \text{for } p_0 \in \partial\Omega\ (j = 1, 2, \ldots, N).$$

Note that

$$\int_{\partial\Omega} \tilde{b}_j(p)\, dS = 0 \qquad \text{for } j = 1, 2, \ldots, N-1. \tag{1.13}$$

Indeed, if we denote by Ω_k and $\partial\Omega_k$ those parts of the domain Ω and of its boundary respectively which are contained in the angle $2\pi(k-1)/N < \varphi \leqslant 2\pi k/N$, it is plain that

$$\int_{\partial\Omega} b_j(p)\,ds = \sum_{k=1}^{N} \int_{\partial\Omega_k} \tilde{b}_j(p)\,dS.$$

But $\Omega_k = S^k_{2\pi/N}\Omega_1$ because of the property $S_{2\pi/N}\Omega = \Omega$. Hence $\Omega_1 = S^{-k}_{2\pi/N}\Omega_k$ and

$$\int_{\partial\Omega} \tilde{b}_j(p)\,dS = \sum_{k=1}^{N} \int_{\partial\Omega_k} \tilde{b}_j(p)\,dS = \int_{\partial\Omega_1} \frac{1}{b_j} \sum_{k=1}^{N} b_{(j)}(S^k_{2\pi/N}p)\,dS$$

$$= \int_{\partial\Omega_1} \frac{1}{b_j} \sum_{k=1}^{N} (\tilde{S}^k_{2\pi/N}\tilde{P}_j b)\,(p)\,dS = \int_{\partial\Omega_1} \frac{1}{b_j} \Big[\sum_{k=1}^{N} \varepsilon^{kj}\Big] (\tilde{P}_j b)\,(p)\,dS$$

$$= \int_{\partial\Omega_1} \frac{1}{b_j} \delta_{N,j} b_{(j)}(p)\,dS = \delta_{N,j} \int_{\partial\Omega_1} \tilde{b}_j(p)\,dS = \begin{cases} 0 & \text{for } j = 1, 2, \ldots, N-1, \\ \int_{\partial\Omega} \tilde{b}_N(p)\,dS & \text{for } j = N, \end{cases}$$

where $\delta_{N,j}$ is the Kronecker symbol, in virtue of Formula (1.6) of Chapter II.

Hence for an arbitrary $b \in \tilde{X}$ we have equalities (1.13) and writing

$$\tilde{b}(p) = \sum_{j=1}^{N} \tilde{b}_j(p),$$

we obtain

$$\int_{\partial\Omega} \tilde{b}(p)\,dS = \int_{\partial\Omega} \tilde{b}_N(p)\,ds. \tag{1.14}$$

Since all the preceding discussion also holds for functions $a_k(p)$ invariant under the rotation $S_{2\pi/N}$, we finally have

THEOREM 1.2. *Let $\Omega \subset \boldsymbol{R}^2$ be a domain invariant with respect to the rotation $S_{2\pi/N}$ with the boundary satisfying the Liapounov condition. Let the functions $a_0(p), \ldots, a_{N-1}(p)$ continuous on $\overline{\Omega}$ be invariant under the rotation $S_{2\pi/N}$, i.e. $a_k(S_{2\pi/N}p) = a_k(p)$ $(k = 0, 1, \ldots, N-1)$, and let*

(i) $b_j(p) = \sum_{k=0}^{N-1} a_k(p)\varepsilon^{kj} \neq 0$ *on $\partial\Omega$, where $\varepsilon = e^{2\pi i/N}$,*

(ii) $\int_{\partial\Omega} b(p)\,dS = 0$, *where $b(p)$ is a continuous function on $\partial\Omega$. Then there exists a solution of equation* (1.3) *satisfying condition* (1.1), *which is of the form*

$$u(p) = \sum_{i=1}^{N} u_i(p),$$

where $u_j(p)$ is a solution of the jth equation (1.7) *satisfying the jth boundary condition* (1.12).

PROOF. The jth equation (1.7) has a solution satysfying the jth condition (1.12) only if the condition

$$\tilde{b}_j(p)\,dS = 0 \qquad (j = 1, 2, \ldots, N),$$

is satisffed.

Our assumptions and Formulae (1.13), (1.14) together imply that this condition is satisfied for $j = 1, 2, \ldots, N$.

In a similar way we can study the equation $u+au = v$ on Ω with the boundary conditions of either (1.4) or (1.11), where $a(p)$ and $v(p)$ are continuous functions for $p \in \tilde{\Omega}$ and $a(S_{2\pi/N}p) = a(p)$.

2. BOUNDARY PROBLEMS FOR THE HEAT EQUATION

We retain the notation of the preceding section and we consider the following boundary problem with rotation of Fourier type:

To find a complex-valued function $u(x, y, t)$ which satisfies 1° the heat equation inside the domain $\Omega \subset \boldsymbol{R}^2$ and for $t > 0$:

$$\Delta u - \frac{\partial u}{\partial t} = 0 \quad \text{for } p = (x, y) \in \Omega \text{ and } t > 0, \tag{2.1}$$

2° an initial condition:

$$\lim_{t \to +0} u(p, t) = v(p) \quad \text{for all } p \in \Omega, \tag{2.2}$$

3° a boundary condition with rotation:

$$\lim_{p \to p_0} \sum_{k=0}^{N-1} a_k(p, t)\, u(S^k_{2\pi/N}p, t) = b(p_0, t) \quad \text{for } t > 0,\ p_0 \in \partial\Omega,\ p \in \Omega. \tag{2.3}$$

The operator $S_{2\pi/N}$ of rotation is defined as in Section 1. The functions $a_k(p, t)$ are invariant with respect to the rotation $S_{2\pi/N}$, i.e.

$$a_k(S_{2\pi/N}p, t) = a_k(p, t) \quad \text{for } p \in \overline{\Omega} \text{ and } t \geqslant 0 \ (k = 1, 2, \ldots, N-1). \tag{2.4}$$

We introduce, as before, the operator $\tilde{S}_{2\pi/N}$ defined by means of the equality

$$(\tilde{S}_{2\pi/N}u)\,(p, t) = u(S_{2\pi/N}p, t) \quad \text{for all } p \in \overline{\Omega} \text{ and } t > 0. \tag{2.5}$$

Since this operator acts only on the variable p, we can treat the time t in our problem as a parameter and solve it in exactly the same manner as in Section 1.

Namely, equation (2.1) with conditions (2.2) and (2.3) is equivalent to a system of N independent heat equations

$$\Delta u_{(j)} - \frac{\partial u_{(j)}}{\partial t} = 0 \quad \text{for } p \in \Omega \text{ and } t > 0 \ (j = 1, \ldots, N), \tag{2.6}$$

with the following initial and boundary conditions for the jth equation (2.6):

$$\lim_{t \to +0} u_{(j)}(p, t) = v_{(j)}(p) \quad \text{for all } p \in \Omega \ (j = 1, 2, \ldots, N), \tag{2.7}$$

$$\lim_{p \to p_0} u_{(j)}(p, t) = \tilde{b}_j(p_0, t) \quad \text{for } t > 0,\ p_0 \in \partial\Omega \ (j = 1, 2, \ldots, N), \tag{2.8}$$

where

$$\tilde{b}_j(p_0, t) = \frac{1}{b_j(p_0, t)} b_{(j)}(p_0, t) \quad \text{and} \quad b_{(j)}(p_0, t) = \sum_{k=0}^{N-1} \varepsilon^{kj} a_k(p_0, t) \neq 0 \tag{2.9}$$

$$\text{for } p_0 \in \partial\Omega,\ t > 0 \ (j = 1, 2, \ldots, N)$$

and for an arbitrary function $u(p)$,

$$u_{(j)}(p, t) = \frac{1}{N} \sum_{k=0}^{N-1} \varepsilon^{-kj} u(S^k_{2\pi/N} p, t) \quad (j = 1, 2, \ldots, N).$$

We obtain the following

THEOREM 2.1. *Let $\Omega \subset \boldsymbol{R}^2$ be a domain invariant with respect to the rotation $S_{2\pi/N}$ and with the boundary $\partial\Omega$ satisfying the Liapounov condition. Let the functions $v(p), b(p, t), a_k(p, t)$ $(k = 0, 1, \ldots, N-1)$ be defined and continuous in the regions $p \in \Omega$; $p \in \partial\Omega$, $t > 0$; $p \in \Omega$, $t > 0$ respectively. Let $b_j(p, t) = \sum_{k=0}^{N-1} a_k(p, t)\varepsilon^{kj} \neq 0$ for $p \in \partial\Omega$, $t > 0$ and let condition* (2.4) *be satisfied.*

Then equation (2.1) *with the initial condition* (2.2) *and with the boundary condition* (2.3) *has a unique solution which is of the form*

$$u(p, t) = \sum_{j=1}^{N} u_j(p, t),$$

where u_j is the unique solution of the jth equation (2.6) *satisfying the jth initial condition* (2.7) *and the jth boundary condition* (2.8).

For the proof it suffices to consider the space X of all functions twice continuously differentiable with respect to the variables x, y and continuously differentiable with respect to t for $(x, y) \in \Omega$ and $t > 0$ and the space $\tilde{X}$ of functions continuous for $(x, y) \in \partial\Omega$ and $t > 0$.

When the distribution of temperature $b(p_0, t)$ on the boundary $\partial\Omega$ changes periodically with period ω in the time, i.e. when

$$b(p_0, t+\omega) = b(p_0, t) \quad \text{for all } p_0 \in \partial\Omega \text{ and } t > 0, \tag{2.10}$$

we can solve the corresponding boundary problem by using multi-involutions. In this case no initial condition will be admitted, because the functions considered will be periodic with respect to the variable t with periods commensurable with ω. Condition (2.3) can be admitted in a more general form:

$$\lim_{p\to p_0} \sum_{k=0}^{N-1} \sum_{j=0}^{M-1} a_{kj}(p,t)\, u\left(S^k_{2\pi/N} p, t-j\frac{\omega}{M}\right) = b(p_0, t),$$

where M is an integer.

3. BOUNDARY PROBLEMS FOR THE WAVE EQUATION

Consider the wave equation with a forcing function $f(p, t)$

$$\Delta u - \frac{1}{c^2}\frac{\partial^2 u}{\partial t^2} = f(p, t), \quad p = (x, y) \in \Omega,\ t > 0, \tag{3.1}$$

where the domain $\Omega \subset \boldsymbol{R}^2$ is invariant with respect to the rotation $S_{2\pi/N}$ (see Section 1).

We are looking for a complex valued function $u(p, t)$ twice continuously differentiable for $p \in \bar{\Omega}$ and $t > 0$, continuous in $\bar{\Omega}$ and for $t > 0$, and satisfying

1° equation (3.1) for $p \in \Omega$, $t > 0$,

2° the initial condition

$$\lim_{t\to+0} u(p, t) = v(p) \quad \text{for } p \in \Omega, \tag{3.2}$$

3° the boundary condition with rotation

$$\lim_{p\to p_0} \sum_{k=0}^{N-1} a_k(p, t)\, u(S^k_{2\pi/N} p, t) = b_0(p_0, t) \quad \text{for } t > 0,\ p_0 \in \partial\Omega. \tag{3.3}$$

We assume that the functions $a_k(p, t)$ are invariant under the rotation $S_{2\pi/N}$, i.e. that

$$a_k(S_{2\pi/N} p, t) = a_k(p, t) \quad \text{for } p \in \bar{\Omega} \text{ and } t > 0. \tag{3.4}$$

Now consider the operator $\tilde{S}_{2\pi/N}$ defined by Formula (2.5). Just as in Sections 1 and 2 we conclude that equation (3.1) is equivalent to N independent wave equations

$$\Delta u_{(j)} - \frac{1}{c^2}\frac{\partial^2 u_{(j)}}{\partial t^2} = f_{(j)}(p, t), \quad p \in \Omega, t > 0\ (j = 1, 2, \ldots, N) \tag{3.5}$$

with the initial condition for the jth equation

$$\lim_{t\to+0} u_{(j)}(p,t) = v_{(j)}(p,t) \quad \text{for } p\in\Omega \ (j=1,2,\ldots,N) \tag{3.6}$$

and with the boundary condition for this equation

$$\lim_{p\to p_0} u_{(j)}(p) = \tilde{b}_j(p_0,t) \quad \text{for } t>0 \text{ and } p_0\in\partial\Omega \ (j=1,2,\ldots,N), \tag{3.7}$$

where we write

$$u_{(j)}(p,t) = \frac{1}{N}\sum_{k=0}^{N-1} \varepsilon^{-kj} u(S^k_{2\pi/N}p,t), \qquad \varepsilon = e^{2\pi i/N} \ (j=1,2,\ldots,N)$$

and

$$\tilde{b}_j(p,t) = \frac{1}{b_j(p,t)} b_{(j)}(p,t), \quad b_j(p,t) = \sum_{k=0}^{N-1} \varepsilon^{kj} a_k(p,t) \neq 0$$

$$\text{for } p\in\partial\Omega,\ t>0 \ (j=1,2,\ldots,N).$$

This yields the following

THEOREM 3.1. *Let $\Omega\subset \boldsymbol{R}^2$ be a domain invariant under the rotation $S_{2\pi/N}$ with the boundary satisfying the Liapounov condition. Let the functions $v(p), b(p,t)$, $a_k(p,t)$ $(k=0,1,\ldots,N-1)$, $f(p,t)$ be defined in the regions $p\in\Omega$; $p\in\partial\Omega$, $t>0$; $p\in\bar{\Omega}$, $t>0$, respectively, and let $b_j(p,t)\neq 0$ for $p_0\in\partial\Omega$, $t>0$. Condition* (3.4) *is assumed to be satisfied. Then equation* (3.1) *with conditions* (3.2), (3.3) *has a unique solution which is of the form*

$$u(p,t) = \sum_{j=1}^{N} u_j(p,t),$$

where $u_j(p,t)$ is the unique solution of the jth equation (3.5) *satisfying the jth initial condition* (3.6) *and the jth boundary condition* (3.7).

In a similar way, using multi-involutions, we can solve a boundary problem with rotation, when the forcing function $f(p,t)$ and the function $b(p,t)$ change periodically in time. In this case we do not assume any initial condition. Moreover, we can admit a boundary condition with shifts of the time t, provided these are commensurable with the periods of $f(p,t)$ and $b(p,t)$.

CHAPTER XI

Integral Equations with Transformed Argument

In this chapter we deal with integral equations with regular kernels and with transformed argument.

1. INTEGRAL AND INTEGRO-DIFFERENTIAL EQUATIONS WITH REFLECTION

Let $\Omega \subset \boldsymbol{R}^n$ be a *domain symmetric* with respect to the origin of coordinates:

(1.1) $$-\Omega = \Omega, \quad \text{where } -\Omega = \{u\colon\ u = -t,\, t \in \Omega\}$$

(Fig. 17).

Consider the integral operator

(1.2) $$(Kx)(t) = \int_{\Omega} K(t, s)x(s)\, ds,$$

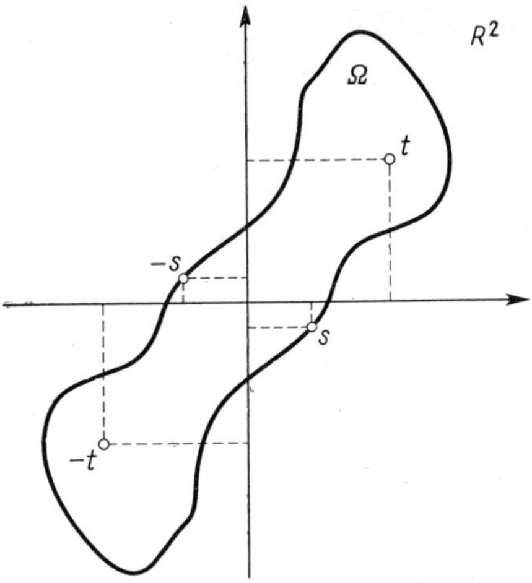

Fig. 17. Two-dimensional symmetric domain

where $K(t, s)$ is a continuous function for $(t, s) \in \Omega \times \Omega$. Let $(Sx)(t) = x(-t)$. The operator S of reflection is an involution on the space $C(\Omega)$ [1]. We have

$$(KSx)(t) = \int_{\Omega} K(t, s) x(-s)\, ds = (-1)^n \int_{\Omega} K(t, -u) x(u)\, du$$

$$= (-1)^n \int_{\Omega} K(t, -u) x(u)\, du,$$

$$(SKx)(t) = \int_{\Omega} K(-t, s) x(s)\, ds.$$

We will consider two cases:

$$K(-t, u) = K(t, u), \quad \text{i.e. } K \text{ is an even function,}$$

and

$$K(-t, u) = -K(t, u), \quad \text{i.e. } K \text{ is an odd function,}$$

since in these two cases we are able to apply our method.

We conclude that

$$(1.3) \quad \begin{aligned} &SK - KS = 0 \quad \text{if either } n \text{ and } K \text{ are even or } n \text{ and } K \text{ are odd,} \\ &SK + KS = 0 \quad \text{if either } n \text{ is odd and } K \text{ is even or } n \text{ is even and } K \text{ is odd.} \end{aligned}$$

Consider the integral equation

$$(1.4) \quad a_0 x(t) + b_0 x(-t) + a_1 \int_{\Omega} K(t, s) x(s)\, ds + b_1 \int_{\Omega} K(-t, s) x(s)\, ds = y(t),$$

where $y \in C(\Omega)$ and a_0, a_1, b_0, b_1 are constants.

In the sequel we will write this equation in the form

$$(1.5) \qquad Ax \equiv [a_0 I + b_0 S + (a_1 I + b_1 S) K] x = y.$$

As the first step we solve this equation in the case where $SK - KS = 0$, i.e. where both n and K are either even or odd. We write, as before,

$$P^+ = \tfrac{1}{2}(I + S), \quad P^- = \tfrac{1}{2}(I - S), \quad x^+ = P^+ x, \quad x^- = P^- x$$

for all $x \in C(\Omega)$,

$$X^+ = C^+(\Omega) = P^+ C(\Omega), \quad X^- = P^-(\Omega) = C(\Omega).$$

Since the operator K commutes with S, we have

$$KX^+ \subset X^+, \quad KX^- \subset X^-,$$

[1] For integral equations with reflection in $L(\Omega)$ see Komjak [1].

Hence by Theorem 5.1 of Chapter II, equation (1.5) is equivalent to two independent integral equations (without reflection):

$$(1.6)\qquad \begin{aligned} (a_0+b_0)x^+ + (a_1+b_1)Kx^+ &= y^+, \\ (a_0-b_0)x^+ + (a_1-b_1)Kx^- &= y^-. \end{aligned}$$

The first equation is considered in the space $C^+(\Omega)$ of all even functions belonging to $C(\Omega)$, the second one—in the space $C^-(\Omega)$ of all odd functions belonging to $C(\Omega)$. If $a_0+b_0 = 0$, $a_0-b_0 = 0$, then both equations are Fredholm equations of the first kind. If $a_0^2-b_0^2 \neq 0$, $a_1^2-b_1^2 \neq 0$, then both equations are Fredholm equations of the second kind. If $a_1^2-b_1^2 = 0$, then equations (1.6) (hence also (1.4)) are not integral equations and their solutions are given by Theorem II.5.2. In other cases one of these equations can be of the first kind and the other one of the second kind.

We thus have

THEOREM 1.1. *Let $\Omega \subset \boldsymbol{R}^n$ be symmetric with respect to the origin of coordinates. Let $K(t, s)$ be continuous for $(t, s) \in \Omega\times\Omega$. If both n and K are either even or odd, then equation* (1.4) *is equivalent to the two independent equations*

$$(a_0+b_0)x^+(t)+(a_1+b_1)\int_\Omega K(t, s)x^+(s)\,ds = -\tfrac{1}{2}[y(t)+y(-t)],$$

$$(a_0-b_0)x^-(t)+(a_1-b_1)\int_\Omega K(t, s)x^-(s)\,ds = -\tfrac{1}{2}[y(t)-y(-t)].$$

If these equations have solutions $x^+(t)$ and $x^-(t)$ respectively, then equation (1.4) *has a solution of the form $x(t) = x^+(t)+x^-(t)$.*

Now consider the case where $SK+KS = 0$, i.e. the case, when either n is even and K is odd or n is odd and K is even. We apply the theorems of Section 1 in Chapter III on operators anticommuting with an involution.

If $a_0^2-b_0^2 \neq 0 \neq a_1^2-b_1^2$, then writing $\lambda = \dfrac{a_0^2-b_0^2}{a_1^2-b_1^2}$, we conclude that any solution of equation (1.5) is of the form

$$x = R_A\tilde{x} + [(a_0-a_1\sqrt{\lambda})I-(b_0+b_1\sqrt{\lambda})S]z_1,$$

where

$$R_A = -(a_1^2-b_1^2)^{-1}[(a_0-b_0S)-(a_1I+b_1S)K],$$

$\tilde{x}$ is a solution of the equation $(K^2-\lambda I)\tilde{x} = y$ and z_1 is an arbitrary solution of the equation $(K-\sqrt{\lambda}I)z_1 = 0$. However,

$$(K^2x)(t) = \int_\Omega K(t, s)\Big[\int_\Omega K(s, u)x(u)\,du\Big]ds = \int_\Omega\Big[\int_\Omega K(t, s)K(s, u)\,ds\Big]x(u)\,du.$$

Hence we obtain the following

THEOREM 1.2. *Let $\Omega \subset R^n$ be symmetric with respect to the origin of coordinates and let $K(t, s)$ be continuous for $(t, s) \in \Omega \times \Omega$. We assume that either n is even and K is odd or n is odd and K is even. If $a_0^2 - b_0^2 \neq 0 \neq a_1^2 - b_1^2$, then equation* (1.4) *has a solution*

$$x(t) = -(a_1^2 - b_1^2)^{-1}\Big[a_0\tilde{x}(t) - b_0\tilde{x}(-t) - a_1\int_\Omega K(t,s)\tilde{x}(s)\,ds - b_1\int_\Omega K(-t,s)\tilde{x}(s)\,ds\Big] +$$
$$+(a_0 - a_1\sqrt{\lambda})z_1(t) - (b_0 + b_1\sqrt{\lambda})z_1(-t),$$

where $\lambda = (a_0^2 - b_0^2)(a_1^2 - b_1^2)^{-1}$, $\tilde{x}(t)$ is a solution (if it exists) of the equation

$$\tilde{x}(t) - \frac{1}{\lambda}\int_\Omega K_1(t,s)\tilde{x}(s)\,ds = -y(t), \quad \text{where } K_1(t,s) = \int_\Omega K(t,u)K(u,s)\,ds$$

and $z_1(t)$ is an arbitrary solution of the equation

$$z_1(t) - \frac{1}{\sqrt{\lambda}}\int_\Omega K(t,s)z_1(s)\,ds = 0.$$

We now restrict ourselves to the case where equation (1.4) is essentially an integral equation, i.e. to the case in which at least one of the numbers $a_0 - b_0$, $a_0 + b_0$ is not equal to zero. The case where $a_1 - b_1 = a_1 + b_1 = 0$ is completely solved by Theorem 1.2 of Chapter VII.

THEOREM 1.3. *Let Ω and $K(t, s)$ satisfy the assumption of Theorem* 1.2. *If $a_0 - b_0 = a_0 + b_0 = 0$ and $a_1 - b_1 \neq 0$, then every solution of equation* (1.4) *satisfies the equation*

$$\int_\Omega K(t,s)x(s)\,ds = y_1(t), \quad \text{where } y_1(t) = (a_1^2 - b_1^2)^{-1}[a_1 y(t) - b_1 y(-t)],$$

and conversely.

THEOREM 1.4. *Let Ω and $K(t, s)$ satisfy the assumptions of Theorem* 1.2. *Let $a_1 + b_1 = 0$ but $a_1 - b_1 \neq 0$. Then*

(a) *if $a_0^2 - b_0^2 \neq 0$, then the solution of equation* (1.4) *is*

$$x(t) = (a_0^2 - b_0^2)^{-1}\Big\{a_0 y(t) - b_0 y(-t) - \tfrac{1}{2}(a_1 - b_1)\int_\Omega [K(t,s) - K(-t,s)]y(s)\,ds\Big\};$$

(b) *if $a_0 + b_0 = 0$ but $a_0 - b_0 \neq 0$, then a solution of equation* (1.4) *exists if and only if $y(t)$ is an odd function. If this condition is satisfied, then*

$$x(t) = x_1(t) - (a_1 - b_1)\int_\Omega K(t,s)x_1(s)\,ds + (a_0 - b_0)^{-1}y(t),$$

where $x_1(t)$ is an arbitrary even function from $C(\Omega)$;

(c) *if* $a_0 - b_0 = 0$ *but* $a_0 + b_0 \neq 0$, *then a solution of equation* (1.4) *exists if and only if*

$$y(t) - y(-t) - (a_0 + b_0)^{-1}(a_1 - b_1) \int_\Omega K(t, s)\,[y(s) + y(-s)]\,ds = 0.$$

If this condition is satisfied, then

$$x(t) = (a_0 + b_0)^{-1}[y(t) + y(-t)] + x_1(t),$$

where $x_1(t)$ *is an arbitrary odd function from* $C(\Omega)$;

(d) *if* $a_0 + b_0 = a_0 - b_0 = 0$, *then a solution exists if and only if* y *is an odd function and if the equation*

$$\int_\Omega K(t, s)\,x^+(s)\,ds = \frac{1}{a_1 - b_1}\,y(t),$$

has an even solution. If these conditions are satisfied, then $x(t) = x^+(t) + x_1(t)$, *where* $x_1(t)$ *is an arbitrary odd function from* $C(\Omega)$.

If $a_1 - b_1 = 0$ but $a_1 + b_1 \neq 0$, we obtain similar results as in Theorem 1.4 if we interchange the roles of $a_1 - b_1$ and $a_1 + b_1$ and of even and odd functions.

Theorems 1.2, 1.3, 1.4 remain true if the coefficients of equation (1.4) are even functions, because even functions are invariant with respect to reflection. If y is a vector valued function and $K(t, s)$ is a square matrix-function satisfying the previous assumptions, then all the theorems of this section remain valid.

2. PERIODIC SOLUTIONS OF LINEAR INTEGRO-DIFFERENTIAL-DIFFERENCE EQUATIONS OF FREDHOLM TYPE

Consider the integral-difference equation of Fredholm type:

$$\sum_{j=0}^{m} a_j(t)\,x(t-\omega_j) + \sum_{j=0}^{m} \int_0^{\omega} K_j(t-s, s)\,x(s-\omega_j)\,ds = y(t), \tag{2.1}$$

where y is an ω_{m+1}-periodic function continuous for $t \in [0, \omega_{m+1}]$. We will assume that $\omega_0 = 0$ and that $\omega_1, \ldots, \omega_{m+1}$ are commensurable. This means that there are integers $n_1, \ldots, n_{m+1}$ and a real $r \neq 0$ such that $\omega_j = n_j r$, and we also put $n_0 = 0$. We assume further that $N > m$ is a common multiple of the numbers $n_1, \ldots, n_{m+1}$ (not necessarily the least one) and that $\omega = Nr$. The functions $a_j(t)$ are supposed to be r-periodic[1] and continuous in the closed interval $[0, r]$. The kernels $K_j(t, s)$ are ω-periodic with respect to the variable t and r-periodic with respect to the variable s and continuous for $0 \leqslant t \leqslant \omega$, $0 \leqslant s \leqslant r$.

---

[1] Equations with ω-periodic coefficients will be considered in the next section.

We are looking for ω-periodic solutions of equation (2.1). Note that this equation could be written in a simpler form. Indeed, since

$$\int_a^{\sigma+a} x(s)\,ds = \int_0^{\sigma} x(s)\,ds \quad \text{for all } \sigma\text{-periodic functions } x(t) \text{ and all } a,$$

we find, using our assumptions, that

$$\int_0^{\omega} K_j(t-s,s)x(s-\omega_j)\,ds = \int_{\omega_j}^{\omega+\omega_j} K(t-u-\omega_j, u+\omega_j)x(u)\,du$$

$$= \int_0^{\omega} K_j(t-u-\omega_j, u+n_j r)x(u)\,du = \int_0^{\omega} K_j(t-u-\omega_j, u)x(u)\,du.$$

Hence if we put

$$K(t, s) = \sum_{j=0}^{m} K_j(t-\omega_j, s),$$

we can rewrite equation (2.1) in the form

$$\sum_{j=0}^{m} a_j(t)x(t-\omega_j) + \int_0^{\omega} K(t-s, s)x(s)\,ds = y(t), \tag{2.2}$$

where the function $K(t, s)$ is ω-periodic with respect to t and r-periodic with respect to s and continuous for $0 \leqslant t \leqslant \omega$, $0 \leqslant s \leqslant r$.

Let C_ω be the space of all ω-periodic functions continuous for $0 \leqslant t \leqslant \omega$ with the norm $\|x\| = \sup_{0 \leqslant t \leqslant \omega} |x(t)|$. This is a Banach space, but it is enough for our purposes merely that C_ω is a linear space. In the space C_ω we consider the shift operator $(Sx)(t) = x(t-r)$, which is an involution of order N on the space C_ω. Indeed, $(S^N x)(t) = x(t-Nr) = x(t-\omega) = x(t)$ for all $x \in C_\omega$.

We show that the operator K defined by means of the integral

$$(Kx)(t) = \int_0^{\omega} K(t-s, s)x(s)\,ds \qquad (x \in C_\omega)$$

commutes with S. Indeed, for all $x \in C_\omega$

$$(SKx)(t) = \int_0^{\omega} K(t-r-s, s)x(s)\,ds = \int_r^{\omega+r} K(t-u, u-r)x(u-r)\,du$$

$$= \int_0^{\omega} K(t-u, u)x(u-r)\,du = (KSx)(t).$$

Hence

$$SK - KS = 0. \tag{2.3}$$

Since S is an involution of order N, the space C_ω is the direct sum of spaces $X_j = P_j C_\omega$, where

$$P_j = \frac{1}{N}\sum_{k=0}^{N-1} \varepsilon^{-kj} S^k \quad \text{for } j = 1, 2, \dots, N, \ \varepsilon = e^{2\pi i/N}. \tag{2.4}$$

Now write: $x_{(j)} = P_j x$ $(j = 1, 2, \dots, N)$ and

$$A_j(t) = \sum_{k=0}^{m} \varepsilon^{jn_k} a_k(t) \quad (j = 1, 2, \dots, N). \tag{2.5}$$

The operators A_j of multiplication by functions $A_j(t)$ commute with S, since the operators a_j of multiplication by r-periodic functions $a_j(t)$ commute with S (cf. Section 2 of Chapter V). We can therefore apply Theorem 5.1 of Chapter II to obtain the following

THEOREM 2.1. *Let the functions $K_j(t, s)$, $K(t, s)$, $a_j(t)$ and the numbers $r, \omega_0, \omega_1, \dots, \omega_{m+1}, \omega$ be as those described above. Then equation* (2.1) (*resp.* (2.2)) *is equivalent in the space C_ω to a system of N independent equations (without shifts)*

$$A_j(t) x_{(j)}(t) + \int_0^\omega K(t-s, s) x_{(j)}(s)\, ds = y_{(j)}(t) \quad (j = 1, 2, \dots, N), \tag{2.6}$$

where the A_j are defined by Formulae (2.5), $y_{(j)}(t) = \dfrac{1}{N}\sum_{k=0}^{N-1} \varepsilon^{-kj} y(t-kr)$ *for $j = 1, 2, \dots, N$ and in the case of equation* (2.1) $K(t, s) = \sum_{j=0}^{m} K_j(t-\omega_j, s)$. *If all these equations have ω-periodic solutions $x_j(t)$, then the ω-periodic solution of equation* (2.1) (*resp.* (2.2)), *which is obviously continuous on the closed interval* $[0, \omega]$, *is of the form*

$$x(t) = \sum_{j=1}^{N} x_j(t).$$

Observe that in the case where $A_j(t) \neq 0$ for $j = 1, 2, \dots, N$ and $0 \leqslant t \leqslant r$, we have instead of system (2.6) a system of independent Fredholm equations of the second kind

$$x_{(j)} + \int_0^\omega \tilde{K}_j(t-s, s) x_{(j)}(s)\, ds = y_j(t) \quad (j = 1, 2, \dots, N), \tag{2.7}$$

where

$$y_j(t) = [A_j(t)]^{-1} y_{(j)}(t) \quad \text{and} \quad \tilde{K}_j(t, s) = [A_j(t+s)]^{-1} K(t, s) \quad (j = 1, 2, \dots, N).$$

The functions $K_j(t, s)$ are r-periodic with respect to s and ω-periodic with respect to t. Indeed,

$$\tilde{K}_j(t, s+r) = [A_j(t+s+r)]^{-1}K(t, s+r) = \Big[\sum_{k=0}^{m} \varepsilon^{jn_k} a_k(t+s+r)\Big]^{-1}K(t, s+r)$$

$$= \Big[\sum_{k=0}^{m} \varepsilon^{jn_k} a_k(t+s)\Big]^{-1}K(t, s) = \tilde{K}_j(t, s), \quad \tilde{K}_j(t+\omega, s)$$

$$= \Big[\sum_{k=0}^{m} \varepsilon^{jn_k} a_k(t+s+rN)\Big]^{-1}K(t+\omega, s) = \Big[\sum_{k=0}^{m} \varepsilon^{jn_k} a_k(t+s)\Big]^{-1}K(t, s) = \tilde{K}_j(t, s).$$

Theorem 2.1 remains valid if x, y are vector valued functions and $a_j(t)$, $K_j(t, s)$ are square matrix-functions with elements satisfying the previous conditions.

Now consider an integro-differential-difference equation[1]

$$\sum_{j=0}^{m} a_j(t)x(t-\omega_j) + \sum_{k=0}^{n} \int_0^{\omega} N_k(t-s, s)x^{(k)}(s)\,ds = y(t), \tag{2.8}$$

where the numbers $\omega_0, \ldots, \omega_{m+1}, \omega$ are as before and $y(t)$ is a given ω_{m+1}-periodic function with the derivative $y^{(n)} \in C[0, \omega_{m+1}]$. Moreover, all the functions $N(t, s)$ and their derivatives with respect to t up to and including the order n are continuous for $0 \leqslant t \leqslant \omega$, $0 \leqslant s \leqslant r$ and ω-periodic with respect to t and r-periodic with respect to s.

It follows from our preceding discussion that the integral operators N_k commute with S. The derivation operator also commutes with S (see Section 1 of Chapter V). Hence equation (2.8) is equivalent to N independent equations without shifts:

$$A_j(t)x_{(j)}(t) + \sum_{k=0}^{N-1} \int_0^{\omega} N_k(t-s, s)x_{(j)}^{(k)}(s)\,ds = y_{(j)} \quad (j = 1, 2, \ldots, N), \tag{2.9}$$

where the A_j are defined by Formula (2.5). Assume now that all the $A_j(t) \neq 0$ for $0 \leqslant t \leqslant r$. Then system (2.9) is equivalent to the system of N independent equations

$$x_{(j)} + \sum_{k=0}^{n} \int_0^{\omega} N_{kj}(t-s, s)x_{(j)}^{(k)}(s)\,ds = y_j(t) \quad (j = 1, 2, \ldots, N), \tag{2.10}$$

where $N_{kj}(t, s) = [A_j(t+s)]^{-1}N_k(t, s)$ $(j = 1, 2, \ldots, N;\ k = 0, 1, \ldots, n)$, and $y_j(t) = [A_j(t)]^{-1}y_{(j)}(t)$ $(j = 1, 2, \ldots, N)$.

---

[1] As has been shown, we can without loss of generality consider an equation analogous to equation (2.2).

It is well known (see for instance Bykov [1]) that the jth equation of system (2.10) has a solution if and only if the following system has a solution:

$$(2.11)\qquad x_{\nu j}(t)+\sum_{k=0}^{n}\int_{0}^{\omega} N_{k\nu}^{(j)}(t-s,\,s)\,x_{\nu k}(s)\,ds = y_{\nu}^{(j)}(t) \qquad (j=0,1,\ldots,n),$$

where $0 \leqslant \nu \leqslant N$ is fixed.

If this system has a solution $(x_{\nu 0},\ldots,x_{\nu n})$, then the solution of the νth equation (2.10) is $x_\nu = x_{\nu 0}$, provided that $x_{\nu k}\in C_\omega$ $(k=0,1,\ldots,n)$. Having found solutions to all equations (2.10) we infer that $x=\sum_{\nu=1}^{n} x_\nu$ is a solution of equation (2.8). Hence we have

THEOREM 2.2. *Let the numbers* $\omega_0,\ldots,\omega_{m+1},\omega$, *the functions* $a_j(t)$ *and* $N_k(t,s)$ *satisfy the same conditions as before. Moreover, let the function* $A_j(t)\neq 0$ *be continuously differentiable up to and including the order* n *for* $0\leqslant t\leqslant r$. *If each of systems* (2.11) *for* $\nu=1,2,\ldots,N$ *has a solution* $(x_{\nu 0},\ldots,x_{\nu n})$ *such that* $x_{\nu k}\in C_\omega$ *for* $k=0,1,\ldots,n$, *then the solution of equation* (2.8) *belonging to* C_ω *is of the form*

$$x(t)=\sum_{\nu=1}^{N} x_{\nu 0}\,.$$

Note that all the results of this section remain valid if we consider kernels of integral operators r-periodic with respect to both variables, because in this case formula (2.3) is also true. Indeed, if $K(t,s)$ is r-periodic with respect to both variables, then

$$(SKx)(t)=\int_0^{\omega} K(t-r,s)x(s)\,ds=\int_r^{\omega+r} K(t-r,u-r)x(u-r)\,du$$

$$=\int_r^{\omega+r} K(t,u)x(u-r)\,du=\int_0^{\omega} K(t,u)x(t-r)\,du=(KSx)(t).$$

Further corresponding proofs are exactly the same.

3. PERIODIC SOLUTIONS OF NON-LINEAR INTEGRO-DIFFERENTIAL-DIFFERENCE EQUATIONS OF FREDHOLM TYPE

Consider an integral-difference equation

$$(3.1)\qquad x(t)=F\big(t,x(t),x(t-\omega_1),\ldots,x(t-\omega_m)\big)+$$
$$+\int_0^{\omega} G\big(t,s,x(s),x(s-\omega_1),\ldots,x(s-\omega_m)\big)ds,$$

where, as before the numbers $\omega_1, \ldots, \omega_m$ are commensurable: $\omega_j = n_j r$, where n_j are integers and $r \neq 0$ is real. We assume that N is a common multiple of integers n_j and that $\omega = Nr$. We assume also that the functions $F(t, z_0, \ldots, z_m)$ and $G(t, s, z_0, \ldots, z_m)$ are ω-periodic with respect to t and s-periodic with respect to s and continuous for $0 \leqslant t \leqslant \omega$, $0 \leqslant s \leqslant r$, $z_j \in \boldsymbol{R}$ $(j = 0, 1, \ldots, m)$.

We look for solutions of equation (3.1) belonging to the space C_ω (defined in Section 2). Since the shift operator S defined in Section 2 is an involution of order N on C_ω, we can act on both sides of equation (3.1) with the operators P_j $(j = 1, 2, \ldots, N)$ to obtain the N equations

$$(3.2)\quad x_{(j)}(t) = \frac{1}{N}\sum_{k=0}^{N-1} \varepsilon^{-kj} F\big(t-kr, x(t-kr), x(t-\omega_1-kr), \ldots, x(t-\omega_m-kr)\big) +$$

$$+\frac{1}{N}\sum_{k=0}^{N-1} \varepsilon^{-kj} \int_0^\omega G\big(t-kr, s, x(s), x(s), x(s-\omega_1), \ldots, x(s-\omega_m)\big) ds$$

$$(j = 1, 2, \ldots, N), \qquad \text{since } x(t) = \sum_{j=1}^{N} x_{(j)}(t), \text{ where } x_{(j)}(t) = (P_j x)(t)$$

$$= \frac{1}{N}\sum_{k=0}^{N-1} \varepsilon^{-kj} x(t-kr) \quad (j = 1, 2, \ldots, N) \text{ and } \varepsilon = e^{2\pi i/N}.$$

We recall (Proposition 2.1 of Chapter V) that $u \in X_j = P_j C_\omega$ if and only if

$$(3.3)\qquad u(t) = e^{-2\pi i j t/\omega} v(t) \quad (j = 1, 2, \ldots, N),$$

where $v(t)$ is an r-periodic function, continuous by definition for $0 \leqslant t \leqslant r$. Therefore, writing

$$(3.4)\qquad \tilde{x}(t) = e^{2\pi i j t/\omega} x_{(j)}(t) \quad (j = 1, 2, \ldots, N),$$

we conclude that $\tilde{x}_j(t)$ are r-periodic functions and that

$$x(t-kr) = \sum_{\nu=1}^{N} x_{(\nu)}(t-kr) = \sum_{\nu=1}^{N} e^{-2\pi i\nu(t-kr)/\omega} \tilde{x}_\nu(t-kr)$$

$$= \sum_{\nu=1}^{N} e^{2\pi i \nu k r/Nr} e^{-2\pi i \nu t/\omega} \tilde{x}_\nu(t) = \sum_{\nu=1}^{N} \varepsilon^{\nu k} e^{-2\pi i \nu t/\omega} \tilde{x}_\nu(t)$$

$$(k = 0, 1, \ldots, N-1).$$

For the same reason

$$x(t-\omega_j-kr) = x(t-n_j r-kr) = \sum_{\nu=1}^{N} \varepsilon^{\nu(k+n_j)} e^{-2\pi i \nu t/\omega} \tilde{x}_\nu(t)$$

$$(j = 0, 1, \ldots, m;\ k = 0, 1, \ldots, N-1).$$

We write

$$(3.5)\quad \tilde{F}_j(t, \tilde{x}_1(t), \ldots, \tilde{x}N(t))$$

$$= e^{2\pi ijt/\omega}\frac{1}{N}\sum_{k=0}^{N-1}\varepsilon^{-kj}F(t-kr, x(t-kr), \ldots, x(t-\omega_m kr))$$

$$= e^{2\pi ijt/\omega}\frac{1}{N}\sum_{k=0}^{N-1}\varepsilon^{-kj}F\Big(t-kr, \sum_{\nu=1}^{N}\varepsilon^{\nu k}e^{-2\pi i\nu t/\omega}\tilde{x}_\nu(t), \ldots$$

$$\ldots, \sum_{\nu=1}^{N}\varepsilon^{(k+n_m)}e^{-2\pi i\nu t/\omega}\tilde{x}_\nu(t)\Big) \qquad (j = 1, 2, \ldots, N)$$

and similarly

$$(3.6)\quad \tilde{G}_j(t, s, \tilde{x}_1(t), \ldots, \tilde{x}_N(t))$$

$$= e^{2\pi ijt/\omega}\frac{1}{N}\sum_{k=0}^{N-1}\varepsilon^{-kj}G\Big(t-kr, s, \sum_{\nu=1}^{N}\varepsilon^{\nu k}e^{-2\pi i\nu t/\omega}\tilde{x}_\nu(s), \ldots, \sum_{k=1}^{N}\varepsilon^{\nu(k+n_m)}e^{-2\pi i\nu t/\omega}\tilde{x}_\nu(s)\Big)$$

$$(j = 1, 2, \ldots, N).$$

The functions $\tilde{F}_j(t, \tilde{x}(t), \ldots, \tilde{x}_N(t))$ and $\tilde{G}_j(t, s, \tilde{x}_1(t), \ldots, \tilde{x}_N(t))$ are r-periodic with respect to t and s. Using the notation of (3.5) and (3.6), we rewrite system (3.2) as follows:

$$(3.7)\qquad \tilde{x}_j(t) = \tilde{F}_j(t, \tilde{x}_1(t), \ldots, \tilde{x}_N(t)) + \int_0^\omega \tilde{G}_j(t, s, \tilde{x}_1(s), \ldots, \tilde{x}_N(s))ds$$

$$(j = 1, 2, \ldots, N).$$

We are looking for r-periodic, solutions, of system (3.7) which are continuous for $0 \leqslant t \leqslant r$, i.e. for solutions $(\tilde{x}_1, \ldots, x_N)$ such that $\tilde{x}_j \in C_r$ for $j = 1, \ldots, N$. If such a solution exists, an ω-periodic solution, continuous for $0 \leqslant t \leqslant \omega$, of equation (3.1) also exists and is of the form

$$x(t) = \sum_{j=1}^{N} e^{-2\pi ijt/\omega}\tilde{x}_j(t).$$

Observe that (as in Section 2 of Chapter V) the method described here enables us to solve a linear equation of the form (2.1) in the case where the coefficients $a_j(t)$ are ω-periodic.

The next observation is that we can study in exactly the same way a system of integral-difference equations, because without any change in the proof we can consider vector valued functions in place of scalar functions.

Using multi-involutions, we obtain the same result if we consider functions of several variables $t = (t_1, \ldots, t_q)$, and we are looking for ω-periodic solutions, where ω is a vector-period as introduced in Section 4 of Chapter V.

We have here considered only integral-difference equations, but in a similar way, to that used in Section 2, we can study integro-differential-difference equations if we make use of property (3.3).

4. PERIODIC SOLUTIONS OF LINEAR INTEGRO-DIFFERENTIAL-DIFFERENCE EQUATIONS OF VOLTERRA TYPE

At the beginning of this section we consider the integral-difference equation of Volterra type

$$\sum_{j=0}^{m} a_j(t)x(t-\omega_j)+\int_0^t K(t-s, s)x(s)ds = y(t), \tag{4.1}$$

where (as in Section 2) y is an ω_{m+1}-periodic function continuous for $0 \leqslant t \leqslant \omega_{m+1}$. We assume that $\omega_0 = 0$ and that $\omega_1, \ldots, \omega_{m+1}$ are commensurable: $\omega_j = n_j r$. Furthermore, we assume that N is a common multiple (not necessarily the least one) of the numbers $n_1, \ldots, n_{m+1}$ and that $\omega = Nr$. The functions $a_j(t)$ are supposed to be r-periodic and continuous in the interval $[0, r]$. The kernel $K(t, s)$ is r-periodic in s, ω-periodic in t and continuous in the triangle $0 \leqslant s \leqslant t$, $0 \leqslant t \leqslant \omega$.

We are looking for ω-periodic solutions of equation (5.1) satisfying the initial condition [1]

$$x(t) = 0 \quad \text{for } t \in [-r, 0] \tag{4.2}$$

(Fig. 18).

Let

$$X_{\omega,r} = \{x \in C_\omega:\ x(t) = 0 \text{ for } -r \leqslant t \leqslant 0\},$$

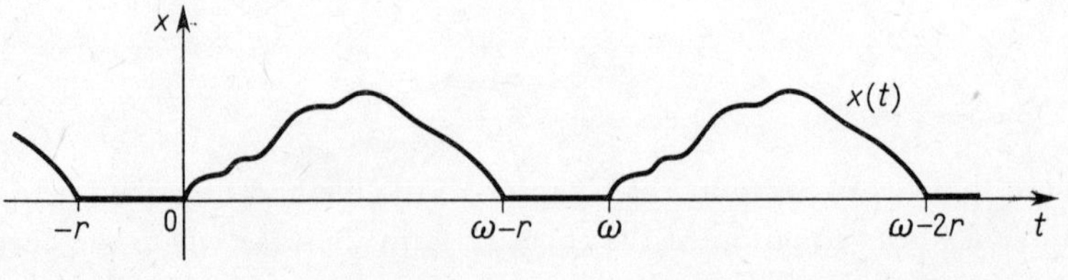

Fig. 18. Form of periodic solution of the Volterra integral-difference equations

[1] This assumption is necessary if the class of ω-periodic functions is to be preserved by the Volterra integral operator.

where C_ω is the space of all ω-periodic functions continuous for $0 \leqslant t \leqslant \omega$ (see Section 2).

In this space the shift operator $(Sx)(t) = x(t-r)$ is an involution of order N, for $(S^N x)(t) = x(t-Nr) = x(t-\omega) = x(t)$.

We show that the operator K defined by means of the Volterra integral

$$(Kx)(t) = \int_0^t K(t-s, s)x(s)\,ds, \quad x \in X_{\omega,r}, \tag{4.3}$$

commutes with S. Indeed, for all $x \in X_{\omega,r}$ condition (4.2) implies that

$$\begin{aligned}(KSx)(t) &= \int_0^t K(t-s, s)x(s-r)\,ds = \int_{-r}^{t-r} K(t-u-r, u+r)x(u)\,du \\ &= \int_{-r}^{t-r} K(t-r-u, u)x(u)\,du = \int_0^{t-r} K(t-r-u, u)x(u)\,du + \\ &+ \int_{-r}^{0} K(t-r-u, u)x(u)\,du = \int_0^{t-r} K(t-r-u, u)x(u)\,du = (SKx)(t).\end{aligned}$$

Hence

$$SK - KS = 0 \quad \text{on } X_{\omega,r}. \tag{4.4}$$

Since S is an involution of order N, the space $X_{\omega,r}$ is the direct sum of the spaces $X_j = P_j X_{\omega,r}$, where $P_j = \frac{1}{N}\sum_{k=0}^{N-1} \varepsilon^{-kj} S^k$ for $j = 1, 2, \ldots, N$, $\varepsilon = e^{2\pi i/N}$.

We write $x_{(j)} = P_j x$ and

$$A_j(t) = \sum_{\nu=0}^{m} \varepsilon^{jn_\nu} a_\nu(t) \quad (j = 1, 2, \ldots, N). \tag{4.5}$$

The operators of multiplication by functions $A_j(t)$ commute with S, since these functions are r-periodic (cf. Section 2 of Chapter V). This enables us to apply Theorem 5.1 of Chapter II, which yields the following

THEOREM 4.1. *Let the functions $a_j(t)$, $K(t, s)$ and the number $\omega_0, \omega_1, \ldots, \omega_{m+1}, \omega$ be as those described above and let $y \in X_{\omega,r}$. Then equation* (4.1) *with the initial condition* (4.2) *is equivalent in the space $X_{\omega,r}$ to a system of N independent equations (without shifts)*

$$A_j(t)x_{(j)}(t) + \int_0^t K(t-s, s)x_{(j)}(s)\,ds = y_{(j)}(t) \quad (j = 1, 2, \ldots, N) \tag{4.6}$$

with the initial condition

$$(4.7) \qquad x_{(j)}(t) = 0 \quad \textit{for} \ -r \leqslant t \leqslant 0 \ (j = 1, 2, \ldots, N),$$

where the $A_j(t)$ *are defined by Formula* (4.5) *and* $y_{(j)}(t) = \frac{1}{N} \sum_{k=0}^{N-1} \varepsilon^{-kj} y(t-kr)$ $(j = 1, 2, \ldots, N)$. *If all these equations have* ω*-periodic solutions satisfying condition* (4.7), *then the* ω*-periodic solution of equation* (4.1) *satisfying condition* (4.2) *is of the form*

$$x(t) = \sum_{j=1}^{N} x_{(j)}(t).$$

Theorem 4.1 remains true if $a_j(t)$, $K(t, s)$ are square matrix-functions whose elements satisfy the previous assumptions and $x(t)$, $y(t)$ are vector valued functions.

Now consider the integro-differential-difference equation of Volterra type:

$$(4.8) \qquad \sum_{j=0}^{m} a_j(t) x(t-\omega_j) + \sum_{k=0}^{n} \int_0^t N_k(t-s, s) x^{(k)}(s) \, ds = y(t),$$

where the numbers $\omega_0, \ldots, \omega_{m+1}, \omega$ are defined as before, and $y(t)$ is an ω_{m+1}-periodic function with the derivative $y^{(n)} \in C[0, \omega_{m+1}]$. Moreover, all the functions $N_k(t, s)$ and their derivatives with respect to t up to and including the order n are continuous in the triangle $0 \leqslant s \leqslant t$, $0 = t = r$ and r-periodic with respect to both variables.

We are looking for ω-periodic n-times continuously differentiable solutions of equation (4.8) satisfying the initial condition

$$(4.9) \qquad x^{(k)}(t) = 0 \quad \text{for} \ -r \leqslant t \leqslant 0 \ (k = 0, 1, \ldots, n).$$

Consider the space

$$X^n_{\omega,r} = \{x \in C_\omega \colon x^{(n)} \in C_\omega \text{ and } x^{(k)}(t) = 0 \text{ for } -r \leqslant t \leqslant 0 \text{ and } k = 0, 1, \ldots, n\}.$$

It follows from our preceding discussion that all the operators in equation (4.8) commute with S in the space $X^n_{\omega,r}$. Therefore, equation (4.8) with the initial condition (4.9) is equivalent in the space $X^n_{\omega,r}$ to the N independent equations

$$(4.10) \quad A_j(t) x^{(j)}(t) + \sum_{k=0}^{N-1} \int_0^t N_k(t-s, s) x_{(j)}(s) \, ds = y_{(j)}(t) \quad (j = 1, 2, \ldots, N)$$

provided that $y \in X^n_{\omega,r}$, where the $A_j(t)$ are defined by Formula (4.5). Arguing as in Section 2, we conclude that the jth equation (4.10) has a solution in the space $X_j = P_j X^n_{\omega,r}$ if and only if the system

$$x_{j\nu}(t) + \sum_{k=0}^{n} \int_0^t N_{kj}^{(\nu)}(t-s, s) x_{jk}(s)\, ds = y_j^{(\nu)}(t) \tag{4.11}$$

has ω-periodic solutions satisfying the condition

$$x_{j\nu}(t) = 0 \quad \text{for } -r \leqslant t \leqslant 0 \ (\nu = 0, 1, \ldots, n;\ 0 \leqslant j \leqslant N \text{ being fixed}), \tag{4.12}$$

where $N_{kj}(t, s) = [A_j(t+s)]^{-1} N_k(t, s)$ and $y_j = A_j^{-1} y_{(j)}$, provided that $A_j \neq 0$ for $j = 1, 2, \ldots, N$ are sufficiently differentiable.

The last system (j being fixed) is a Volterra system of integral equations which under our assumptions has a solution that is unique (see Pogorzelski [1], also Section 5 of Chapter I) but which does not necessarily satisfy condition (4.12). Hence we have

THEOREM 4.2. *Let the numbers $\omega_0, \ldots, \omega_{m+1}, \omega$, and the functions $N_k(t, s)$, $a_j(t)$ satisfy the same conditions as before. Let the functions $A_j(t) \neq 0$ defined by Formula* (4.5) *be continuously differentiable up to and including the order n for $0 \leqslant t \leqslant r$. If each of systems* (4.11) *for $j = 1, 2, \ldots, N$ has a solution $(x_{j0}, \ldots, x_{jn})$ such that $x_{jk} \in C_\omega$ and satisfies condition* (4.12) $(k = 0, 1, \ldots, n)$, *then equation* (4.8) *has an ω-periodic solution which is n-times differentiable and which satisfies condition* (4.9), *which is of the form $x(t) = \sum_{j=1}^{N} x_{j0}(t)$.*

Theorems 4.1 and 4.2 remain valid if, instead of the kernels $K(t-s, s)$ and $N_k(t-s, S)$ we consider kernels $K(t, s)$ and $N_k(t, s)$, r-periodic with respect to both variables. Indeed, Formula (4.4) holds, because

$$\begin{aligned}(KSx)(t) &= \int_0^t K(t, s) x(s-r)\, ds = \int_{-r}^{t-r} K(t, u+r) x(u)\, du \\ &= \int_{-r}^{t-r} K(t, u) x(u)\, du = \int_0^{t-r} K(t, u) x(u)\, du + \int_{-r}^{0} K(t, u) x(u)\, du \\ &= \int_0^{t-r} K(t, u) x(u)\, du = \int_0^{t-r} K(t-r, u) x(u)\, du = (SKx)(t).\end{aligned} \tag{4.13}$$

Further corresponding proofs are exactly the same.

In the particular case where $K(t, s) = \text{const}$, it might be convenient to assume, instead of condition (4.2), the weaker condition

$$\int_{-r}^{0} x(u)\,du = 0,$$

which also implies the commutativity of the operators K and S.

The same results can be obtained in the case where x, y are vector valued functions and a_j, $K(t, s)$, $N_k(t, s)$ are square matrix-functions whose elements satisfy the previous assumptions. The same proofs remain valid.

5. EXPONENTIAL-PERIODIC SOLUTIONS OF INTEGRO-DIFFERENTIAL-DIFFERENCE EQUATIONS OF FREDHOLM TYPE

Consider the following linear integral-difference equation of Fredholm type

$$(5.1)\qquad \sum_{i=0}^{m} a_j(t)x(t-\omega_j) + \sum_{i=0}^{m}\int_0^{\omega} K_j(t-s, s)x(s-\omega_j)\,ds = y(t) \qquad (\omega_0 = 0).$$

We assume that all the numbers ω_j are commensurable, i.e. that there exist a real $r \neq 0$ and integers $n_1, \ldots, n_m$ such that $\omega_j = n_j r$. Let $n_0 = 0$ and let N be a common multiple (not necessarily the least one) of the numbers n_j and let $\omega = Nr$. The functions $a_j(t)$ are supposed to be r-periodic and continuous in the closed interval $[0, r]$.

We consider the space $E_\omega(\lambda) = \{x(t) = e^{\lambda t}\xi(t),\ \xi(t) \in E_\omega\}$, where E_ω is the space of all ω-periodic functions on the real line (cf. with Section 1 of Chapter VI and also Section 3 of Chapter IX).

We assume that $K_j(t, s) = e^{\lambda t}\, q_j(t, s)$, where $q_j(t, s)$ are ω-periodic with respect to the variable t and r-periodic with respect to the variable s, and continuous for $0 \leqslant t \leqslant \omega$, $0 \leqslant s \leqslant r$.

Let $y \in E_\omega(\lambda)$. We are looking for solutions of equation (5.1) belonging to the space $E_\omega(\lambda)$.

Note that this equation can be written in a simpler form. Indeed, since $\int_a^{\sigma+a} x(s)\,ds = \int_0^{\sigma} x(s)\,ds$ for all σ-periodic functions and all a, we find from our assumptions for $x(t) = e^{\lambda t}\xi(t)$, $\xi(t) \in E_\omega$, that

$$\int_0^{\omega} K_j(t-s, s)x(s-\omega_j)\,ds = \int_0^{\omega} e^{\lambda(t-s)}q_j(t-s, s)e^{\lambda(s-\omega_j)}\xi(s-\omega_j)\,ds$$

$$= \int_{-\omega}^{\omega-\omega_j} e^{\lambda t}q_j(t-u-\omega_j,\ u+\omega_j)e^{-\lambda\omega_j}\xi(u)\,du$$

$$= e^{\lambda(t-\omega_j)} \int_{-\omega_j}^{\omega-\omega_j} q_j(t-u-\omega_j, u)\xi(u)\,du$$

$$= e^{\lambda(t-\omega_j)} \int_0^{\omega} q_j(t-u-\omega_j, u)\xi(u)\,du$$

$$= \int_0^{\omega} e^{\lambda(t-\omega_j-u)} q_j(t-u-\omega_j, u) e^{\lambda u}\xi(u)\,du$$

$$= \int_0^{\omega} K_j(t-u-\omega_j, u)x(u)\,du.$$

Hence if we put

$$K(t, s) = \sum_{j=0}^{m} K_j(t-\omega_j, s),$$

we can rewrite equation (5.1) in the form

$$\sum_{j=0}^{m} a_j(t)x(t-\omega_j) + \int_0^{\omega} K(t-s, s)x(s)\,ds = y(t), \tag{5.2}$$

where

$$K(t, s) = \sum_{j=0}^{m} e^{\lambda(t-\omega_j)} q_j(t-\omega_j, s) = e^{\lambda t}\sum_{j=0}^{m} e^{-\lambda\omega_j} q_j(t-\omega_j, s) = e^{\lambda t} q(t, s),$$

and the function $q(t, s)$ is ω-periodic with respect to t and r-periodic with respect to s and continuous for $0 \leqslant t \leqslant \omega$, $0 \leqslant s \leqslant r$.

Let X be the space of all functions $x(t)$ belonging to $E_\omega(\lambda)$ and continuous for $0 \leqslant t \leqslant \omega$. In the space X we consider the shift operator $(Sx)(t) = x(t-r)$. S is an algebraic operator of order N as the following identity is satisfied: $(S^N s)(t) = e^{-\lambda\omega} x(t)$ for all $x \in X$. We show that the operator K defined by means of the integral $(Kx)(t) = \int_0^{\omega} K(t-s, s)x(s)\,ds$ for $x \in X$ commutes with S. Indeed, for $x \in X$

$$(SKx)(t) = \int_0^{\omega} K(t-r-s, s)x(s)\,ds = \int_r^{\omega+r} K(t-u, u-r)x(u-r)\,du$$

$$= \int_r^{\omega+r} e^{\lambda(t-u)} q(t-u, u-r) e^{\lambda(u-r)}\xi(u-r)\,du = e^{\lambda(t-r)} \int_0^{\omega} q(t-u, u)\xi(u-r)\,du$$

$$= \int_0^{\omega} e^{\lambda(t-u)} q(t-u, u) e^{\lambda(u-r)}\xi(u-r)\,du = \int_0^{\omega} K(t-u, u)x(u-r)\,du = (KSx)(t).$$

Hence $SK - KS = 0$.

Since S is an algebraic operator of order N, the space X is the direct sum of the spaces $X_j = P_jX$, where

$$(5.3)\qquad P_j = \frac{1}{N}\sum_{k=0}^{N-1} e^{\lambda kr}\varepsilon^{-kj}S^k \quad \text{for } j = 1, 2, \ldots, N \text{ and } \varepsilon = e^{2\pi i/N}.$$

Now write $x_{(j)} = P_jx$ $(j = 1, 2, \ldots, N)$ and

$$(5.4)\qquad A_\nu(t) = \sum_{j=0}^{m} e^{-\lambda\omega_j}\varepsilon^{\nu n_j}a_j(t) \quad (\nu = 1, 2, \ldots, N).$$

The operators A_ν of multiplication by functions $A_\nu(t)$ commute with S because the operators of multiplication by functions $a_j(t)$, which are r-periodic, commute with S (cf. Section 2 of Chapter V).

We can therefore apply Corollary 5.1 of Chapter II to obtain

THEOREM 5.1 (Włodarska-Dymitruk [4]). *Let the functions $K_j(t, s)$ (resp. $K(t, s)$) $a_j(t)$, $y(t)$ and the numbers $r, \omega_0, \ldots, \omega_m, \omega$ satisfy the assumptions listed under equation* (5.1). *Then equation* (5.1) (*resp.* (5.2)) *is equivalent in the space X to a system of N independent equations (without shifts)*

$$(5.5)\qquad A_j(t)x_{(j)}(t) + \int^{\omega} K(t-s, s)x_{(j)}(s)\,ds = y_{(j)}(t) \quad (j = 1, 2, \ldots, N),$$

where the A_j are defined by Formula (5.4), $y_{(j)} = \frac{1}{N}\sum_{k=0}^{N-1} e^{\lambda kr}\varepsilon^{-kj}y(t-kr)$ $(j = 1, 2, \ldots, N)$ *and, in the case of equation* (5.1) $K(t, s) = \sum_{j=0}^{m} K_j(t-\omega_j, s)$.

If all these equations have solutions $x_j \in X$, then the solution of equation (5.1) (*resp.* (5.2)) *belonging to X is of the form* $x(t) = \sum_{j=1}^{N} x_{(j)}(t)$.

Observe that in the case where $A_j(t) \neq 0$ for $j = 1, 2, \ldots, N$ and $0 \leqslant t \leqslant r$, we have instead of system (5.5) a system of independent Fredholm equations of the second kind

$$(5.6)\qquad x_{(j)}(t) + \int_0^{\omega} \tilde{K}_j(t-s, s)x_{(j)}(s)\,ds = y_j(t) \quad (j = 1, 2, \ldots, N),$$

where

$$y_j = A_j^{-1}y_{(j)}, \quad \tilde{K}_j(t, s) = [A_j(t+s)]^{-1}K(t, s) \quad (j = 1, 2, \ldots, N).$$

Moreover, $\tilde{K}_j(t, s)$ is an r-periodic function with respect to s and $\tilde{K}_j(t+\omega, s) = e^{\lambda\omega}\tilde{K}_j(t, s)$ $(j = 1, 2, \ldots, N)$. Indeed,

$$\tilde{K}_j(t, s+r) = [A_j(t+s+r)]^{-1}K(t, s+r) = \Big[\sum_{j=0}^{m} e^{-\lambda\omega_j}\varepsilon^{jn_k}a_k(t+s+r)\Big]^{-1}K(t, s)$$

$$= \Big[\sum_{k=0}^{m} e^{-\lambda\omega_j}\varepsilon^{jn_k}a_k(t+s)\Big]^{-1}K(t, s) = \tilde{K}_j(t, s),$$

$$\tilde{K}_j(t+\omega, s) = [A_j(t+\omega, s)]^{-1}K(t+\omega, s) = \Big[\sum_{k=0}^{m} e^{-\lambda\omega_j}\varepsilon^{jn_k}a_k(t+s+rN)\Big]^{-1}e^{\lambda t}q(t, s)e^{\lambda\omega}$$

$$= \Big[\sum_{k=0}^{m} e^{-\lambda\omega_j}\varepsilon^{jn_k}a_k(t+s)\Big]^{-1}K(t, s)e^{\lambda\omega} = e^{\lambda\omega}\tilde{K}_j(t, s).$$

Theorem 5.1 remains true if x, y are vector valued functions and a_j, $K_j(t, s)$, $K(t, s)$ are square matrix functions, all with elements satisfying the previous conditions.

Now consider an integro-differential-difference equation [1]

$$\sum_{j=0}^{m} a_j(t)x(t-\omega_j) + \sum_{k=0}^{n}\int_0^{\omega} N_k(t-s, s)x^{(k)}(s)\,ds = y(t), \tag{5.7}$$

where the numbers $\omega_0, \ldots, \omega_m, \omega$ are as before, and $y(t)$ is a given function belonging to the space $E_\omega(\lambda)$ with the nth derivative continuous on $[0, \omega]$. Moreover, all the functions $N_k(t, s)$ and their derivatives with respect to t up to and including the order n are continuous for $0 \leqslant t \leqslant \omega$, $0 \leqslant s \leqslant r$ and $N_k(t, s) = e^{\lambda t}p_k(t, s)$, where the functions $p_k(t, s)$ are ω-periodic with respect to t and r-periodic with respect to s.

It follows from our preceding discussion that the integral operators N_k commute with S, as well as with derivation. Hence equation (5.7) is equivalent to the N independent equations without shifts

$$A_j(t)x_{(j)}(t) + \sum_{k=0}^{n}\int_0^{\omega} N_k(t-s, s)x_{(j)}^{(k)}(s)\,ds = y_{(j)}(t) \quad (j = 1, 2, \ldots, N), \tag{5.8}$$

[1] As before, we can show that without loss of generality it is enough to consider an equation analogous to equation (5.2).

where the A_j are defined by Formula (5.4). Now assume that all the $A_j(t) \neq 0$ for $0 \leqslant t \leqslant r$. Then system (5.8) is equivalent to the system of N idependent equations

$$(5.9) \qquad x_{(j}(t) + \sum_{k=0}^{n} \int_0^{\omega} N_{kj}(t-s, s) x_{(j)}^{(k)}(s)\, ds = y_j(t) \qquad (j = 1, 2, \ldots, N),$$

where $N_{kj}(t, s) = [A_j(t+s)]^{-1} N_k(t, s)$, $y_j = A_j^{-1} y_{(j)}$.

As in the preceding sections we obtain the following

THEOREM 5.2 (Włodarska-Dymitruk [4]). *Let the numbers $\omega_0, \ldots, \omega_m, \omega$ the functions $a_j(t)$ and $N_k(t, s)$ satisfy the same assumptions as before. Moreover, let the functions $A_j(t) \neq 0$ be continuously differentiable up to and including the order n for $0 \leqslant t \leqslant r$ $(j = 1, 2, \ldots, N)$. If for each fixed ν $(\nu = 1, 2, \ldots, N)$ the system*

$$(5.10) \qquad x_{\nu j}(t) + \sum_{k=0}^{n} \int_0^{\omega} N_k^{(j)}(t-s, s) x_{\nu k}(s)\, ds = y_\nu^{(j)}(t) \qquad (j = 0, 1, \ldots, n)$$

has a solution $(x_{\nu 0}, \ldots, x_{\nu n})$ such that $x_{\nu k} \in E_\omega(\lambda)$ have continuous nth derivatives for $k = 0, 1, \ldots, n$, then equation (5.7) has an n-times continuously differentiable solution belonging to $E_\omega(\lambda)$ which is of the form: $x = \sum_{j=1}^{N} x_{j0}$.

CHAPTER XII

Singular Integral Equations of one Variable

In this chapter we assemble those results on singular integral equations which can easily be obtained by our method.

1. CAUCHY AND HILBERT SINGULAR INTEGRALS

We recall that a *regular open arc* in the complex plane is a set of points of the form

$$L = \{z\colon\ z = z(t),\ \ \alpha < t < \beta\},$$

where

1. the function $z(t)$ is one-to-one and has a continuous non-vanishing derivative at each point $t \in (\alpha, \beta)$,

2. $\lim_{t\to\alpha+0} z'(t) \neq 0$ and $\lim_{t\to\beta-0} z'(t) \neq 0$,

3. the extension of the function $z(t)$ to the closed interval $[\alpha, \beta]$ remains a one-to-one map (cf. also Section 1 of Chapter VIII).

A *regular closed arc* in the complex plane is a set of points of the form

$$L = \{z\colon\ z = z(t),\ \ \alpha \leqslant t \leqslant \beta,\ \ z(\alpha) = z(\beta)\},$$

where

1. the function $z(t)$ is one-to-one for $t \neq \alpha, \beta$ and has a continuous non-vanishing derivative at each point $t \in (\alpha, \beta)$,

2. $\lim_{t\to\alpha+0} z'(t) = \lim_{t\to\beta-0} z'(t) \neq 0$.

If L is a regular arc, then there is a positive constant χ such that

$$\chi \leqslant \frac{t_1 - t_2}{s_{1,2}} \leqslant 1 \qquad \text{for all } t_1 t_2 \in L,$$

where $s_{1,2}$ denotes the length of that part of the arc L which lies between the points t_1 and t_2.

Let $x(t)$ be a function (real valued or complex valued) defined and integrable on the arc L. The *Cauchy integral* of the function $x(t)$ is the function

$$\Phi(z) = \int_L \frac{x(t)}{t-z}\,dt. \tag{1.1}$$

This integral is well-defined for all points $z \notin \bar{L}$. Moreover, if $z \notin \bar{L}$, the function $\Phi(z)$ is analytic and it is easy to verify that

$$\Phi'(z) = \int_L \frac{x(t)}{(t-z)^2}\,dt. \tag{1.2}$$

This definition can easily be extended to the case where L is the union of a finite number of pairwise disjoints arc L_j and $z \notin \bigcup \bar{L}_j$. Evidently, the set $L = \bigcup L_j$ divides the plane into a finite number of connected domains. In each of these domains the function $\Phi(z)$ is analytic. Moreover, we conclude from Formula (1.1) that $\Phi(z) = O(1/|z|)$ as $z \to 0$.

If $z \in L$, then the Cauchy integral does not exist in general. We therefore ntroduce a new notion. The *principal value in the sense of Cauchy* of integral (1.1) is defined to be the limit

$$\lim_{\varepsilon\to 0} \int_{L/\tilde{L}_\varepsilon} \frac{x(\tau)\,d\tau}{\tau - t} = \text{V.P.} \int_L \frac{x(\tau)}{\tau-t}\,d\tau,$$

where

$$t \in L \quad \text{and} \quad \tilde{L}_\varepsilon = \{z\colon |z-t| < \varepsilon\} \cap L.$$

If the function $x(t)$ is integrable, then the integral $\int_{L/\tilde{L}_\varepsilon} \frac{x(\tau)}{\tau-t}\,d\tau$ always exists. However, the principal value of integral (1.1) does not necessarily exist, even in the case where the function $x(t)$ is continuous. If the principal value of integral (1.1) exists, then the following formula holds:

$$\text{V.P.} \int_L \frac{x(\tau)}{\tau-t}\,d\tau = \text{V.P.} \int_L \frac{x(\tau)-x(t)}{\tau-t}\,d\tau + \pi i x(t) + x(t) \log\frac{b-t}{a-t}, \tag{1.3}$$

where $a = z(\alpha)$ and $b = z(\beta)$ denote the two end-points of the arc L. In particular, if the arc L is closed, then the last term in Formula (1.3) vanishes.

In the sequel we write briefly $\int_L \frac{x(\tau)}{\tau-t}\,d\tau$ instead of V.P. $\int_L \frac{x(\tau)}{\tau-t}\,d\tau$ and we call this integral the *singular integral* in the *Cauchy sense*, or simply: the *singular integral* on the arc L.

It follows immediately from Formula (1.3) that a singular integral of a function satisfying the Hölder condition (compare Example 4.3 of Chapter 1) always exists. Moreover, the integral on the right-hand side of Formula (1.3) is an ordinary Riemann integral.

Observe that the singular integral $\int_L \frac{x(\tau)}{\tau - t} d\tau$ is a function of the variable t. Hence the map

$$x(t) \xrightarrow{S} y(t) = \frac{1}{\pi i} \int_L \frac{x(\tau)}{\tau - t} d\tau$$

is a linear operator. Our subsequent discussion will be based on the following important theorem:

THEOREM 1.1 (Plemelj [1], Privalov [1]). *If L is a finite system of pairwise disjoint regular closed arcs, then*

$$S \in B(H^{\mu}(L)) \quad \textit{if } 0 < \mu < 1\,^{1},$$
$$S \in B(H^{1}(L)) \to (H^{1-\varepsilon}(L)) \quad \textit{for arbitrary positive } \varepsilon < 1.$$

We do not give the proof of this theorem here; the reader can find it in Pogorzelski's monograph [1], § 1, Chapter XV. The formulation of the theorem given there is not the same as that given above. However, Theorem 1.1 is an immediate consequence of the fact that the constant C appearing in the classical formulation depends only on the line L.

The following relations hold between the boundary values of the function $\Phi(z)$ defined by Formula (1.1) and the function $x(t)$.

THEOREM 1.2 (Plemelj [1]). *If a function $x(t)$ satisfies the Hölder condition on a regular arc L, then the function* $\Phi(z) = \frac{1}{2\pi i} \int_L \frac{x(\tau)}{\tau - z} d\tau$ *can be extended continuously on both sides of the arc L. If we denote the respective boundary values by $\Phi^+(t)$ and $\Phi^-(t)$, then for every $t \in L$ which is not an end-point of L*:

$$(1.4) \quad \begin{aligned} \Phi^+(t) &= \tfrac{1}{2}x(t) + \frac{1}{2\pi i} \int_L \frac{x(\tau)}{\tau - t} d\tau, \\ \Phi^-(t) &= -\tfrac{1}{2}x(t) + \frac{1}{2\pi i} \int_L \frac{x(\tau)}{\tau - t} d\tau. \end{aligned}$$

The proof of this theorem can be found in Pogorzelski's monograph [1], § 3, Chapter XV.

---

[1] For the definition of $B(X \to Y)$ see Section 4 of Chapter I.

Since the function $\Phi(z)$ is analytic for $z \notin \overline{L}$, Plemelj's Formulae (1.4) also hold for finite unions of pairwise disjoint arcs.

It also follows from Formulae (1.4) that every function satisfying the Hölder condition on a regular closed arc can be represented as the difference of two functions

$$x(t) = \Phi^+(t) - \Phi^-(t), \tag{1.5}$$

where $\Phi^+(t)$ is the boundary value of a certain function holomorphic in a domain whose boundary is L, and $\Phi^-(t)$ is the boundary value of a function holomorphic outside the domain with boundary L and bounded at infinity. This representation is unique. From the Cauchy integral formula we immediately obtain

$$S\Phi^+(t) = \Phi^+(t) \quad \text{and} \quad S\Phi^-(t) = -\Phi^-(t), \tag{1.6}$$

where S denotes the operator $(Sx)(t) = \frac{1}{\pi i}\int_L \frac{x(\tau)}{\tau - t}\,d\tau$, as before. Hence we obtain

THEOREM 1.3. *If a regular arc L is closed, then the operator S in an involution on the space $H^\mu(L)$, $0 < \mu < 1$: $S^2 = I$.*

Suppose we are given a finite system of pairwise disjoint closed regular arcs $L_1, \ldots, L_n$. These arcs divide the plane into components $\Omega_0, \ldots, \Omega_n$. We associate the sign "$-$" with the component Ω_0 containing the point ∞, and the sign "$+$" with the components which have a common boundary with Ω_0. We next associate the sign "$-$" with the components which have a common boundary with the components having the sign "$+$", but not with Ω_0, and so on. We orient the arcs in such a manner that on the left of each of then lies a domain with the sign "$+$", and on the right, a domain with the sign "$-$". Such a system is called an *oriented system* (see Fig. 19).

Integral (1.1) induces a function $\Phi(z)$ analytic in each of the domains $\Omega_0, \ldots, \Omega_n$ separately. We denote the boundary values of this function in passing from a domain with the sign "$+$" by $\Phi^+(t)$, and in passing from a domain with the sign "$-$" by $\Phi^-(t)$. The Plemelj Formulae (1.4) then also hold for an oriented system. Arguing just as in the case of one arc, we obtain

COROLLARY 1.4. *If a system L is oriented, then $S^2 = I$ on $H^\mu(L)$, $0 < \mu < 1$.*

For open arcs Theorem 1.3 does not hold, although in this case we have an inversion formula for the operator S. This was done in a very general way by Pogorzelski [1] (Chapter XX, § 3) in the class $\mathfrak{h}_\mu^\alpha$ introduced by him in [3].

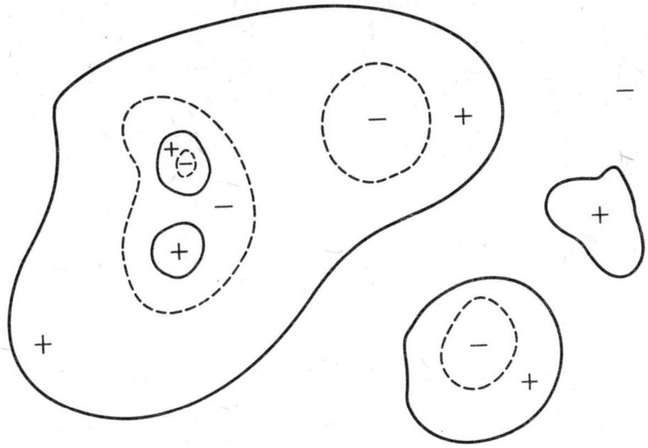

Fig. 19. Oriented system

Similar theorems can be obtained for singular operators in the L^p-spaces, $p > 1$. Singular operators in L^p-spaces were studied by Calderón and Zygmund [1], Zygmund [1], [2], Tricomi [2], [4], Hvedelidze [1], Widom [1] and others.

Suppose that in the complex plane completed by the point at infinity we are given an arbitrary fixed point $z_0 = x_0 + iy_0$. The map h_{z_0}: $z_0 \to w$, where

$$w = \frac{1}{z - z_0} \qquad (w = \xi + i\eta) \tag{1.7}$$

is a *homographic transformation* for all arcs L not containing z_0.

An *arc L closed by the point at infinity* is said to be *regular* if the homographic map (1.7), where $z_0 \notin L$, transforms L into a regular closed arc L (in the usual sense). We call L briefly, a *regular arc closed at infinity*. For example, the real axis is a regular arc closed at infinity, because every homographic map with $\operatorname{im} z_0 \neq 0$ maps it onto a circle.

On a regular arc closed at infinity we can define the Cauchy integral as a limit:

$$\int_L \frac{x(\tau)}{\tau - z}\, d\tau = \lim_{N\to\infty} \int_{L \cap \{z: |z| \leqslant N\}} \frac{x(\tau)}{\tau - z}\, d\tau.$$

This improper integral exists if $x(t)$ is continuous on L and if

(1.8)

1. $\lim_{t\to\infty} x(t) = x(\infty)$ exists and is $< +\infty$,

2. $|x(t) - x(\infty)| < \dfrac{c}{|t|^{\alpha}}$, where c, α are positive constants.

If L is a regular arc closed at infinity, denote by $H^{\mu}_{\infty}(L)$ the set of all complex valued functions $x(t)$ defined for $t \in L$ such that limit (1.8) exists and the following inequality holds:

$$(1.9) \qquad |x(t)-x(t')| < c\left|\frac{1}{t-z_0}-\frac{1}{t'-z_0}\right|^{\mu}, \qquad 0<\mu\leqslant 1 \text{ and } z_0 \notin L.$$

It is easy to check that the homographic map (1.7) transforms $H^{\mu}_{\infty}(L)$ into $H^{\mu}(L^*)$, where L^* is the image of L by this map.

Denote by S_{∞} the operator defined by means of a singular integral on a regular arc closed at infinity:

$$(1.10) \qquad (S_{\infty}x)(t) = \frac{1}{\pi i}\int_L \frac{x(\tau)}{\tau - t}\,d\tau, \qquad t \in L.$$

This mapping is called a *Hilbert transform.* Theorems 1.1, 1.2, 1.3 also hold in the case of S_{∞} considered on the space $H^{\mu}_{\infty}(L)$, where L is a regular arc closed at infinity. Moreover, these theorems remain true if L is an oriented system containing a regular arc closed at infinity (Fig. 20)[1].

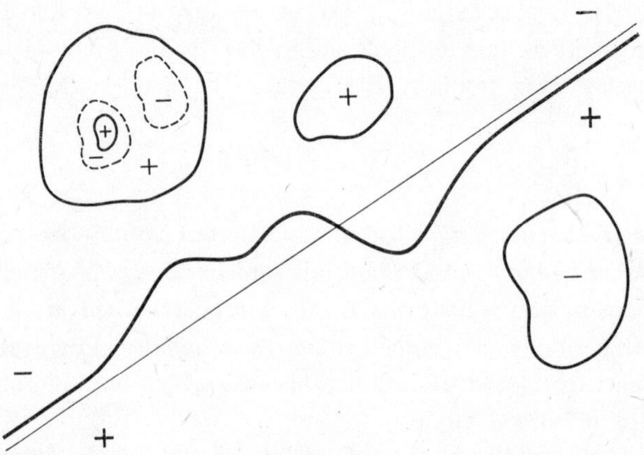

Fig. 20. Oriented system containing an arc closed at infinity

The Hilbert transform, in the case where L is the real line, is usually studied in the spaces $L^p(\boldsymbol{R}), p \geqslant 1$. In this case we have, analogously to the Plemelj–Privalov Theorem 1.1:

---

[1] The properties of a singular integral on the real line in the class $H^{\mu}_{\infty}(L)$ were given in Muskhelishvili's book [1], p. 263–273, and on a regular arc closed at infinity in the present author's paper [14] in 1960. These last results were repeated (with mistakes) by Motkin [1] in 1967.

THEOREM 1.5 (M. Riesz [1]). *For $p > 1$ the Hilbert transform defined by means of the singular integral*

$$(1.11) \qquad (Hx)(t) = \frac{1}{\pi} \int_{-\infty}^{+\infty} \frac{x(\tau)}{\tau - t}\, d\tau, \qquad t \in \boldsymbol{R},$$

is a bounded operator mapping $L^p(\boldsymbol{R})$ onto itself.

Hence the operator $S_\infty = \frac{1}{i} H \in B(L^p(\boldsymbol{R}))$. In this case we also have an involution, i.e. $S_\infty^2 = I$ on $L^p(\boldsymbol{R})$.

When $p = 1$, the range of the Hilbert transform is not the whole of $L(\boldsymbol{R})$ and we will not consider this case (see Zygmund [2]). Multi-dimensional Hilbert transforms were studied by Calderón and Zygmund [1] (see also a survey: Zygmund [2]).

If $x \in L^2(\boldsymbol{R})$, then $x = x^+ + x^-$, where x^+ and x^- are boundary values (understood here as limits in the norm of the space $L^2(\boldsymbol{R})$) of analytic functions regular for $\operatorname{im} z > 0$ and $\operatorname{im} z < 0$, respectively (Titchmarsh [2]).

In the theory of singular integral equations another type of Hilbert transform is also considered. Namely, denote by $H_0(0, 2\pi)$ the space of all 2π-periodic functions defined on the real line which satisfy the Hölder condition with the exponent μ. In this space we now define a singular integral operator by the equality:

$$(1.12) \qquad (S_0 x)(s) = \frac{1}{2\pi} \int_0^{2\pi} x(\sigma) \cot \frac{\sigma - s}{2}\, d\sigma, \qquad s \in [0, 2],$$

where the integral is understood as a Cauchy principal value [1]. If we change both the variables simultaneously by the substitutions

$$t = e^{is}, \qquad \tau = e^{i\sigma}, \qquad 0 \leqslant \sigma, s \leqslant 2\pi,$$

we obtain

$$\frac{1}{\pi i} \frac{d\tau}{\tau - t} = \frac{1}{2\pi} \left(\cot \frac{\sigma - s}{2} - \pi i \right),$$

where t, τ belong to the unit circle $L = \{\tau\colon |\tau| = 1\}$.

The following identity holds (see Mikhlin [1]):

$$(1.13) \qquad I + S_0^2 = K_0, \qquad \text{where } (K_0 x)(s) = \frac{1}{2\pi} \int_0^{2\pi} x(\sigma)\, d\sigma.$$

[1] Singular integral equations with the cotangent kernel were studied by Poincaré [1], F. Noether [1]. For details see for instance Mikhlin [1], Schmeidler [1].

Since the operator K_0 is one-dimensional, the operator $\frac{1}{i}S_0$ is an almost involution (see Section 4 of Chapter II).

2. INVOLUTIONARY CASES OF SINGULAR INTEGRAL EQUATIONS

Suppose we are given an oriented system L. We consider the singular integral equation

$$A_0(t)x(t)+\frac{1}{\pi i}\int_L \frac{K(t,\tau)x(\tau)}{\tau-t}d\tau = f(t). \tag{2.1}$$

We suppose that the function $K(t,\tau)$ satisfies the Hölder condition with an exponent μ on L:

$$|K(t,\tau)-K(t',\tau')| \leqslant C(|t-t'|^{\mu}+|\tau-\tau'|^{\mu}).$$

Similarly we suppose that $A_0(t)\in H^{\mu}(L)$. The assumptions concerning the function $f(t)$ will be formulated later.

We rewrite equation (2.1) as

$$A_0(t)x(t)+\frac{A_1(t)}{\pi i}\int_L \frac{x(\tau)}{\tau-t}d\tau+\int_L T(t,\tau)x(\tau)d\tau = f(t),$$

where

$$A_1(t) = K(t,t) \quad \text{and} \quad T(t,\tau) = \frac{K(t,\tau)-K(t,t)}{\tau-t}\cdot\frac{1}{\pi i}.$$

The assumption that the function $K(t,\tau)$ satisfies the Hölder condition with the exponent μ immediately implies that $|T(t,\tau)| < \frac{C}{|\tau-t|^{1-\mu}}$, i.e. that the function $T(t,\tau)$ is a weakly singular kernel.

It follows from Theorem 1.1 that the operator S preserves the spaces $H^{\mu}(L)$ for all $\mu<1$ and for all those L which are oriented systems. In order to discover those spaces in which equation (2.1) ought to be studied, we apply the following theorems (Proofs of Theorem 2.1 and 2.3 can be found in the book [6] by the present author and Rolewicz, part D, Chapter 2):

THEOREM 2.1. *If a function $M(t)$ satisfies the Hölder condition with an exponent μ on the set L, then the operator of multiplication by the function $M(t)$ belongs to $B(H^{\alpha}(L))$ for all $\alpha\leqslant\mu$.*

THEOREM 2.2 (Pogorzelski [1]). *If the function $K(t,\tau)$ satisfies the Hölder condition with an exponent μ with respect to the variable $t,\tau\in L$ and if the integral* *operator T is defined by means of the kernel $T(t, \tau) = [K(t, \tau) - K(t, t)]/(\tau - t)$, then $T \in B(C(L) \to H^{\alpha}(L))$ for $\alpha < \mu/2$.*

The proof is given in Pogorzelski's monograph [1], Lemma 2 on p. 495.

THEOREM 2.3. *If the conditions of Theorem 2.2 are satisfied, then $T \in B(H^{\alpha}(L))$ for all $\alpha < 1/2$ and T is compact in $H^{\alpha}(L)$.*

As a corollary to Theorem 2.3 we obtain

COROLLARY 2.1. *Let M denote the operator of multiplication by a function $M(t)$ satisfying the Hölder condition with an exponent μ on an oriented system L, and let $(Sx)(t) = \frac{1}{\pi i}\int_L \frac{x(\tau)}{\tau - t}\, d\tau$ $(t \in L)$. Then the commutator $SM - MS$ belongs to $B(H^{\alpha}(L))$ for $\alpha < \mu/2$ and is compact in this space.*

It follows from these theorems and Corollary 2.1 that equation (2.1) should be considered in the space $H^{\alpha}(L)$, $\alpha < \mu/2$. Hence a natural assumption is that $f(t) \in H^{\alpha}(L)$. Recall that S is an involution on the space $H^{\alpha}(L)$ (Corollary 1.4). Hence we can apply to equation (2.1) the method of regularization given in Section 4 of Chapter II (see also Section 3 of this chapter).

As an immediate consequence of Corollary 2.1 and Theorem 2.2 of this chapter and Theorem 4.2 of Chapter II we obtain

THEOREM 2.4. *Let L be an oriented system. Let the functions $A_0(t)$ and $K(t, \tau)$ satisfy the Hölder condition with an exponent μ for $t, \tau \in L$. Let*

$$T(t, \tau) = \frac{1}{\pi i} \cdot \frac{K(t, \tau) - K(t, t)}{\tau - t}, \qquad (Tx)(t) = \int_L T(t, \tau) x(\tau)\, d\tau,$$

$$A_1(t) = K(t, t), \qquad A(S) = A_0 + A_1 S,$$

where A_0 and A_1 are operators of multiplication by functions $A_0(t)$ and $A_1(t)$ respectively. Finally, let $(Sx)(t) = \frac{1}{\pi i}\int_L \frac{x(\tau)}{\tau - t}\, d\tau$.

If $A_0^2 - A_1^2 \neq 0$ for $t \in L$, then the operator $A(S) + T \in B(H^{\alpha}(L))$, $\alpha < \mu/2$, and the operator $R_{A(S)} = (A_0^2 - A_1^2)^{-1}(A_0 - A_1 S)$ is a simple regularizer of the operator $A(S) + T$ to the ideal $T(H^{\alpha}(L))$ of compact operators.

Hence by Theorem 4.3 of Chapter II we have

THEOREM 2.5. *If the conditions of Theorem 2.4 are satisfied, then the operator $A(S) + T$ has a finite d-characteristic and $\varkappa_{A(S)+T} = \varkappa_{A(S)}$.*

We usually investigate, not the codimension of the image of this operator in the space $H^\alpha(L)$, but the nullity of the conjugate operator defined on the space Ξ_0 of all functionals of the form

$$\xi(x) = \int_L \xi(t)x(t)\,dt, \quad \text{where } \xi(t) \in H^\alpha(L)$$

(see Section 3 of Chapter II).

This space will be identified with the space $H^\alpha(L)$, and the space of functionals of the form

$$\xi(x) = \int_L \xi(t)x(t), \quad \text{where } \xi(t) \text{ is a continuous function,}$$

will be identified with the space $C(L)$. Note that, with this convention, we have $A(S) \in L_0(H^\alpha(L), \Xi_0) = L_0(H^\alpha(L), H^\alpha(L))$. On the other hand, if

$$R_{A(S)}A(S) = I+T_1, \quad A(S)R_{A(S)} = I+T_2,$$

then the integral operators T_1, T_2 are weakly singular (i.e. given by weakly singular kernels) and can be extended to operators $\tilde{T}_1$ and $\tilde{T}_2$ defined on the whole space $C(L)$. The operators $I+\tilde{T}_1, I+\tilde{T}_2$ belong to $L_0(C(L) \to C(L))$. Hence Theorem 3.3 of Chapter *I* implies that the operator $A(S)$, and hence also the operator $A(S)+T$, are $\Phi_{H^\alpha(L)}$-operators. This means that the dimension of the null space of the conjugate operator is equal to the codimension of the image of the given operator.

Observe, that the operator conjugate to $A(S)$ (in the sense of the space $H^\alpha(L)$) is of the form: $A'(S) = A_0 - SA_1$. We thus have

THEOREM 2.6. *Let the conditions of Theorem* 2.4 *be satisfied and let the following equations be given*:

(2.2) $$A_0(t)x(t) + \frac{A_1(t)}{\pi i}\int_L \frac{x(\tau)}{\tau - t}\,d\tau = f(t),$$

(2.3) $$A_0(t)y(t) - \frac{1}{\pi i}\int_L \frac{A_1(\tau)y(\tau)}{\tau - t}\,d\tau = g(t),$$

where $f, g \in H^\alpha(L)$, $\alpha < \mu/2$. *Then*

(i) *both the homogeneous equations (i.e.* $f \equiv 0$ *and* $g \equiv 0$*) have a finite number (not necessarily the same for both) of linearly independent solutions satisfying Hölder conditions and all these solutions belong to the space* $H^\alpha(L)$;

(ii) *a necessary and sufficient condition for equation* (2.2) (*resp.* (2.3)) *to have a solution is that*

$$\int_L f(t)y(t)\,dt = 0 \qquad \Big(\int_L g(t)x(t)\,dt = 0 \text{ resp.}\Big)$$

for every solution $y(t)$ *of the homogeneous equation* (2.3) (*resp. for every solution* $x(t)$ *of the homogeneous equation* (2.2)).

Note, that from (i) we conclude that (ii) in fact gives only a finite number of conditions for the function $f(t)$ (resp. $g(t)$).

THEOREM 2.5 immediately implies

THEOREM 2.7. *Theorem* 2.6 *is true for the equations*

$$A_0(t)x(t)+\frac{1}{\pi i}\int_L \frac{K(t,\tau)x(\tau)}{\tau-t}\,d\tau = f(t), \tag{2.4}$$

$$A_0(t)y(t)-\frac{1}{\pi i}\int_L \frac{K(\tau,t)y(\tau)}{\tau-t}\,d\tau = g(t), \tag{2.5}$$

in place of equations (2.2) *and* (2.3).

Theorem 2.5 implies also that equations (2.2) and (2.4) both have the same index. To calculate the index we can make use of the following

THEOREM 2.8. *Under the assumptions of Theorem* 2.4, *the index of equation* (2.2), *and hence also of* (2.4), *is*

$$\varkappa_{A(S)} = \frac{1}{2\pi}\int_L d_t\left(\frac{A_0(t)-A_1(t)}{A_0(t)+A_1(t)}\right),$$

i.e. it is equal to the increment in the argument of the function $(A_0-A_1)/(A_0+A_1)$, *as* t *moves along* L *in the positive direction.*

The proof can be found in the book [6] by the present author and Rolewicz, Part D, Chapter II, Section 5.

Let L denote, as before, an oriented system. We now consider the system of singular integral equations

$$A_0(t)x(t)+\frac{1}{\pi i}\int_L \frac{K(t,\tau)}{\tau-t}x(t)s\,d\tau = f(t), \tag{2.6}$$

where $A_0(t) = (A_0^{i,j}(t))$ and $K(t,\tau) = (K^{i,j}(t,\tau))$ are square $n\times n$ matrix-functions. All functions $A_0^{i,j}(t)$, $K^{(i,j)}(t,\tau)$ satisfy the Hölder condition with an exponent μ, and $f(t)$ is an n-dimensional vector valued functions whose components belonging to a space $H^\alpha(L)$, $\alpha < \mu/2$.

We denote by $H_n^\mu(L)$ the linear space of all n-dimensional vector valued functions with components belonging to $H^\mu(L)$. Arguing as in the one-dimensional case, we write equation (2.6) as

$$A_0(t)x(t)+\frac{A_1(t)}{\pi i}\int_L \frac{x(\tau)}{\tau-t}\,d\tau+\frac{1}{\pi i}\int_L \frac{K(t,\tau)-K(t,t)}{\tau-t}\,x(\tau)\,d\tau = f(t),$$

where $A_1(t) = K(t,t)$. Note that the last integral is weakly singular. We consider the matrix

$$S_{(n)} = \begin{pmatrix} S & 0 & \dots & 0 \\ 0 & S & \dots & 0 \\ \cdot & \cdot & \cdot\,\cdot & \cdot \\ 0 & 0 & \dots & S \end{pmatrix},$$

where $(Sx)(t) = \dfrac{1}{\pi i}\displaystyle\int_L \frac{x(\tau)}{\tau-t}\,d\tau$ for $x \in H^\alpha(L)$ and $t \in L$. It is easily verified that

$$S_{(n)}^2 = I_{(n)}, \qquad \text{where } I_{(n)} = (\delta_{ij} I),$$

and the unit matrix $I_{(n)}$ is the identity operator in the space $H_n^\alpha(L)$. Moreover, the commutators $A_j S_{(n)} - S_{(n)} A_j$ ($j = 0$ or 1) are compact operators in the space $H_n^\alpha(L)$. Hence applying once more the theorems of Section 4 in Chapter II we obtain,

THEOREM 2.9. *If the matrix* $A_0^2 - A_1^2$ *is invertible for* $t \in L$, *then the operator* $A(S_{(n)})x = A_0 x + A_1 S_{(n)} x + Tx \equiv A_0(t)x(t) + \dfrac{1}{\pi i}\displaystyle\int_L \frac{K(t,\tau)x(\tau)}{\tau-t}\,d\tau$ *has a simple regularizer*

$$R_{A(S_{(n)})+T} = R_{A(S_{(n)})} = (A_0^2 - A_1^2)^{-1}(A_0 - A_1 S_{(n)})$$

to the ideal of compact operators $T(H_n^\alpha(L))$, $\alpha < \mu/2$.

COROLLARY 2.2. *The operator* $A(S_{(n)})+T$ *has a finite d-characteristic and* $\varkappa_{A(S_{(n)})+T} = \varkappa_{A(S_{(n)})}$.

THEOREM 2.10. *The operator* $A(S_{(n)})+T$ *is a* $\Phi_{H_\alpha^n(L)}$*-operator for* $\alpha < \mu/2$.

We finally obtain

THEOREM 2.11. *Suppose that L is an oriented system and that we are given equations* (2.6) *and*

$$A_0^*(t)y(t)+\frac{1}{\pi i}\int_L \frac{K^*(t,\tau)y(\tau)}{\tau-t}\,d\tau = g(t), \tag{2.7}$$

where the elements of the square $n \times n$ *matrix-functions* $A_0(t)$ *and* $K(t, \tau)$ *satisfy the Hölder condition with the exponent* μ, $A_1(t) = K(t, t)$, *the matrix* $A_0^2 - A_1^2$ *is invertible for every* $t \in L$, $f(t), g(t) \in H_n^\alpha(L)$, $\alpha < \mu/2$ *and* $A_0^*(t), K^*(t, \tau)$ *denote the matrices conjugate to* $A_0(t)$ *and* $K(t, \tau)$, *respectively. Then*

(i) *the homogeneous equations (i.e.* $f \equiv 0$, $g \equiv 0$) *both have a finite (but not necessarily the same for both) number of linearly independent solutions satisfying the Hölder condition and all these solutions belong to* $H_n^\alpha(L)$;

(ii) *equation* (2.6) (*resp.* (2.7)) *has a solution if and only if*

$$\int_L f(t)y(t)\,dt = 0 \quad (\textit{resp. } \int_L g(t)x(t)\,dt = 0) \tag{2.8}$$

for every solution $y(t)$ *of the homogeneous equation* (2.7) (*resp. for every solution* $x(t)$ *of the homogeneous equation* (2.6)).

THEOREM 2.12. *If the conditions of Theorem* 2.11 *are satisfied, then*

$$\varkappa_{A(S_{(n)})} = \frac{1}{2\pi}\int_L d_t \arg D(t), \quad \textit{where } D(t) = \det[A_0(t) - A_1(t)]/\det[A_0(t) + A_1(t)].$$

The proof can be found in the book [6] by the present author and Rolewicz, Part D, Chapter II, Section 6.

All the preceding theorems remain true if, instead of the operator S in the space $H^\mu(L)$, we consider the Hilbert transform S_∞ defined by Formula (1.10) in the space $H_\infty^\mu(\tilde{L})$, where $\tilde{L}$ is an oriented system containing a regular arc closed at infinity. The proofs are just the same; this follows from the discussion of Section 1.

Consider now the equation

$$A_0(t)x(t) + \frac{1}{\pi i}\int_{-\infty}^{+\infty} \frac{K(t,\tau)}{\tau - t} x(\tau)\,d\tau = y(t) \tag{2.9}$$

in the space $L^2(\boldsymbol{R})$. Writing this equation in the form

$$A(S_\infty) + Tx = y, \quad \text{where } A(S_\infty) = A_0 + A_1 S,\ A_1(t) = K(t, t), \tag{2.10}$$

$$(Tx)(t) = \frac{1}{\pi i}\int_{-\infty}^{+\infty} \frac{K(t, \tau) - K(t, t)}{\tau - t} x(\tau)\,d\tau,$$

we obtain

THEOREM 2.13. *If the functions* $A_0(t), K(t, \tau)$ *are continuous and bounded for* $t, \tau \in \boldsymbol{R}$ *and satisfy the conditions*

$$(1) \qquad \int_{-\infty}^{+\infty}\left[\int_{-\infty}^{+\infty}\frac{A_0(\tau)-A_0(t)}{\tau-t}d\tau\right]dt < +\infty,$$

$$\int_{-\infty}^{+\infty}\left[\int_{-\infty}^{+\infty}\frac{K(t,\tau)-K(t,t)}{\tau-t}d\tau\right]dt < +\infty,$$

$$\int_{-\infty}^{+\infty}\left[\int_{-\infty}^{+\infty}\frac{A_1(\tau)-A_1(t)}{\tau-t}d\tau\right]dt < +\infty, \qquad \text{where } A_1(t) = K(t,t),$$

$$(2) \qquad \inf_{t\in R}|A_0^2-A_1^2| > 0 \quad \text{and} \quad (A_0^2-A_1^2)^{-1} \in L^2(\boldsymbol{R}),$$

then the operator $A(S_\infty)+T$ *has a finite d-characteristic, and* $\varkappa_{A(S)+T} = \varkappa_{A(S_\infty)}$. *Moreover, it is a* Φ_Ξ*-operator with respect to the family*

$$\Xi = \left\{\xi\colon \xi(x) = \int_{-\infty}^{+\infty}\xi(t)x(t)dt;\ \xi(t)\in L^2(\boldsymbol{R})\right\}.$$

The proof is similar to those of the preceding theorems, if we observe that conditions (1) imply the compactness of the operators $S_\infty A_0 - A_0 S_\infty$, $S_\infty A_1 - A_1 S_\infty$ and T and that the product of a function belonging to $L^2(\boldsymbol{R})$ with a continuous and bounded function is again in $L^2(\boldsymbol{R})$.

As a corollary we obtain the necessary and sufficient condition for the solvability of equation (2.9), if $y \in L^2(\boldsymbol{R})$.

Let $L_k^2(\boldsymbol{R})$ $(k = 1, 2, \ldots)$ denote the Hilbert space consisting of all k-dimensional vector valued functions whose components belong to the space $L_k^2(\boldsymbol{R})$, with the norm $\|x\| = \left(\sum_{j=1}^{k}\|x_j\|_2^2\right)^{1/2}$. For systems of singular integral equations with Hilbert transform in the space $L_k^2(\boldsymbol{R})$ we obtain results similar to Theorems 2.9, 2.10, 2.11, 2.12, provided that the elements of the square matrices $A_0(t)$ and $K(t, \tau)$ satisfy conditions (1) and (2) of Theorem 2.13.

We now indicate other involutionary singular integral equations for which similar results can be proved (see Samko [6]).

Let L be a regular closed arc. In the sequel we assume that

1) $M(t, \tau) \in C(L\times L)$, 2) $|M(t,\tau)-M(t,t)| < A|t-\tau|^\chi$ for some $\chi > 0$ and $t, \tau \in L$.

Let X denote either the space $L^p(L)$, $p > 1$ or the space $H^\mu(L)$, $0 < \mu < 1$. Let

$$K(t,\tau) = \frac{M(t,\tau)-M(t,t)}{\tau-t}, \qquad (Kx)(t) = \frac{1}{\pi i}\int_L K(t,\tau)x(\tau)\,d\tau,$$

(2.11)

$$(S_M x)(t) = \frac{1}{\pi i}\int_L \frac{M(t,\tau)}{\tau-t}x(\tau)\,d\tau.$$

LEMMA. *If the function $M(t,\tau)$ can be extended to the domain Ω bounded by L in such a manner that $M(t,\tau)$ is analytic in both variables in Ω and continuous in $\bar{\Omega} = \Omega \cup L$, and if $M(t,t) \equiv 1$, then the operator S_M is an involution on the space X: $S_M^2 = I$ on X.*

We obtain the proof of the lemma by means of a decomposition of X into the direct sum $X = X^+ \oplus X^-$ for $x \in X$ using the projectors $(I+S)/2$ and $(I-S)/2$, where $(Sx)(t) = \frac{1}{\pi i}\int_L \frac{x(t)}{\tau-t}d\tau$. It follows from the conditions imposed on the function $M(t,\tau)$ that $Kx \in X^+$ for $x \in X$ and $Kx^+ = 0$ for $x^+ \in X^+$, where $x^+ = (I+S)x/2$, and so on. Therefore $(I-S)K = 0$ and $K(I+S) = 0$. Hence

$$KS = -K \quad \text{and} \quad SK = K.$$

Hence, if we recall that $S^2 = I$, we obtain $K^2 = KS^2K = (KS)(SK) = -K^2$ which implies $K^2 = 0$. Since $M(t,t) \equiv 1$, we obtain $S_M = K+S$, and then $S_M^2 = S^2+SK+KS+K^2 = S^2 = I$.

This enables us to solve the equation

$$a(t)x(t)+\frac{1}{\pi i}\int_L \frac{M(t,\tau)x(\tau)}{\tau-t}x(\tau)\,d\tau = y(t), \tag{2.12}$$

i.e. the equation $(aI+S_M)x = y$, by the same method as has already been used in this section. In particular if $a(t) = a = \text{const}$, then from Theorem 5.1 of Chapter II we immediately find that:

1) if $a^2-1 \neq 0$, then equation (2.12) has a unique solution in the space X, namely $x(t) = (a^2-1)^{-1}\left[ay(t)-\frac{1}{\pi i}\int_L \frac{M(t,\tau)}{\tau-t}y(t)\,dt\right]$.

2) If $a+1 = 0$ but $a-1 \neq 0$, then $x(t) = \frac{1}{2}y(t)+z(t)+\frac{1}{\pi i}\int_L \frac{M(t,\tau)z(\tau)}{\tau-t}d\tau$ if and only if $y(t)-\frac{1}{\pi i}\int_L \frac{M(t,\tau)y(\tau)}{\tau-t}d\tau$, where z is an arbitrary function belonging to X.

3) Interchanging the roles of the signs "+" and "−" we obtain the corresponding result in the case where $a+1 \neq 0$ but $a-1 = 0$.

If $M(t,t) \neq 1$ on Ω, we put $M'(t,\tau) = M(t,\tau)/M(t,t)$ and $b(t) = M(t,t)$. Since $M'(t,t) \equiv 1$, we have $S_{M'}^2 = I$ and we can solve equation (2.12) by rewriting it as

$$ax + bS_{M'}x = y, \quad \text{where } S_{M'}^2 = I.$$

and then applying the preceding argument.

3. ALMOST INVOLUTIONARY CASES OF SINGULAR INTEGRAL EQUATIONS

Let L be an oriented system. We have studied singular operators of the form $\dfrac{1}{\pi i}\displaystyle\int_L \frac{K(t,\tau)}{\tau - t} x(\tau)\,d\tau$. However, in some cases, it happens that in place of the function $\tau - t$ appears the function $h(\tau - t)$, where $h(u)$ is a continuous function, $h(u) = u + o(u)$, and the function $h(u)/u$ satisfies the Hölder condition with an exponent μ. We write

$$S_h x = \frac{1}{\pi i}\int_L \frac{x(\tau)}{h(\tau - t)}\,d\tau.$$

THEOREM 3.1. *If the function $h(u)$ satisfies the above conditions, then the operator $H = S_h - S$ belongs to $B(C(L) \to H^\alpha(L))$, $\alpha < \mu/2$ and is compact in the space $H^\alpha(L)$.*

(For the proof see the present author and Rolewicz [6], part D, Chapter II, Section 6, also the present author [7].)

COROLLARY 3.1. *The operator S_h belongs to $B(H^\alpha(L))$ and it is an almost involution, i.e. it satisfies the identity $S_h^2 = I + T$, where $T \in B(H^\alpha(L))$ and is compact in $H^\alpha(L)$, $\alpha < \mu/2$.*

Indeed, since $S_h = H + S$, we have $S_h^2 = (H+S)^2 = H^2 + HS + SH + S^2$. But $S^2 = I$, and the operator H is compact. Hence the operator $T = H^2 + SH + HS$ is also compact.

COROLLARY 3.2. *If B is the operator of multiplication by a function $B(t)$ satisfying the Hölder condition with an exponent μ, then the commutator $BS_h - S_h B$ is compact in the space $H^\alpha(L)$, $\alpha < \mu/2$.*

Indeed, $BS_h - S_h B = B(S+H) - (S+H)B = BS - SB + BH - HB$ is compact by Theorem 2.1 and Theorem 3.1.

THEOREM 3.2. *If L is an oriented system and*

$$(T_h x)(t) = \frac{1}{\pi i}\int_L \frac{K(t,\tau)-K(t,t)}{\tau-t}\,x(\tau)\,d\tau,$$

where the function $K(t,\tau)$ satisfies the Hölder condition with an exponent μ for $t,\tau\in L$, then the operator T_h belongs to $B\big(C(L)-H^{\alpha}(L)\big)$, $\alpha<\mu/2$ and is compact in the space $H^{\alpha}(L)$.

For the proof see the present author and Rolewicz [6], Part D, Chapter II, Section 6, and the present author [7].

This Theorem and Corollary 3.3 of Chapter II immediately imply

THEOREM 3.3. *Let L, $h(u)$ and $K(t,\tau)$ satisfy the conditions of Theorem 3.2. If the function $A_0(t)$ satisfies the Hölder condition and if $A_0^2-A_1^2\neq 0$, where $A_1(t)=K(t,t)$, then the operator*

$$(A_h x)(t) = A_0(t)+\frac{1}{\pi i}\int_L \frac{K(t,\tau)}{h(\tau-t)}\,x(\tau)\,d\tau \tag{3.2}$$

has a simple regularizer of the form

$$R_{A_h} = (A_0^2-A_1^2)^{-1}(A_0-A_1 S_h)$$

to the ideal of compact operators in $H^{\alpha}(L)$, $\alpha<\mu/2$.

COROLLARY 3.3. *Suppose that the conditions of Theorem 3.3 are satisfied. Then the operator A_h defined by Formula (3.2) has a finite d-characteristic and $\varkappa_{A_h} = \varkappa_{A_0+A_1 S_h} = \varkappa_{A_0+A_1 S}$, where the last index is given by Formula (2.8).*

Indeed, $A_h = A_0+A_1 S_h+T_h = A_0+A_1(S+H)+T_h = A_0+A_1 S+(A_1 H+T_h)$. Moreover, the operator $A_1 H+T_h$ is compact by assumptions, and therefore does not affect the index.

If we study conjugate operators, we can obtain theorems fully analogous to Theorem 2.7, if we observe that the function $h^*(u)=h(-u)$ defining the kernel of the conjugate operator has the same properties as the function $h(u)$.

These arguments can be extended without any essential change to systems of singular integral equations with kernels defined by the function $h(u)$.

We now consider singular integral equations with a cotangent kernel:

$$A_0(s)x(s)+\frac{1}{2\pi}\int_0^{2\pi} K(s,\sigma)\cot\frac{\sigma-s}{2}\,x(\sigma)\,d\sigma = f(s). \tag{3.3}$$

We suppose that $A_0(s)$ and $K(s,\sigma)$ are real-valued 2π-periodic functions satisfying the Hölder condition with an exponent μ. We are looking for solutions of equation (3.3) in the class of real-valued 2π-periodic functions, which belong to the space $H^{\alpha}(0, 2\pi)$, $\alpha < \mu/2$. If the above conditions are satisfied, then we obtain (for the proofs see the present author and Rolewicz [6], and also the present author [7]).

THEOREM 3.4. *The operator*

$$(T_c x)(t) = \frac{1}{2\pi}\int_0^{2\pi} [K(s, \sigma) - K(s, s)] \cot\frac{\sigma - s}{2} x(\sigma)\, d\sigma,$$

belongs to $B(H^{\alpha}(L))$, $\alpha < \mu/2$ *and is compact in* $H^{\alpha}(L)$.

THEOREM 3.5. *If* $A_0^2 + K^2(s, s) \neq 0$, *then the operator*

$$(A_c x)\,(t) = A_0(s)x(s) + \frac{1}{2\pi}\int_0^{2\pi} K(s, \sigma) \cot\frac{\sigma - s}{2}\, d\sigma,$$

belongs to $B(H^{\alpha}(L))$, $\alpha < \mu/2$, *has finite d-characteristic and is a* Φ_{Ξ}-*operator with respect to the family* $\Xi = \left\{\xi\colon \xi(x) = \int_0^{2\pi} \xi(t)x(t)\,dt,\ x \in H_0^{\alpha}[0, 2\pi]\right\}$.

This gives the respective conditions for the solvability of the equation $A_0 x = y$, where $y \in H_0^{\alpha}[0, 2\pi]$, $\alpha < \mu/2$.

As before, this theorem can be generalized to systems of singular integral equations with cotangent kernel.

As in the case of singular integral equations with the kernel $1/(\tau - t)$ we can extend all the theorems concerning a cotangent kernel to equations with the kernel $h(\sigma - s)$, where $h(u)$ is of the form $h(u) = u + o(u)$ and the function $h(u)\cot\frac{u}{2}$ satisfies the Hölder condition with an exponent μ (see the present author [7]).

4. SINGULAR INTEGRAL EQUATIONS IN THE CASE $A_0^2 - A_1^2 = 0$

So far we have considered singular integral equations in the case where $A_0^2 - A_1^2 \neq 0$ throughout the range of integration. We now give some results for the case, where $A_0^2 - A_1^2 = 0$ at one point at least. To begin with we examine an arbitrary equation with involution. The method already given does not work in this case.

Let S be an involution on a linear space X and let $A(S) = A_0 + A_1 S$, where A_0, A_1 are linear operators transforming X into itself. We can write (see Section 4 of Chapter II):

$$A(S) = A_+ P^+ + A_- P^-, \quad \text{where } A_+ = A(1) = A_0 + A_1, A_- = A(-1) = A_0 - A_1,$$
$$P^+ = \tfrac{1}{2}(I+S), \quad P^- = \tfrac{1}{2}(I-S).$$

THEOREM 4.1. *Let S be an involution on the linear space X and let $A_0, A_1 \in L_0(X)$. Let $A(S) = A_0 + A_1 S$. Then*

(1) *If $A_+ = 0$ and A_- is invertible, then the equation*

$$A(S)x = y, \quad y \in X, \tag{4.1}$$

has a solution $x = z_+ + 2A_0^{-1}y$, where $z^+ \in X^+ = P^+X$ is arbitrary, under the necessary and sufficient condition that $P^+ A_-^{-1} y = 0$, i.e. that $y \in A_0 X^-$.

(2) *If $A_- = 0$ and A_+ is invertible, then equation* (4.1) *has a solution $x = 2A_0^{-1}y + z^-$, where $z^- \in X^- = P^-X$ is arbitrary, under the necessary and sufficient condition that $P^{-1}A_0^{-1}y = 0$, i.e. that $y \in A_0 X^+$.*

PROOF. Suppose that $A_+ = 0$ and A_- is invertible. Then equation (4.1) can be written in the form $A_- P^- x = y$, or in the equivalent (by assumption) form $P^- x = 2A_0 y$. Hence the first part of our theorem is an immediate consequence of Theorem 5.2 of Chapter II. Moreover, the condition $P^+ A_0^{-1} y = 0$ is equivalent to the condition that $y \in A_0 X^-$. Interchanging the roles of A_+ and A_-, we obtain the second part of our theorem

COROLLARY 4.1. *If the conditions of Theorem* 4.1 *are satisfied, then*

$$\alpha_{A(S)} = \begin{cases} \dim X^+ & \text{if } A_+ = 0, \\ \dim X^- & \text{if } A_- = 0, \end{cases} \qquad \beta_{A(S)} = \alpha_{A(S)}.$$

Hence the operator $A(S)$ has finite d-characteristic if and only if $\dim X < +\infty$ *if $A_+ = 0$ and* $\dim X^- < +\infty$ *if $A_- = 0$. Under this condition $\varkappa_{A(S)} = 0$.*

Indeed, if $A_+ = 0$, then $\beta_{A(S)} = \operatorname{codim} A(S)X = \operatorname{codim} A_- P^- X = \operatorname{codim} A_- X^- = \dim X / A_- X^- = \dim X^+$, since $\dim A_- X^- = \dim X^-$, $\dim X = \dim X^+ + \dim X^-$. Similarly we find $\beta_{A(S)} = \dim X^-$, when $A_- = 0$.

Since all the spaces of functions under consideration have the property that $\dim X^+ = \dim X^- = +\infty$, no singular integral equations have a finite d-characteristic, or even a semi-finite d-characteristic if either $A_0 - A_1 = 0$ or $A_0 + A_1 = 0$. Nevertheless, we can solve these equations by using the basic Theorem 4.1. We immediately obtain

THEOREM 4.2. *Let L be an oriented system and let the functions $A_0(t)$ and $A_1(t) \in H^\mu(L)$. Let*

$$[A(S)x](t) = A_0(t)x(t) + \frac{A_1(t)}{\pi i} \int_L \frac{x(\tau)}{\tau - t} d\tau = y(t), \tag{4.2}$$

where $y \in H^\mu(L)$. Then

(1) *If $A_+ = A_0 + A_1 \equiv 0$ on L and $A_- = A_0 - A_1 = 2A_0$ is invertible in $H^\mu(L)$ (i.e. $A_0 \neq 0$ on L), then equation (5.2) has a solution*

$$x(t) = \tfrac{1}{2}\left[z(t) + \frac{1}{\pi i} \int_L \frac{z(\tau)}{\tau - t} d\tau\right] + 2[A_0(t)]^{-1} y(t)$$

where $z \in H^\mu(L)$ is arbitrary, under the necessary and sufficient condition that $y \in A_0 X^-$, i.e.

$$[A_0(t)]^{-1} y(t) + \frac{1}{\pi i} \int_L \frac{[A_0(\tau)]^{-1} y(\tau)}{\tau - t} d\tau = 0.$$

(2) *If $A_- \equiv 0$ and A_+ is invertible, then equation (5.2) has a solution*

$$x(t) = \tfrac{1}{2}\left[z(t) - \frac{1}{\pi i} \int_L \frac{z(\tau)}{\tau - t} d\tau\right] + 2[A_0(t)]^{-1} y(t),$$

$z \in H^\mu(L)$ being arbitrary, under the necessary and sufficient condition that

$$[A_0(t)]^{-1} y(t) - \frac{1}{\pi i} \int_L \frac{[A_0(\tau)]^{-1} y(\tau)}{\tau - t} d\tau = 0.$$

This theorem is also true when y is a vector valued function belonging to $H_n^\mu(L)$ and $A_0(t)$, $A_1(t)$ are square $n \times n$ matrices whose elements belong to $H^\mu(L)$.

Theorem 4.2 remains valid if, instead of the space $H^\mu(L)$, we consider the space $H_\infty^\mu(\tilde{L})$, where $\tilde{L}$ is an oriented system containing an arc closed at infinity (see Sections 1 and 2).

The next possibility is, where $X = L^2(\boldsymbol{R})$ and the singular operator in equation (4.2) is the Hilbert transform S_∞, provided that the coefficients A_0 and A_1 are continuous and bounded on $\boldsymbol{R}$. For vector valued functions we can study equation (4.2) with the Hilbert transform S_∞ in the space $L_k^2(\boldsymbol{R})$, where k is the dimension of the vectors (see Section 2).

A very different situation occurs when one of the functions A_+, A_- is equal to zero, but not identically, and the second one is invertible. The following should be noted

PROPERTY 4.1. *Let M be the operator of multiplication by a function $M(t) \in L^p(\Omega, \Sigma, \mu)$, $p > 1$* (see Section 4 of Chapter I, Example 4.4). *Let*

$$Z_M = \{t \in \Omega: M(t) = 0\}. \tag{4.3}$$

Then

(1) *if $\mu(Z_M) = 0$, then the operator M is invertible,*

(2) *if $\mu(Z_M) > 0$, then $\alpha_M = +\infty$, $\beta_M = +\infty$.*

PROPERTY 4.2. *Let M be the operator of multiplication by a function $M(t) \in H^\mu(L)$, where L is an oriented system, $0 < \mu \leqslant 1$. If the set Z_M defined in the preceding property is not empty but contains only a finite number of points, then the operator M has the semifinite d-characteristic $(0, +\infty)$.*

The same is true for the space $H^\mu_\infty(L)$.

Theorems 2.2 and 3.1 of Chapter I imply the following

PROPOSITION 4.1. *If S is an involution on a linear space X and the operators A_0, A_1 transforming X into itself have the following properties:*

(1) *$A_0 S - SA_0$, $A_1 S - SA_1$ belong to a Fredholm ideal $J \subset L_0(X)$,*

(2) *one of the operators $A_+ = A_0 + A_1$, $A_- = A_0 - A_1$ is invertible and the second one has the semi-finite d-characteristic $(0, +\infty)$; then the operator $A(S) + T = A_0 + A_1 S + T$ has the semi-finite d-characteristic $(0, +\infty)$ for all $T \in J$.*

PROOF. We have (see Section 3 of Chapter II)

$$(A_0 - A_1 S)(A_0 + A_1 S + T) = A_0^2 - A_1^2 + T_1,$$

where

$$T_1 = A_1 S(SA_1 - A_1 S) - A_1(SA_0 - A_0 S) + (A_0 - A_1 S)T \in J.$$

Now assume that $A_0 + A_1$ is invertible. Then $\alpha_{A_0+A_1} = \beta_{A_0+A_1} = 0$, $\alpha_{A_0^2-A_1^2} = \alpha_{A_0-A_1}$ and the operator $(A_0+A_1)^{-1}T \in J$. Since all the operators belonging to J are perturbations of operators with a semi-finite d-characteristic, which preserve the index, we obtain

$$\alpha_{A_0+A_1S+T} \leqslant \alpha_{(A_0-A_1S)(A_0+A_1S+T_1)} = \alpha_{A_0^2-A_1^2+T_1} = \alpha_{A_0^2-A_1^2} = \alpha_{A_0-A_1} = 0$$

and $\alpha_{A_0+A_1S+T} = 0$. Since the operator $A_0 - A_1 S$ is a left regularizer of an operator with the d-characteristic $(0, +\infty)$ and the superposition also has zero nullity, we have $\alpha_{A_0-A_1S} = +\infty$, $\beta_{A_0-A_1S} = 0$ and $\varkappa_{A_0-A_1S} = -\infty$.

Hence

$$\beta_{A_0+A_1S+T} = \varkappa_{A_0+A_1S+T_1} = \varkappa_{A_0^2-A_1^2+T_1} - \varkappa_{A_0-A_1S} = \beta_{A_0^2-A_1^2+T} - \alpha_{A_0^2-A_1^2+T_1} - \varkappa_{A_0-A_1S}$$
$$= \beta_{A_0^2-A_1^2} - \varkappa_{A_0-A_1S} = +\infty.$$

The proof is similar when $A_0 - A_1$ is invertible.

PROPOSITION 4.2. *If the conditions of either Property* 4.1 *or of Property* 4.2 *are satisfied, then every compact operator T acting in the appropriate space is a perturbation of the operator M preserving the d-characteristic*: $\alpha_{M+T} = \alpha_M = \beta_{M+T} = \beta_M$.

We thus have

THEOREM 4.3. *Let L be an oriented system and let the functions* $A_0(t)$ *and* $K(t, \tau)$ *satisfy the Hölder condition with an exponent* μ *for* $t, \tau \in L$. *If one of the functions* $A_0 + A_1$, $A_0 - A_1$, *where* $A_1(t) = K(t, t)$, *vanishes on a finite set* $L_0 \subset L$ *and the second one is invertible, then the operator*

$$[A(S)+T]x(t) = A_0(t)x(t) + \frac{1}{\pi i}\int_L \frac{K(t, \tau)x(\tau)}{\tau - t}\,d\tau \tag{4.4}$$

has the semi-finite d-characteristic $(0\ {+\infty})$ *for* $\alpha < \mu/2$.

The proof follows immediately from Propositions 4.1, 4.2 and from Theorem 2.2, since the ideal of compact operators in $H^\alpha(L)$ is a Fredholm ideal (see also Prössdorf [1], [2]).

From Proposition 4.1 it also follows that every solution of the equation $[A(S)+T]x = y$, $y \in H^\mu(L)$, is a solution of an equation of Fredholm type:

$$(I+T_1)x = A_0 - A_1 Sy, \tag{4.5}$$

where T_1 is defined by Formula (4.3).

The last theorem is also true if, instead of the space $H^\mu(L)$, we consider the space $H^\mu_\infty(L)$ and $L^2(\mathbf{R})$ with the Hilbert transform under the conditions asserting the commutativity of commutators, as was done in Section 2.

We now present without proof a theorem of a similar kind, obtained by Laĭterer on the basis of Gohberg results [1] and [2].

Let $L^2_k(K)$ denote the space of all k-dimensional vector valued functions with components belonging to $L^2(K)$, where K is the unit circle. We consider the equation $[A(S)+T]x = y$, where the operator $A(S)+T$ is defined by Formula (4.4), where we put $L = K$ and $y \in L^2_m(K)$.

THEOREM 4.4 (Laĭterer [1]). *If the* $n \times m$ *matrix-functions* $A_0(t)+A_1(t)$, $A_0(t) - A_1(t)$, *where* $A_1(t) = K(t, t)$, *are continuous on the unit circle K, then a necessary and sufficient condition for the operator* $A(S) = A_0 - A_1 S \in B(L^2_n(K) \to L^2_m(K))$ *to be either a* Φ_+*-operator or a* Φ*-operator* (see Section 4 of Chapter II) *is that*

$$\operatorname{rank}[A_0(t)+A_1(t)] = \operatorname{rank}[A_0(t)-A_1(t)] = \min(m, n) \quad \textit{for all } t \in K.$$

Under this condition, if $m < n$, *then* $A(S)$ *is a* Φ_+*-operator which is not a* Φ*-operator, if* $m > n$, *it is a* Φ_+*-operator which is not a* Φ*-operator, and if* $m = n$, $A(S)$ *is a* Φ*-operator.*

CHAPTER XIII

Singular Integral Equations in one Variable with Transformed Argument

In this chapter we study singular integral equations in one variable when the argument of the unknown function is transformed by reflection, by rotation and by a Carleman function.

1. SINGULAR INTEGRAL EQUATIONS WITH REFLECTION

Consider singular integral equations with the Hilbert transform (see Section 1 of Chapter XI) and with reflection of the argument:

$$(1.1)\qquad a_0 x(t)+b_0 x(-t)+\frac{a_1}{\pi i}\int_{-\infty}^{+\infty}\frac{x(s)}{s-t}ds+\frac{b_1}{\pi i}\int_{-\infty}^{+\infty}\frac{x(s)}{s+t}ds = y(t),$$

where a_0, b_0, a_1, b_1 are constants complex and $y \in L^2(\boldsymbol{R})$.

If we denote the Hilbert transform, as before, by

$$(S_\infty x)(t) = \frac{1}{\pi i}\int_{-\infty}^{+\infty}\frac{x(s)}{s-t}\,ds,$$

and the reflection by S: $(Sx)(t) = x(-t)$, we can write equation (1.1) as

$$(1.2)\qquad Ax \equiv [a_0 I+b_0 S+(a_1 I+b_1 S)S_\infty] = y,$$

where $S^2 = I$, $S_\infty^2 = I$ on $L^2(\boldsymbol{R})$ (Section 1 of Chapter XI). Moreover, we observe that S_∞ anticommutes with S:

$$(1.3)\qquad SS_\infty+S_\infty S = 0.$$

Indeed, for all $x \in L^2(\boldsymbol{R})$,

$$(SS_\infty x)(t) = \frac{1}{\pi i}\int_{-\infty}^{+\infty} \frac{x(s)}{s+t}\,ds$$

and

$$(S_\infty Sx)(t) = \frac{1}{\pi i}\int_{-\infty}^{+\infty} \frac{x(-s)\,ds}{s-t} = -\frac{1}{\pi i}\int_{-\infty}^{+\infty} \frac{x(u)}{-u-t}\,du = -\frac{1}{\pi i}\int_{-\infty}^{+\infty} \frac{x(u)}{u+t}$$
$$= -(SS_\infty x)(t).$$

We therefore can apply to equation (1.1) all the arguments of Section 1 in Chapter III concerning equations with two anticommuting involutions. We study a few particular cases.

I. Let $a_0^2-b_0^2 \neq 0 \neq a_1^2-b_1^2$ and let $\lambda = (a_0^2-b_0^2)/(a_1^2-b_1^2) = 1$.

(a) Let $b_1 = b_0 \neq 0$. If $a_1 = a_0$, we conclude that a necessary and sufficient condition for equation (1.1) to have a solution is $Ay = 0$ and the solution is then of the form

$$x = \tfrac{1}{2}(a_0^2-b_0^2)^{-1}(a_0 I - b_0 S)y + (I - S_\infty)z, \qquad \text{where } z \text{ is arbitrary,}$$

i.e. explicitly

$$x(t) = \tfrac{1}{2}(a_0^2-b_0^2)^{-1}[a_0 y(t) - b_0 y(-t)] + z(t) - \frac{1}{\pi i}\int_{-\infty}^{+\infty} \frac{z(s)}{s-t}\,ds,$$

where $z \in L^2(\boldsymbol{R})$ is arbitrary.

If $a_1 = -a_0$, then a necessary and sufficient condition for equation (1.1) to have a solution is that $(I+S_\infty)y = 0$, i.e.

$$y(t) + \frac{1}{\pi i}\int_{-\infty}^{+\infty} \frac{y(s)}{s-t}\,ds = 0,$$

and the solution is of the form

$$x(t) = (a_0^2-b_0^2)^{-1}\left\{a_0\left[\tfrac{1}{2}y(t) + z(t) + \frac{1}{\pi i}\int_{-\infty}^{+\infty} \frac{z(s)}{s-t}\,ds\right] - \right.$$
$$\left. -b_0\left[\tfrac{1}{2}y(-t) + z(-t) + \frac{1}{\pi i}\int_{-\infty}^{+\infty} \frac{z(s)}{s+t}\,ds\right]\right\},$$

where $z \in L^2(\boldsymbol{R})$ is arbitrary.

(b) $b_1 = -b_0$. If $a_1 = a_0$, then a necessary and sufficient condition for equation (1.1) to have a solution is

$$y(t) - \frac{1}{\pi i} \int_{-\infty}^{+\infty} \frac{y(s)}{s-t} ds = 0$$

and the solution is then

$$x(t) = (a_0^2 - b_0^2)^{-1} \left\{ a_0 \left[\tfrac{1}{2} y(t) + z(t) - \frac{1}{\pi i} \int_{-\infty}^{+\infty} \frac{z(s)}{s-t} ds \right] - \right.$$

$$\left. - b_0 \left[\tfrac{1}{2} y(-t) + z(-t) - \frac{1}{\pi i} \int_{-\infty}^{+\infty} \frac{z(s)}{s+t} ds \right] \right\}.$$

If $a_1 = -a_0$, then a necessary and sufficient condition for equation (1.1) to have a solution is $Ay = 0$ and the solution is then of the form

$$x(t) = \tfrac{1}{2}(a_0^2 - b_0^2)^{-1} [a_0 y(t) - b_0 y(-t)] + z(t) + \frac{1}{\pi i} \int_{-\infty}^{+\infty} \frac{z(s)}{s+t} ds.$$

(c) Let $b_1 = b_0 = 0$ and $a_1 = a_0 \neq 0$. A necessary and sufficient condition for equation (1.1) to have a solution is

$$y(t) - \frac{1}{\pi i} \int_{-\infty}^{+\infty} \frac{y(s)}{s-t} ds = 0$$

and the solution is of the form

$$x(t) = \frac{1}{2a_0} y(t) + z(t) - \frac{1}{\pi i} \int_{-\infty}^{+\infty} \frac{z(s)}{s-t} ds,$$

where $z \in L^2(\boldsymbol{R})$ is arbitrary.

(d) Let $b_1 = b_0$ and $a_1 = -a_0 \neq 0$. A necessary and sufficient condition for equation (1.1) to have a solution is

$$y(t) + \frac{1}{\pi i} \int_{-\infty}^{+\infty} \frac{y(s)}{s-t} ds = 0$$

and the solution is of the form

$$x(t) = \frac{1}{2a_0} y(t) + z(t) + \frac{1}{\pi i} \int_{-\infty}^{+\infty} \frac{z(s)}{s-t} ds.$$

II. Let $a_0^2-b_0^2 \neq 0 \neq a_1^2-b_1^2$ and let $\lambda = (a_0^2-b_0^2)/(a_1^2-b_1^2) \neq 1$. Then equation (1.1) has the unique solution

$$x = \frac{1}{1-\lambda} R_A y, \quad \text{where } R_A = -(a_1^2-b_1^2)^{-1}[a_0 I - b_0 S - (a_1 I + b_1 S) S_\infty].$$

Hence

$$x = \frac{-1}{a_1^2-b_1^2-a_0^2+b_0^2}[a_0 I - b_0 S - (a_1+b_1 S) S_\infty] y$$

and

$$x(t) = \frac{1}{a_0^2-b_0^2-a_1^2+b_1^2}\left[a_0 y(t) - b_0 y(-t) - \frac{a_1}{\pi i}\int_{-\infty}^{+\infty}\frac{y(s)}{s-t}ds - \frac{b_1}{\pi i}\int_{-\infty}^{+\infty}\frac{y(s)}{s+t}ds\right]$$

$$= \frac{1}{a_0^2-b_0^2-a_1^2+b_1^2}\left\{a_0 y(t) - b_0 y(-t) - \frac{1}{\pi i}\int_{-\infty}^{+\infty}\left[\frac{a_1}{s-t}+\frac{b_1}{s+t}\right]y(s)\,ds\right\}$$

$$= \frac{1}{a_0^2-b_0^2-a_1^2+b_1^2}\left[a_0 y(t) - b_0 y(-t) - \frac{1}{\pi i}\int_{-\infty}^{+\infty}\frac{(a_1+b_1)s+(a_1-b_1)t}{s^2-t^2}y(s)\,ds\right].$$

III. Let $a_1+b_1 = a_1-b_1 = 0$. The solution is given by Theorem 1.2 in Chapter VIII.

IV. If $a_0+b_0 = a_0-b_0 = 0$, but $a_1^2-b_1^2 \neq 0$, then equation (1.2) has the unique solution

$$x(t) = (a_1^2-b_1^2)^{-1}[(a_1 I + b_1 S) S_\infty y](t)$$

$$= (a_1^2-b_1^2)^{-1}\frac{1}{\pi i}\int_{-\infty}^{+\infty}\left[\frac{a_1}{s-t}+\frac{b_1}{s+t}\right]y(s)\,ds$$

$$= (a_1^2-b_1^2)^{-1}\frac{1}{\pi i}\int_{-\infty}^{+\infty}\frac{(a_1+b_1)s+(a_1-b_1)t}{s^2-t^2}y(s)\,ds.$$

V. Let $a_1+b_1 = 0$ but $a_1-b_1 \neq 0$.

(a) Let $a_0^2-b_0^2 \neq 0$. Then the unique solution is

$$x(t) = (a_0^2-b_0^2)^{-1}\left[(a_0 I - b_0 S) - \tfrac{1}{2}(a_1-b_1)(I-S) S_\infty y\right](t)$$

$$= (a_0^2-b_0^2)^{-1}\left\{a_0 y(t) - b_0 y(-t) - \frac{1}{2\pi i}\left[\int_{-\infty}^{+\infty}\frac{y(s)}{s-t}ds - \int_{-\infty}^{+\infty}\frac{y(s)}{s+t}ds\right]\right\}$$

$$= (a_0^2-b_0^2)^{-1}\left[a_0 y(t) - b_0 y(-t) - \frac{a_1-b_1}{\pi i}\int_{-\infty}^{+\infty}\frac{y(s)}{s^2-t^2}ds\right].$$

(b) Let $a_0+b_0 = 0$, but $a_0-b_0 \neq 0$. A necessary and sufficient condition for equation (1.1) to have a solution is $[(I+S)y](t) = y(t)+y(-t) = 0$, which means that $y(t)$ must be an odd function. If this condition is satisfied, then

$x = (I+S)z+(a_0-b_0)^{-1}y-(a_1-b_1)S_\infty(I+S)z,$ where $z \in L^2(\boldsymbol{R})$ is arbitrary.

If we observe that for all $z \in L^2(\boldsymbol{R})$ the function $(I+S)z$ is an even function, this can be written as

$$x(t) = x_1(t)-\frac{a_1-b_1}{\pi i}\int_{-\infty}^{+\infty}\frac{x_1(s)}{s-t}ds+(a_0-b_0)^{-1}y(t),$$

where $x_1(t) \in L^2(\boldsymbol{R})$ is an arbitrary even function.

(c) Let $a_0-b_0 = 0$ but $a_0+b_0 \neq 0$. Then a necessary and sufficient condition for equation (1.1) to have a solution is

$$(I-S)[I-(a_0+b_0)^{-1}(a_1-b_1)S_\infty]y = 0,$$

which is written explicitly

$$y(t)-y(-t)-(a_0+b_0)^{-1}(a_1-b_1)\frac{t}{\pi i}\int_{-\infty}^{+\infty}\frac{y(s)}{s^2-t^2}ds = 0.$$

The solution is of the form

$x = \frac{1}{2}(a_0+b_0)^{-1}(I+S)y+\frac{1}{2}(I-S)z,$ where $z \in L^2(\boldsymbol{R})$ is arbitrary.

Since $(I-S)z$ is an odd function, we finally obtain

$$x(t) = \tfrac{1}{2}(a_0+b_0)^{-1}[y(t)+y(-t)]+x_1(t),$$

where $x_1 \in L^2(\boldsymbol{R})$ is an arbitrary odd function.

(d) $a_0-b_0 = a_0+b_0 = 0$. A necessary and sufficient condition for equation (1.1) to have a solution is $y \equiv 0$, and $x = S_\infty(I+S)z$, where $z \in L^2(\boldsymbol{R})$ is arbitrary. Since $(I+S)z$ is an even function, we find

$x(t) = \dfrac{1}{\pi i}\displaystyle\int_{-\infty}^{+\infty}\frac{x_1(s)}{s-t}ds,$ where $x_1 \in L^2(\boldsymbol{R})$ is an arbitrary even function.

VI. Let $a_1-b_1 = 0$, but $a_1+b_1 \neq 0$. Interchanging the roles of a_1-b_1 and of a_1+b_1 and of odd and even functions, we obtain results analogous to those of V.

Observe that all these results remain true if the coefficients of equation (1.1) are even functions, since the operator of multiplication by an even function commutes with S. It is enough to assume that the respective combinations of coefficients vanish identically. Only the final transformation of solutions calls for attention.

An oriented system L (see Section 1 of Chapter XII) is said to be *symmetric* if the condition $t \in L$ implies that $-t \in L$ (Fig. 21). In the space $H^{\mu}(L)$, where L is an oriented symmetric system, we consider the equation

$$a_0 x(t) + b_0 x(-t) + \frac{a_1}{\pi i} \int_L \frac{x(\tau)}{\tau - t} d\tau + \frac{b_1}{\pi i} \int_L \frac{x(\tau)}{\tau + t} d\tau = y(t), \tag{1.4}$$

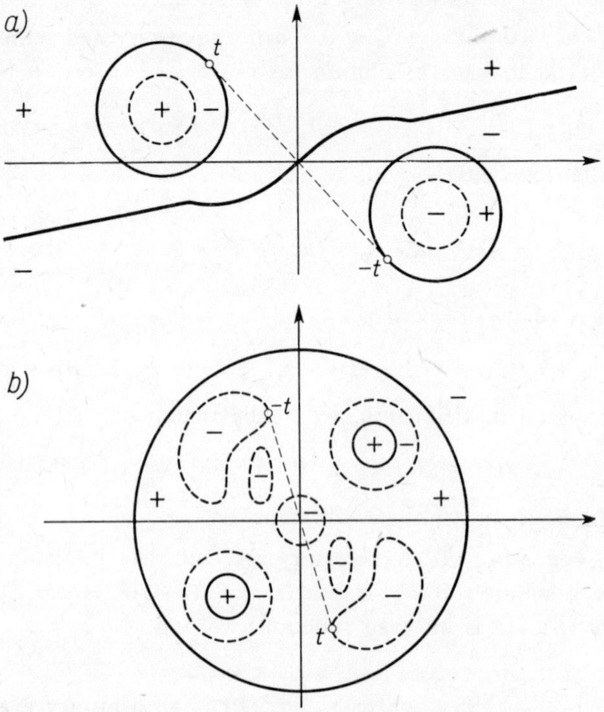

Fig. 21. Symmetric oriented systems

where $y \in H^{\mu}(L)$ and a_0, b_0, a_1, b_1 are constant or even functions. In exactly the same manner as in the preceding case, we obtain solutions of equation (1.4). The same is true in the space $H^{\mu}_{\infty}(L)$, where L is a symmetric oriented system containing an arc closed at infinity. The argument can also be generalized to systems of equations with scalar coefficients considered in the appropriate spaces.

2. SINGULAR INTEGRAL EQUATIONS WITH ROTATION

Let L be an oriented system. We write

(2.1) $\alpha L = \{\alpha t\colon t \in L\}$, where α is an arbitrary complex number.

An oriented system L is said to be *invariant with respect to rotation* through an angle $2\pi/N$, where N is an arbitrary positive integer, if

(2.2) $$\varepsilon L = L, \quad \text{where } \varepsilon = e^{2\pi i/N}$$

(see Fig. 22, Fig. 23).

Note, that equality (2.2) implies

(2.3) $$\varepsilon^{N-1} L = L.$$

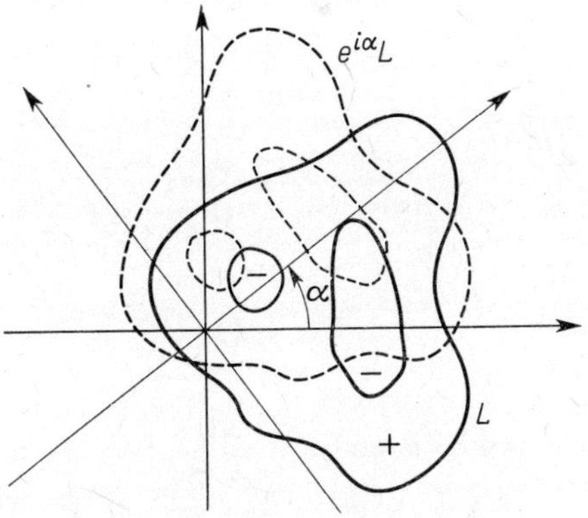

Fig. 22. Rotated oriented system

Indeed, $\varepsilon^{N-1}L = \varepsilon^{N-1}\varepsilon L = \varepsilon^N L = L$.

Now consider the following operators:

(2.4) $$(S_0 x)(t) = \frac{1}{\pi i}\int_L \frac{x(\tau)}{\tau - t}\,d\tau, \qquad Sx = \varepsilon x, \qquad \text{where } \varepsilon = e^{2\pi i/N}.$$

If L is an oriented system, then $S_0^2 = I$ on the space $H^\mu(L)$ (Corollary 1.4 in Chapter XII). Since $\varepsilon^N = 1$, the rotation operator is an involution of order N: $S^N = I$ on this space. We need the following

PROPOSITION 2.1. *If L is an oriented system invariant with respect to rotation through $2\pi/N$, then*

(2.5) $$S_0 S - S S_0 = 0.$$

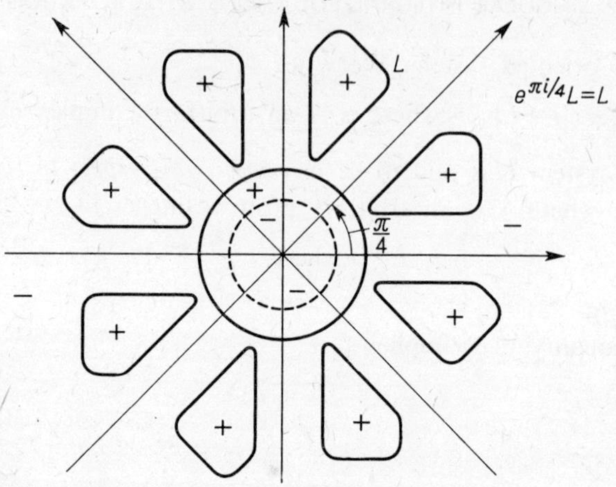

Fig. 23. Oriented system invariant with respect to rotation

Indeed, for all $x \in H^{\mu}(L)$

$$(S_0 Sx)(t) = \frac{1}{\pi i}\int_L \frac{x(\varepsilon\tau)}{\tau - t} d\tau,$$

and

$$(SS_0 x)(t) = \frac{1}{\pi i}\int_L \frac{x(\tau)}{\tau - \varepsilon t} d\tau.$$

Put $u = \varepsilon\tau$. Then $\tau = \varepsilon^{N-1}u$ and

$$(S_0 Sx)(t) = \int_{\varepsilon^{N-1}L} \frac{x(u)}{\varepsilon^{N-1}u - t}\varepsilon^{N-1} du = \frac{1}{\pi i}\int_L \frac{x(u)}{\varepsilon^N u - \varepsilon t}\varepsilon^N du$$

$$= \frac{1}{\pi i}\int_L \frac{x(u)}{u - \varepsilon t} du = (SS_0 x)(t)$$

since $\varepsilon^{N-1}L = L$ (Formula (2.3)).

Now consider the following singular integral equation with rotation:

$$\sum_{k=0}^{N-1}\left[a_k x(\varepsilon^k t) + \frac{b_k}{\pi i}\int_L \frac{x(\tau)}{\tau - \varepsilon^k t} d\tau\right] = y(t), \tag{2.6}$$

where $\varepsilon = e^{2\pi i/N}$, L is an oriented system invariant with respect to rotation through $2\pi/N$, $y \in H^{\mu}(L)$, $0 < \mu < 1$ and the coefficients are complex constants. Using the notation of (2.4), we can rewrite equation (2.6) in the form:

$$\sum_{k=0}^{N-1}(a_k S^k + b_k S^k S_0)x = y,$$

or, if we apply Proposition 2.1, in the form

$$\sum_{k=0}^{N-1} (a_k I + b_k S_0) S^k x = y. \tag{2.7}$$

Since S and S_0 commute, we can apply Theorem 5.3 of Chapter II to obtain the following

THEOREM 2.1. *If L is an oriented system invariant with respect to rotation through $2\pi/N$ and if $[a(\varepsilon^m)]^2 - [b(\varepsilon^m)]^2 \neq 0$ for $m = 1, 2, \ldots, N$, where $a(t) = \sum_{k=0}^{N-1} a_k t^k$, $b(t) = \sum_{k=0}^{N-1} b_k t^k$, then equation (2.6) has the unique solution*

$$x(t) = \frac{1}{N} \sum_{m=1}^{N} \frac{1}{[a(\varepsilon^m)]^2 - [b(\varepsilon^m)]^2} \left[a(\varepsilon^m) \sum_{k=0}^{N-1} \varepsilon^{-km} y(\varepsilon^k t) - \right.$$
$$\left. - \frac{b(\varepsilon^m)}{\pi i} \sum_{k=0}^{N-1} \int \frac{y(\varepsilon^k \tau)}{\tau - t} \varepsilon^{-km} d\tau \right]. \tag{2.8}$$

PROOF. Write

$$A(t) = \sum_{k=0}^{N-1} (a_k I + b_k S_0) t^k.$$

Then

$$A(\varepsilon^m) = \sum_{k=0}^{N-1} (a_k I + b_k S_0) \varepsilon^{km} = a(\varepsilon^m) I + b(\varepsilon^m) S_0$$

The operators $A(\varepsilon^m)$ are invertible, as $[a(\varepsilon^m)]^2 - [b(\varepsilon^m)]^2 \neq 0$ and

$$[A(\varepsilon^m)]^{-1} = [a^2(\varepsilon^m) - b^2(\varepsilon^m)]^{-1} [a(\varepsilon^m) I - b(\varepsilon^m) S_0].$$

Hence by Theorem 5.3 of Chapter II, the solution is

$$x(t) = \sum_{m=1}^{N} [A(\varepsilon^m)]^{-1} P_m y$$
$$= \sum_{m=1}^{N} [a^2(\varepsilon^m) - b^2(\varepsilon^m)]^{-1} [a(\varepsilon^m) I - b(\varepsilon^m) S_0] \frac{1}{N} \sum_{k=0}^{N-1} \varepsilon^{-km} S^k y,$$

which gives Formula (2.8) in an explicit form.

It is plain that we can study in the same way equation (2.6) in the space $L^2(L)$. This method also works in the case of singular integral equations with a cotangent kernel (see Sections 1 and 3 of Chapter XII), as the cotangent function is periodic (compare Flaĭšer [1]).

3. SINGULAR INTEGRAL EQUATIONS OF THE CARLEMAN TYPE

Let L be an oriented system. Suppose we are given on L a Carleman function $g(t) \not\equiv t$ of order n (see Section 1 of Chapter VIII), i.e. a function such that

(3.1) $\quad g_n(t) \equiv t \quad$ on L, $\quad$ where $g_0(t) = t$ and $g_{k+1}(t) = g_k(g(t))$ $(k = 0, 1, \ldots)$.

We recall that on every set homeomorphic with the real line is no Carleman function of order $n > 2$ exists apart from the function identically equal to t (Theorem 1.2 of Chapter VIII). The operator S_g defined by means of a Carleman function of order n

$$(S_g x)(t) = x(g(t)) \quad \text{on } L \tag{3.2}$$

is a multiplicative involution of order n:

$$(S_g x)(t) = I \quad \text{on } H^\mu(L) \ (0 < \mu < 1). \tag{3.3}$$

In the sequel we will also assume that $g' \in H^\mu(L)$ (compare Section 2 of Chapter VIII). It follows from the definition that S_g maps $H^\mu(L)$ onto itself and is bounded in $H^\mu(L)$. Indeed, the function $g(t)$, as a differentiable function, satisfies the Lipschitz condition. Therefore, for all $t, s \in L$

$$|x(g(t)) - x(g(s))| \leqslant \text{const } |g(t) - g(s)|^\mu \leqslant \text{const } |t-s|^\mu. \tag{3.4}$$

We now write

$$gL = \{u\colon u = g(t), t \in L\}.$$

It follows from the definition and from Proposition 1.1 of Chapter VIII that

$$gL = L. \tag{3.5}$$

Consider the singular integral operator

$$(Sx)(t) = \frac{1}{\pi i} \int_L \frac{x(\tau)}{\tau - t} d\tau, \qquad x \in H^\mu(L),$$

which is an involution: $S^2 = L$ (Theorem 1.3 of Chapter XII). We prove the following

THEOREM 3.1. *If L is an oriented system and g is a Carleman function of order n such that $g' \in H^\mu(L)$, then the commutator $T_g = SS_g - S_g S$ is compact in the space $H^\mu(L)$, $0 < \mu < 1$.*

PROOF. Observe that

$$T_g = SS_g - S_g S = SS_g - S_g S S_g^{n-1} S_g = (S - S_g S S_p^{n-1}) S_g. \tag{3.6}$$

Since S_g is a bounded operator in $H^\mu(L)$, it is enough to prove that the operator $T_1 = S - S_g S S_g^{n-1}$ is compact in $H^\mu(L)$. But

$$(SS_g^{n-1}x)(t) = \frac{1}{\pi i}\int_L \frac{x(g_{n-1}(\tau))}{\tau - t}\,d\tau.$$

If we put $u = g_{n-1}(\tau)$, then by definition $g(u) = g(g_{n-1}(\tau)) = g_{n-1}(g(\tau)) \equiv \tau$ and

$$(SS_g^{n-1}x)(t) = \frac{1}{\pi i}\int_{gL} \frac{x(u)}{g(u)-t}\,g'(u)\,du = \frac{1}{\pi i}\int_L \frac{x(u)}{g(u)-t}\,g'(u)\,du,$$

because $gL = L$ (Formula 3.5). Hence

$$(T_1 x)(t) = (S - S_g S S_g^{n-1}x)(t) = \frac{1}{\pi i}\int_L \frac{x(u)}{u-t}\,du - \frac{1}{\pi i}\int_L \frac{x(u)}{g(u)-g(t)}\,g'(u)\,du$$

$$= \frac{1}{\pi i}\int_L \left[1 - \frac{u-t}{g(u)-g(t)}\,g'(u)\right]\frac{x(u)}{u-t}\,du.$$

Now write

$$K(t, u) = \frac{u-t}{g(u)-g(t)}\,g'(u).$$

Since $\lim_{u\to t} K(t, u)$ exists and $= 1$, we can put $K(t, t) = 1$. Hence the kernel of the integral operator T_1 is of the form

$$-\frac{K(t, u) - K(t, t)}{u-t}.$$

Moreover, the function $K(t, u)$ satisfies the Hölder condition with the exponent μ with respect to both variables. Indeed, since $g' \in H^\mu(L)$ by assumption, we have only to show that $\dfrac{u-t}{g(u)-g(t)}$ satisfies the Hölder condition. But from our assumptions together with Proposition 1.7 of Chapter VIII and Formula (3.4) it follows that $g' \not\equiv 0$ on L and that

$$\frac{u-t}{g(u)-g(t)} = \frac{1}{g'[g(t+\theta(u-t))]} = g'[g(t+\theta(u-t))], \qquad 0 < \theta < 1$$

satisfies the Hölder condition with the exponent μ. Hence Theorem 2.2 of Chapter XII implies that the operator T_1 is compact in the space $H^\mu(L)$, which was to be proved.

Now consider the singular integral equation of Carleman type

$$(3.7)\qquad \sum_{j=0}^{m-1} a_j(t)x(g_j(t)) + \frac{1}{\pi i}\int_L \frac{K_j(\tau,t)x(\tau)}{\tau-g_j(t)}\,d\tau = y(t),$$

where L and g satisfy the conditions of Theorem 3.1, $y \in H^\mu(L)$, $K_j(\tau,t)$, $a_j(t)$ satisfy the Hölder condition with the exponent μ for $t, \tau \in L$ and, moreover, the functions $a_j(t)$ and $b_j(t) = K_j(t,t)$ are *invariant* with respect to the transformation S_g, i.e.

$$(3.8)\qquad a_j(g(t)) = a_j(t),\quad b_j(g(t)) = b_j(t) \quad \text{for all } t \in L \ (j = 1, 2, \ldots, n-1).$$

Linear equations of the type (3.7) were considered by Litvinčuk [1]–[3], Litvinčuk and Habasov [1]–[5], Litvinčuk and Zverovič [1], Kordzadze [1], Karimov [1], Paradoksova [1], Sosunov [1], and others and also Samko [1] in the case of automorphic kernels. The method used here is different. We write

$$(3.9)\qquad K_j^*(\tau,t) = K_j(\tau, g_{n-j}(t)) \quad (j = 0, 1, \ldots, n-1).$$

This means that

$$K_j^*(\tau, g_j(t)) = K_j\big(\tau, g_j(g_{n-j}(t))\big) = K_j(\tau, g_n(t)) = K_j(\tau,t) \quad (j = 0, 1, \ldots, n-1).$$

By (3.8) we can write equation (3.7) as follows:

$$\sum_{j=0}^{n-1} a_j(t)x(g_j(t)) + \frac{b_j(g_j(t))}{\pi i}\int_L \frac{x(\tau)}{\tau-g_j(t)}\,d\tau + {} + \frac{1}{\pi i}\int_L \frac{K_j^*(\tau,g_j(t)) - K_j^*(t,t)}{\tau-g_j(t)}x(\tau)\,d\tau = y(t).$$

Using notation of (3.2), we can write this equation as follows

$$(3.10)\qquad \sum_{j=0}^{n-1} (a_j S_g^j + b_j S_g^j S)x = y,$$

where

$$T = \sum_{j=0}^{n-1} S_g^j T_j, \qquad (T_j x)(t) = \frac{1}{\pi i}\int_L \frac{K_j^*(\tau, g_j(t)) - K_j^*(t,t)}{\tau - t}x(\tau)\,d\tau.$$

The operator T is compact in $H^\mu(L)$, since all the operators T_j are compact in $H^\mu(L)$ which in turn follows from our assumptions and from Theorem 2.2 of Chapter XII.

Consider the operator

$$(3.11)\qquad A(S) = A_0 + A_1 S, \quad \text{where } A_0 = \sum_{j=0}^{n-1} a_j S_g^j, \; A_1 = \sum_{j=0}^{n-1} b_j S_g^j.$$

Since the commutator $T_g = SS_g - S_g S$ is compact in $H^\mu(L)$ (Theorem 3.1), all the commutators $SS_g^j - S_g^j S$ are compact in $H^\mu(L)$. Condition (3.8) asserts that the operators of multiplication by functions a_j and b_j commute with S_g, and so the commutators $A_0 S - SA_0$ and $A_1 S - SA_1$ are compact in $H^\mu(L)$.

Since S_g is an involution of order n, the operators $A_0 + A_1$ and $A_0 - A_1$ are invertible if (Section 5 of Chapter II)

$$(3.12)\qquad d_+(\varepsilon^m) = \sum_{j=0}^{n-1} \varepsilon^{mj}(a_j + b_j) \neq 0, \quad d_-(\varepsilon^m) = \sum_{j=0}^{n-1} \varepsilon^{mj}(a_j - b_j) \neq 0,$$

$$\varepsilon = e^{2\pi i/n} \quad (m = 1, 2, \ldots, n)$$

respectively, and in this case

$$(A_0 + A_1)^{-1} = \sum_{m=1}^{n} d_+(\varepsilon^m)^{-1} P_m^g; \quad (A_0 - A_1)^{-1} = \sum_{m=1}^{n} d_-(\varepsilon^m)^{-1} P_m^g,$$

where

$$P_m^g = \frac{1}{n} \sum_{k=0}^{n-1} \varepsilon^{-km} S_g^k \quad (m = 1, 2, \ldots, n).$$

Thus, if condition (3.12) is satisfied, the operator $A(S)$ has a simple regularizer $R_{A(S)} = (A_0^2 - A_1^2)^{-1}(A_0 - A_1 S)$ to the ideal $T(H^\mu(L))$ of compact operators in $H^\mu(L)$. Using Theorem 4.3 and Corollary 4.2 from Chapter II, we therefore obtain

THEOREM 3.2. *Let L be an oriented system and let $g(t)$ be a differentiable Carleman function of order n on L such that $g' \in H^\mu(L)$, $0 < \mu < 1$. Let the functions $a_j(t)$, $b_j(t)$, $K_j(t, \tau)$ $(j = 0, 1, \ldots, n-1)$ satisfy the Hölder condition on L with the exponent μ and, moreover,*

(1) $a_j(g(t)) = a_j(t)$, $b_j(g(t)) = b_j(t)$, *where* $b_j(t) = K_j(t, t)$ $\quad (j = 0, 1, \ldots, n-1)$,

(2) $d_+(\varepsilon^m) \neq 0$, $d_-(\varepsilon^m) \neq 0$ *on* L $(m = 1, 2, \ldots, n)$,

where d_+ and d_- are defined by Formula (3.12). *Then the operator $A(S)+T$ defined by the left-hand side of equation* (3.7) (*or equation* (3.10)) *has a finite d-characteristic and $\varkappa_{A(S)+T} = \varkappa_{A(S)}$. Moreover, it is a Φ_Ξ-operator with respect to the family $\Xi = \{\xi\colon \xi(x) = \int_L \xi(t) x(t)\, dt,\ \xi \in H^\mu(L)\}$.*

If A_0+A_1 and A_0-A_1 are not simultaneously invertible, we can apply Theorem 4.1 and Proposition 4.1 of Chapter XII. For example we obtain the following

THEOREM 3.3. *Let all the assumptions of Theorem 3.2 except (2) be satisfied. If $d_+(\varepsilon^m) = 0$ and $d_-(\varepsilon^m) \neq 0$ on L for $m = 1, 2, \ldots, n$, then the equation $A(S)x = y$ has a solution $x = \frac{1}{2}(I+S)z+2A_0^{-1}y$ if and only if $y \in A_0(I+S)H^\mu(L)$, where $z \in H^\mu(L)$ is arbitrary. If $d_+(\varepsilon^m) \neq 0$ and $d_-(\varepsilon^m) = 0$ on L for $m = 1, 2, \ldots$ $\ldots, n$, then the equation $A(S)x = y$ has a solution $x = \frac{1}{2}(I+S)z+2A_0^{-1}y$ if and only if $y \in A_0(I-S)H^\mu(L)$, where $z \in H^\mu(L)$ is arbitrary.*

Similar results can be obtained for the space $H^\mu_\infty(L)$, where L is an oriented system containing an arc closed at infinity, and for the space $H^\mu_0[0, 2\pi]$ for equations with the cotangent Hilbert transform.

CHAPTER XIV

Singular Integral Equations in Several Variables

Singular integral equations in several variables can be solved by the method of algebraic operators in two cases only: either the domain of integration or the kernel of the singular operator under consideration must be of a very special form. We here present these cases.

1. SINGULAR INTEGRAL EQUATIONS IN THE POLYCYLINDRICAL DOMAINS

In this section we will consider singular integral equations where the domain of integration of the form

(1.1) $$L = L_1 \times L_2 \times \ldots \times L_q,$$

where $L_1, \ldots, L_q$ are regular closed arcs (or, more generally, oriented systems). The solution for the case $q = 2$ has been given by Kakičev [1]–[6].

We here use as a tool the notion of a multi-involution (see Section 7 of Chapter II). We write for $t, \tau \in L$

$$t = (t_1, \ldots, t_q), \quad \tau = (\tau_1, \ldots, \tau_q), \quad t_j, \tau_j \in L_j \ (j = 1, 2, \ldots, q)$$

and let

(1.2) $$(S_j x)(t) = \frac{1}{\pi i} \int_{L_j} \frac{x(t_1, \ldots, t_{j-1}, \tau_j, t_{j+1}, \ldots, t_q)}{\tau_j - t_j} d\tau_j \quad (j = 1, 2, \ldots, q),$$

(1.3) $$S = S_1 \ldots S_q.$$

By definition all the operators $S_1, \ldots, S_q$ commute, with one another. The operator S can also be written in the form

(1.4) $$(Sx)(t) = \frac{1}{(\pi i)^q} \int_L \frac{x(\tau)}{(\tau_1 - t_1) \ldots (\tau_q - t_q)} d\tau_1 \ldots d\tau_q.$$

We denote by $H_q^\mu(L), 0 < \mu < 1$, the space of all functions satisfying the Hölder condition with respect to each coordinate $t_j \in L_j$. The norm in the space $H_q^\mu(L)$ can be defined as follows

$$||x||_{H_q^\mu L} = \sup_L |x(t)| + \sum_{j=1}^{q} \sup_{t_j, t_j' \in L_j} \frac{|x(t)-x(t_1, \ldots, t_{j-1}, t_j', t_{j+1}, \ldots, t_q)|}{|t_j - t_j'|^\mu}.$$

The space $H_q^\mu(L)$ with this norm is a Banach space.

All the operators $S_1, \ldots, S_q$ are involutions in the space $H_q^\mu(L)$ because each of them is an involution on the corresponding space $H^\mu(L)$ (see Section 1 of Chapter XII). Hence the operator $S = S_1 \ldots S_q$ is a multi-involution of order $N = (2)_q = (2, \ldots, 2)$. We write

(1.5) $$P_{p,1} = \tfrac{1}{2}(I - S_p); \quad P_{p,2} = \tfrac{1}{2}(I + S_p) \quad (p = 1, 2, \ldots, q),$$

(1.6) $$P_j = P_{1,j_1} \ldots P_{q,j_q} \quad j = (j_1, \ldots, j_q) \text{ and } (1)_q \leqslant j \leqslant (2)_q.$$

By the considerations of Section 7 in Chapter II, we have

(1.7) $$\sum_{(1)_q \leqslant j \leqslant (2)_q} P_j = I; \quad P_j P_m = \delta_{jm} I \quad \text{for } (1)_q \leqslant j, m \leqslant (2)_q$$

and

(1.8) $$S^m P_j = (-1)^{mj} P_j \quad \text{for } (1)_q \leqslant j, m \leqslant (2)_q.$$

The space $H_q^\mu(L)$ is a direct sum:

(1.9) $$H_q^\mu(L) = \bigoplus_{(1)_q \leqslant j \leqslant (2)_q} X_j, \quad \text{where } X_j = P_j H_q^\mu(L).$$

Consider now a singular integral equation with constant coefficients:

(1.10) $$\sum_{(0)_q \leqslant j \leqslant (1)_q} a_k S^k x = y, \quad y \in H_q^\mu(L),$$

where $S^k = S_1^{k_1} \ldots S_q^{k_q}$, $a_k = a_{k_1, \ldots, k_q}$ and the sum runs over all system of q numbers, each of which is either 0 or 1; there are thus 2^q terms in the sum.

Equation (1.10) is equivalent to a system of 2 independent equations:

(1.11) $$\Big(\sum_{(0)_q \leqslant j \leqslant (1)_q} (-1)^{kj} a_k \Big) x_{(j)} = y_{(j)}, \quad (1)_q \leqslant j \leqslant (2)_q,$$

where $y_{(j)} = P_j y$, $x_{(j)} = P_j x$.

As in Theorem 5.2 of Chapter II we obtain

THEOREM 1.1. *Equation* (1.10) *is equivalent to the* 2^q *independent equations* (1.11). *A necessary and sufficient condition for equation* (1.10) *to have a solution is that* $P_j y = 0$ *for all* j *such that* $b_j = \sum_{(0)_q \leqslant k \leqslant (1)_q} (-1)^q a_{kj} = 0$, *where* $(1)_q \leqslant \leqslant j \leqslant (2)_q$. *If this condition is satisfied, then the solution is of the form*

$$x = \sum_{(1)_q \leqslant j \leqslant (2)_q : b_j \neq 0} b_j^{-1} P_j y + \sum_{(1)_q \leqslant j \leqslant (2)_q : b_j = 0} P_j z, \tag{1.12}$$

where $z \in H_q^\mu(L)$ *is arbitrary and the first sum runs over all* j *such that* $b_j \neq 0$, *and the second sum runs over all* j *such that* $b_j = 0$, $(1)_q \leqslant j \leqslant (2)_q$.

Now consider a singular integral equation in a more general form

$$\sum_{(0)_q \leqslant k \leqslant (1)_q} A_k S^k x + Tx = y, \tag{1.13}$$

where the functions $A_k(t), y(t) \in H_q^\mu(L)$ and the operator T is compact in $H_q^\mu(L)$ Since the operators $A_k S_p - S_k A_p$ are compact for $p = 1, 2, \ldots, q$ and $(0)_q \leqslant k \leqslant (1)_q$ (Theorem 2.1 of Chapter XII), we also have

(1.14) $A_k S - S A_k$ are compact in $H_q^\mu(L)$ for $(0)_q \leqslant k \leqslant (1)_q$.

Now write

$$A(t) = \sum_{(0)_q \leqslant k \leqslant (1)_q} A_k t^k \quad (t^k = t_1^{k_1} \ldots t_q^{k_q}).$$

Then the simple regularizer of the operator $A(S)$ is of the form

$$R_{A(S)} = \sum_{(1)_q \leqslant j \leqslant (2)_q} A[((-1)^{|j|})]^{-1} P_j$$

(compare Section 4 of Chapter II) provided that $A((-1)^{|j|}) \neq 0$ on L.

Hence just as in Section 2 of Chapter XII, we obtain the following

THEOREM 1.2. *Let* $A_k(t), y(t) \in H_q^\mu(L)$, $(0)_q \leqslant k \leqslant (1)_q$ *and suppose that* $A(t) \neq 0$ *for* $t = (-1)^{|j|}$, $(1)_q \leqslant j \leqslant (2)_q$. *Then the operator* $A(S)$ *has finite d-characteristic and* $\varkappa_{A(S)+T} = \varkappa_{A(S)}$ *for all* T *compact in* $H_q^\mu(L)$. *Moreover, the operator* $A(S)+T$ *is a* Φ_Ξ*-operator with respect to the family*

$$\Xi = \{\xi \colon \xi(x) = \int_L \xi(t) x(t)\, dt, \ \xi(t) \in H_q^\mu(L)\}.$$

As in Section 4 of Chapter XII, if $A(t) \neq 0$, where $t = (-1)^{|j|}$ for all j except for one multi-index j^0 and $A(t) = 0$ for $t = (-1)^{|j^0|}$, we can still obtain solutions of equation (1.13). In other cases it is much more difficult to find solutions.

2. SINGULAR INTEGRAL EQUATIONS WITH THE BICADZE OPERATOR

A surface S is said to be of the *class* L if $S = \bigcup_{j=1}^{n} \bar{S}_j$, where the S_j are open (i.e. not compact) Liapounov surfaces [1] which have no common interior points ($\bar{S}_j$ denotes the closure of S_j in R^{n+1}) and, moreover, the angle between two surfaces S_i and S_j having common boundary points is not greater than some constant angle $\alpha < \pi$ (Gordadze [1]).

Let S be a closed two-dimensional surface of the class L and let Ω be the domain bounded by S and not containing the point at infinity. Denote by $\alpha_\tau, \beta_\tau, \gamma_\tau$ the direction cosines of the normal to S at the point $\tau \in S$, exterior with respect to Ω. Let

$$(2.1)\qquad \begin{gathered} r_i(t-\tau) = (t_i-\tau_i)/|t-\tau| \quad \text{for } i = 1, 2, 3,\ t = (t_1, t_2, t_3), \\ \tau = (\tau_1, \tau_2, \tau_3), \\ r(t, \tau) = \alpha_\tau r_1(t-\tau)+\beta_\tau r_2(t-\tau)+\gamma_\tau r_3(t-\tau). \end{gathered}$$

We define a matrix-function $M(t, \tau)$ by

$$M(t, \tau) = |t-\tau|^{-2} \begin{Vmatrix} r(t, \tau) & \gamma_\tau r_2(t-\tau)-\beta_\tau r_3(t-\tau) & \alpha_\tau r_3(t-\tau)-\gamma_\tau r_1(t-\tau) & \beta_\tau r_1(t-\tau)-\alpha_\tau r_2(t-\tau) \\ \beta_\tau r_3(t-\tau)-\gamma_\tau r_2(t-\tau) & r(t, \tau) & \beta_\tau r_1(t-\tau)-\alpha_\tau r_2(t-\tau) & \gamma_\tau r_1(t-\tau)-\alpha_\tau r_3(t-\tau) \\ \gamma_\tau r_1(t-\tau)-\alpha_\tau r_3(t-\tau) & \alpha_\tau r_2(t-\tau)-\beta_\tau r_1(t-\tau) & r(t, \tau) & \gamma_\tau r_2(t-\tau)-\beta_\tau r_3(t-\tau) \\ \alpha_\tau r_2(t-\tau)-\beta_\tau r_1(t-\tau) & \alpha_\tau r_3(t-\tau)-\gamma_\tau r_1(t-\tau) & \beta_\tau r_3(t-\tau)-\gamma_\tau r_2(t-\tau) & r(t, \tau) \end{Vmatrix}.$$

We define the *Bicadze operator* (see Bicadze [1], Gegelia [1]) by means of the matrix-function $M(t, \tau)$:

$$(2.2)\qquad (Bx)(t) = \int_S M(t, \tau)x(\tau)d_\tau S,$$

where $x(t)$ is a four-dimensional vector valued function. The integral is understood to be a principal value in the sense of Cauchy, i.e.

$$\int_S M(t, \tau)x(\tau)dS = \lim_{\varepsilon\to 0} \int_{S\setminus S_\varepsilon} M(t, \tau)x(\tau)d_\tau S,$$

where $S_\varepsilon = \{\tau\colon |\tau-t| \leqslant \varepsilon\} \cap S$.

---

[1] See Section 1 of Chapter X.

If S belongs to the class L, then the Bicadze operator B is bounded in $L^p(S)$, $p > 1$ and is an involution:

$$B^2 = I \text{ on } L^p(S), \quad p > 1 \tag{2.3}$$

(see Bicadze [1], Gegelia [1]).

This enables us to obtain

THEOREM 2.1. *Let S be a closed surface of the class* **L**, *and let B be the Bicadze operator defined by Formula* (2.2). *Then the equation*

$$(a_0 I + a_1 B)x = y, \quad y \in L^p(S), \quad p > 1, \tag{2.4}$$

where a_0, b_0 are real, has, in the case $a_0^2 - b_0^2 \neq 0$, a unique solution

$$x = (a_0^2 - a_1^2)^{-1}(a_0 - a_1 B).$$

If $a_0 + a_1 = 0$, $a_0 - a_1 \neq 0$, then a necessary and sufficient condition for equation (2.4) to have a solution is $(I+B)y = 0$ and the solution is then

$$x = \frac{1}{2a_0} y + (I+B)z, \quad \text{where } z \in L^p(S) \text{ is arbitrary.}$$

If $a_0 + a_1 \neq 0$, $a_0 - a_1 = 0$, a necessary and sufficient condition for equation (2.4) to have a solution is $(I-B)y = 0$ and this solution is

$$x = \frac{1}{2a_0} y + (I-B)z, \quad \text{where } z \in L^p(S) \text{ is arbitrary.}$$

If the coefficients are not constant, we obtain theorems on the solvability of a singular equation with the Bicadze operator by means of a regularization analogous to those already encountered.

CHAPTER XV

Wiener–Hopf Equations, Generalized Abel Equations and Equations with the Fourier Transform and Similar Transformations

In this chapter we consider Wiener–Hopf equations which appear in many problems of economics and engineering, for instance in queueing theory. The method for solving these equations used here is based on the idea of Čerskiĭ [1] introduced in 1957: Acting on both sides of such an equation with the Fourier transform and using some operator identities we obtain a singular integral equation with the Hilbert transform. The first section contains basic properties of the Fourier transform. The second one is devoted to Wiener–Hopf equations, integro-differential as well as integral. In Section 3 a generalized Abel equation is studied which is also solved by means of a reduction to a singular integral equation. Section 4 deals with equations containing the Fourier transform and associated transforms.

1. THE FOURIER TRANSFORM IN $L^2(\boldsymbol{R}^n)$

In this Section we assemble the basic properties of the Fourier transform needed in our subsequent studies. We here omit the proofs which can be found in the following monographs: Titchmarsh [2], Zygmund [3], Bochner and Chandrasekharan [1] and Bracewell [1].

Let $x \in L^2(\boldsymbol{R})$ and let

(1.1) $$(Fx)(t) = \hat{x}(t) = \underset{a \to \infty}{\text{l.i.m.}} \frac{1}{\sqrt{2\pi}} \int_{-a}^{a} x(s) e^{-ist} ds,$$

where "l.i.m." denotes convergence in the $L^2(\boldsymbol{R})$-norm (see Chapter I, Example

4.4 regarding the spaces $L^p(\boldsymbol{R}^n)$). This limit always exists and $\hat{x} \in L^2(\boldsymbol{R})$. Moreover, F is a transformation of $L^2(\boldsymbol{R})$ onto itself. The operator F is called the *Fourier transform* and the function $\hat{x} = Fx$ is said to be the *Fourier transform of the function* x.

We define the *conjugate Fourier transform* (inverse Fourier transform) in a similar manner:

$$(F^*x)(t) = \underset{a\to\infty}{\text{l.i.m.}} \frac{1}{\sqrt{2\pi}} \int_{-a}^{a} x(s)e^{ist}ds \quad \text{for } x \in L^2(\boldsymbol{R}). \tag{1.2}$$

As before, we have $F^*x \in L^2(\boldsymbol{R})$ and F^* is also a mapping onto. We have the following *Parseval–Plancherel theorem*:

$$(Fx, Fy) = (F^*x, F^*y) = (x, y) \quad \text{for all } x, y \in L^2(\boldsymbol{R}), \tag{1.3}$$

where (x, y) denotes the inner product in $L^2(\boldsymbol{R})$, i.e. $(x, y) = \int_{\boldsymbol{R}} x(t)y(t)dt$. This formula shows that F and F^* both are unitary transformations on $L^2(\boldsymbol{R})$. Furthermore we obtain the so-called *Parseval equality*:

$$\|Fx\|_2 = \|F^*x\|_2 = \|x\|_2 \quad \text{for all } x \in L^2(\boldsymbol{R}), \tag{1.4}$$

which tell us that F and F^* both are isometries in $L^2(\boldsymbol{R})$.

Since the following identity holds

$$F^*F = FF^* = I, \tag{1.5}$$

we conclude that the Fourier transform and its conjugate both are invertible and

$$F^{-1} = F^*, \quad (F^*)^{-1} = F. \tag{1.6}$$

In the sequel we therefore denote the conjugate Fourier transform by F^{-1}.

If $x \in L^p(\boldsymbol{R})$, $p > 1$, then $Fx \in L^q(\boldsymbol{R})$, where $1/p+1/q = 1$ (the limit is to be understood as a limit in norm).

We define the *convolution* of two functions x and y by means of the integral

$$(x * y)(t) = \frac{1}{\sqrt{2\pi}} \int_{-\infty}^{+\infty} x(s)y(t-s)ds. \tag{1.7}$$

The convolution satisfies the following relations:

$$x * y = y * x, \quad x * (y * z) = (x * y) * z.$$

If $x \in L^2(\boldsymbol{R})$, $y \in L^1(\boldsymbol{R})$, then $x * y$ and $(Fx)(Fy)$ belong to $L^2(\boldsymbol{R})$ and

$$F(x * y) = (Fx)(Fy), \tag{1.8}$$

provided all these convolutions exist. We have the following *Titchmarsh theorem*:

If $x \in L^2(\boldsymbol{R})$, $y \in L^1(\boldsymbol{R})$, then $x * y$ and $(Fx)(Fy)$ belong to $L^2(\boldsymbol{R})$ and

$$F(x * y) = (Fx)(Fy). \tag{1.8}$$

This implies under the same assumption that xy and $(F^{-1}x) * (F^{-1}y)$ belong to $L^2(\boldsymbol{R})$ and that

$$(F^{-1}x) * (F^{-1}y) = F^{-1}(xy). \tag{1.9}$$

Hence if x and y belong to $X = L^2(\boldsymbol{R}) \cap L^1(\boldsymbol{R})$, then xy, $x * y$, and their Fourier transforms all belong to X.

If $x \in L^p(\boldsymbol{R})$, $y \in L^r(\boldsymbol{R})$ for p and r such that $1/p + 1/r > 1$ [1], then $x * y \in L^s(\boldsymbol{R})$, where $1/s = 1/p + 1/r - 1$ and $||x * y||_s \leqslant ||x||_p ||y||_r$. Hence, if $p = 2$, $r = 1$, we have $s = 2$ and $||x * y||_2 \leqslant ||x||_2 ||y||_1$.

All the previous theorems are also true for the Fourier transform in $\boldsymbol{R}^n$. Namely, if $s, t \in \boldsymbol{R}^n$, then writing $(s, t) = \sum_{i=1}^{n} s_i t_i$, where $s = (s_1, \ldots, s_n)$ and $t = (t_1, \ldots, t_n)$, we have

$$(Fx)(t) = (2\pi)^{n/2} \int_{R^n} x(s) e^{-i(t,s)} ds, \qquad (F^{-1}x)(t) = (2\pi)^{-n/2} \int_{R^n} x(s) e^{i(t,s)} ds,$$

$$(x * y)(t) = (2\pi)^{-n/2} \int_{R^n} x(s) y(t-s)\, ds.$$

Since

$$(F^{-1}x)(-t) = (Fx)(t) \quad \text{for all } x \in L^2(\boldsymbol{R}^n), \tag{1.10}$$

then

$$(F^2x)(t) = F[(Fx)(t)] = F(F^{-1}x)(-t) = x(-t)$$

and

$$(F^2x)(t) = x(-t) \quad \text{for } x \in L^2(\boldsymbol{R}^n). \tag{1.11}$$

Therefore

$$x(t) = x(-(t)) = (-1)^n (F^2x)(-t) = (-1)^{2n} (F^4x)(t) = (F^4x)(t).$$

This implies that the Fourier transform is an involution of order 4:

$$F^4 = I \quad \text{on } L^2(\boldsymbol{R}^n). \tag{1.12}$$

Moreover, from this identity and Formula (1.6) we find

$$F^* = F^{-1} = F^3, \tag{1.13}$$

which implies that F^* is also an involution of order 4.

[1] This last inequality does not hold if $p = r = 2$.

Formula (1.11) shows that the Fourier transform is an involution on the subspace of even functions belonging to $L^2(\boldsymbol{R}^n)$ and that the Fourier transform multiplied by i is an involution on the subspace of odd functions belonging to $L^2(\boldsymbol{R}^n)$. Since the characteristic roots are i, -1, $-i$, 1, we have 4 disjoint projectors giving a partition of unity (compare Section 4 of Chapter II):

$$P_1 = \tfrac{1}{4}(I+iF-F^2-iF^3) = \tfrac{1}{4}(I+iF)(I-F^2),$$

$$P_2 = \tfrac{1}{4}(I-F+F^2-F^3) = \tfrac{1}{4}(I-F)(I+F^2),$$

$$P_3 = \tfrac{1}{4}(I-iF-F^2+iF^3) = \tfrac{1}{4}(I-iF)(I-F^2),$$

$$P_4 = \tfrac{1}{4}(I+F+F^2+F^3) = \tfrac{1}{4}(I+F)(I+F^2).$$

Since we can express every function $x(t)$ as the sum of an even and an odd function in a unique manner

$$x = x^+ + x^-, \quad \text{where } x^+(t) = \frac{x(t)+x(-t)}{2}, \; x^-(t) = \frac{x(t)-x(-t)}{2},$$

and $x^+(-t) = x^+(t)$, $x^-(-t) = -x^-(t)$, we have for $x \in L^2(\boldsymbol{R}^n)$

$$F^2x^+ = x^+, \quad F^2x^- = -x^- \quad \text{and} \quad x^+ = \tfrac{1}{2}(I+F^2)x, \quad x^- = \tfrac{1}{2}(I-F^2)x,$$

$$P_1x^+ = 0, \quad P_2x^+ = \tfrac{1}{2}(I-F)x^+, \quad P_3x^+ = 0, \quad P_4x^+ = \tfrac{1}{2}(I+F),$$

$$P_1x^- = \tfrac{1}{2}(I+iF)x^-, \quad P_2x^- = 0, \quad P_3x^- = \tfrac{1}{2}(I-iF)x^-, \quad P_4x^- = 0.$$

Therefore for all $x \in L^2(\boldsymbol{R}^n)$

$$P_1x = \tfrac{1}{2}(I+iF)x^-, \quad P_2x = \tfrac{1}{2}(I-F)x^+,$$

$$P_3x = \tfrac{1}{2}(I-iF)x^-, \quad P_4 = \tfrac{1}{2}(I+F)x^+.$$

Now put $n = 1$ and let us consider the Hilbert transform in $L^2(\boldsymbol{R})$:

$$(S_\infty x)(t) = \frac{1}{\pi i}\int_{-\infty}^{+\infty} \frac{x(s)}{s-t}\,ds,$$

(see Chapter XII, Section 1). Define the operator S_H by the equality

(1.14) $$(S_Hx)(t) = x(t)\operatorname{sign} t, \quad \text{where } \operatorname{sign} t = \begin{cases} 1 & \text{for } t > 0, \\ 0 & \text{for } t = 0, \\ -1 & \text{for } t < 0 \end{cases}$$

(see Fig. 24).

We observe that $S_H^2 = I$. We have the following relations between the Hilbert transform S_∞ and the operator S_H:

(1.15) $$S_\infty F = -FS_H$$

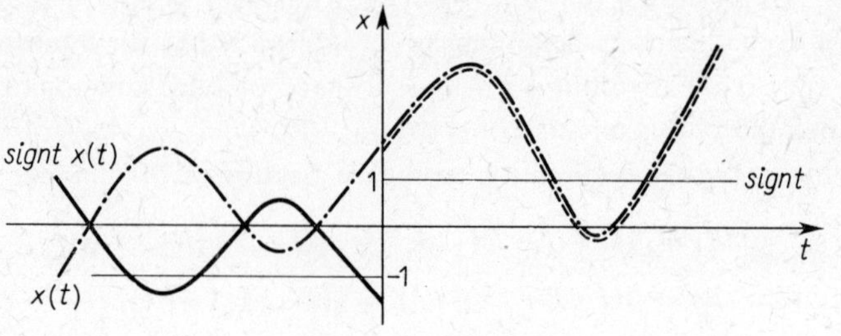

Fig. 24. Graph of the function sign t $x(t)$

which implies

$$S_H = -F^{-1}S_\infty F. \tag{1.16}$$

Note that the following identities hold in $L^2(\boldsymbol{R}^n)$:

$$\hat{x}(t+h) = (2\pi)^{-n/2} \int_{\boldsymbol{R}^n} x(s)e^{-i(t,s)-i(h,s)}ds = \widehat{x(s)e^{-i(h,s)}}, \tag{1.17}$$

$$e^{i(t,h)}\hat{x}(t) = (2\pi)^{-n/2} \int_{\boldsymbol{R}^n} x(s+h)e^{-i(t,s)}ds = \widehat{x(s+h)}, \tag{1.18}$$

where $\hat{x} = Fx$. Hence if we write $(\Delta_h x)(t) = x(t+h)-x(t)$, we have

$$\left.\begin{aligned} &(1.19)\quad \Delta_h F = FN, \quad \text{where } (Nx)(s) = (e^{-i(h,s)}-1)x(s) \\ &(1.20)\quad F\Delta_h = N^*F, \quad \text{where } (N^*x)(t) = (e^{i(t,h)}-1)x(t) \end{aligned}\right\} \text{ for all } x \in L^2(\boldsymbol{R}).$$

We denote by M_j the operator of multiplication by t_j, where t_j is the jth coordinate of the point $t \in \boldsymbol{R}^n$, and

$$M^k = i^{|k|}M_1^{k_1} \dots M_n^{k_n}, \qquad D^k = \frac{\partial^{k_1}}{\partial t_1^{k_1}} \cdots \frac{\partial^{k_n}}{\partial t_n^{k_n}}, \tag{1.21}$$

where $k = (k_1, \dots, k_n)$ is a multi-index (see Chapter II, Section 7), $|k| = k_1 + \dots + k_n$ and $k_1, \dots, k_n$ are non-negative integers. For $k = n = 1$ we have the fundamental relations:

$$FMx = -DFx \quad \text{for all } x \in L^2(\boldsymbol{R}) \text{ such that } itx \in L^2(\boldsymbol{R}), \tag{1.22}$$

$$FDx = MFx \quad \text{for all } x \in L^2(\boldsymbol{R}) \text{ such that } x' \in L^2(\boldsymbol{R}). \tag{1.23}$$

This can be generalized to derivatives of an arbitrary order if we use the following theorem: If $x \in L^2(\boldsymbol{R})$, x has derivatives up to and including the order k and $x^{(k)} \in L^2(\boldsymbol{R})$, then $x^{(j)} \in L^2(\boldsymbol{R})$ for $0 < j < k$. Thus we now assume that together with $x \in L^2(\boldsymbol{R})$ we have $(it)^k x$ and $x^{(k)}$ also belong to $L^2(\boldsymbol{R})$. Under this condition, if we write $X_k = \{x \in L^2(\boldsymbol{R})\colon (it)^k x \in L^2(\boldsymbol{R}), x^{(k)} \in L^2(\boldsymbol{R})\}$, we find

$$FM^k x = (-1)^k D^k F, \qquad FD^k = M^k F \quad \text{on } X_k. \tag{1.24}$$

This immediately implies

$$F(M^{2k}+D^{2k}) = (-1)^{2k}M^{2k}F+D^{2k}F = (M^{2k}+D^{2k})F \quad \text{on } X_{2k}, \tag{1.25}$$

i.e. the operators $M^{2k}+D^{2k}$ commute with the Fourier transform on the space X_{2k}. Relations (1.22), (1.23), (1.24), (1.25) are also true for $n > 1$, if we use notation (1.21).

Now consider the following operators in $L^2(\boldsymbol{R}^n)$:

$$F_c = \tfrac{1}{2}(F^{-1}+F), \qquad F_s = \frac{1}{2i}(F^{-1}-F). \tag{1.26}$$

Since for all $x \in L^2(\boldsymbol{R}^n)$ we have

$$\begin{aligned} (F_c x)(t) &= (2\pi)^{-n/2} \int_{R^n} x(s)\cos(t, s)\,ds, \\ (F_s x)(t) &= (2\pi)^{-n/2} \int_{R^n} x(s)\sin(t, s)\,ds, \end{aligned} \tag{1.27}$$

these transformations are called the *cosine* and *sine Fourier transform* respectively.

Since $F^4 = I$ and $F^{-1} = F^3$ (Formulae (1.12) and (1.13)), we find

$$\begin{aligned} F_c^3 &= \tfrac{1}{8}(F^{-1}+F)^3 = \tfrac{1}{8}(F^3+F)^3 = \tfrac{1}{8}(F^2+I)^3F^3 = \tfrac{1}{8}(F^6+3F^4+3F^2+I)F^3 \\ &= \tfrac{1}{8}(F^2+3I+3F^2+I)F^3 = \tfrac{1}{2}(F^2+I)F^3 = \tfrac{1}{2}(F^5+F^3) = \tfrac{1}{2}(F^{-1}+F) = F_c \end{aligned}$$

and similarly $F_s^3 = F_s$. Hence

$$F_c^3 - F_c = 0 \quad \text{and} \quad F_s^3 - F_s = 0 \quad \text{on } L^2(\boldsymbol{R}^n)$$

and F_c and F_s both are algebraic operators with the characteristic polynomial $P(t) = t^3 - t$ and with the characteristic roots $-1, 0, 1$. As for the Fourier transform, if we write $x = x^+ + x^-$, where x^+ and x^- are respectively the even and the odd part of x, we obtain

$$\begin{aligned} F_c x^+ &= \tfrac{1}{2}(F^{-1}+F)x^+ = \tfrac{1}{2}(F^3+F)x^+ = \tfrac{1}{2}F(F^2+I)x^+ = Fx^+, \\ F_c x^- &= \tfrac{1}{2}F(F^2+I)x^- = 0, \\ F_s x^+ &= \frac{1}{2i}(F^{-1}-F)x^+ = \frac{1}{2i}F(F^2-I)x^+ = 0, \\ F_s x^- &= \frac{1}{2i}F(F^2-I)x^- = -\frac{1}{i}Fx^- = iFx^-. \end{aligned}$$

Hence $F_c x = Fx^+$, $F_s x = iFx^-$. On the other hand, if we write

(1.28) $$L^2_+(\boldsymbol{R}^n) = \{x^+ : x \in L^2(\boldsymbol{R}^n)\}, \quad L^2_-(\boldsymbol{R}^n) = \{x^- : x \in L^2(\boldsymbol{R}^n)\},$$

then we conclude that the kernels and ranges of the operators F_c and F_s are

$$Z_{F_c} = L^2_-(\boldsymbol{R}^n), \quad Z_{F_s} = L^2_+(\boldsymbol{R}^n), \quad F_c L^2(\boldsymbol{R}^n) = FL^2_+(\boldsymbol{R}^n); \quad F_s L^2(\boldsymbol{R}^n) = iFL^2_-(\boldsymbol{R}^n).$$

A function $x(t)$ defined for $t \in \boldsymbol{R}^n$ is said to be *radial* if it depends only on the distance of t from the origin, i.e. on $|t| = (t_1^2 + \ldots + t_n^2)^{1/2}$. The Fourier transform of a radial function belonging to $L^2(\boldsymbol{R}^n)$ is again a radial function. Every radial function is an even function. If we denote

$$L^2_0(\boldsymbol{R}^n) = \{x \in L^2(\boldsymbol{R}^n) \colon x(t) = x(|t|)\},$$

we have $FL^2_0(\boldsymbol{R}^n) = L^2_0(\boldsymbol{R}^n)$. If x is a radial function, then $x \in L^2(\boldsymbol{R}^n)$ if and only if $x(|t|)\,t^{n-1} \in L^2(\boldsymbol{R}^n)$.

If $x \in L^2_0(\boldsymbol{R}^n)$, then we can express its Fourier transform as

(1.29) $$\hat{x}(t) = (2\pi)^{-n/2} t^{1-n/2} \int_0^\infty \tilde{x}(r) r^{n/2} J_{(n-2)/2}(tr)\,dr,$$

where $r = |t|$, $\tilde{x}(r) = x(t)$ and J_k denotes the *Bessel function* of the first kind and of order k. The last integral defines the so-called *Hankel transform*. Namely,

(1.30) $$(H_0 x)(t) = \int_0^\infty x(s) s^{n-1} V_{(n-2)/2}(st)\,ds,$$

where $V_k(t) = J_k(t)/t^k$. The Hankel transform is an involution on the space of all continuous functions on $(0, \infty)$ such that

$$\int_0^\infty |x(s)|\, s^{n-1} ds < +\infty$$

(Bochner and Chandrasekharan [1], p. 175).

For $n = 1$ we have

$$(H_0 x)(t) = 2 \int_0^\infty x(s) \cos ts\, ds,$$

hence H_0 is the cosine transform on the space of even functions.

2. THE WIENER–HOPF INTEGRAL AND INTEGRO-DIFFERENTIAL EQUATIONS

The classical *Wiener–Hopf equation* is of the form

(2.1) $$\lambda x(t) + \frac{1}{\sqrt{2\pi}} \int_{-\infty}^{+\infty} k(t-s) x(s)\,ds = y(t), \quad y \in L^2(\boldsymbol{R}).$$

The integral operator in this equation is of the convolution type. We consider an equation of a more general form, namely an integro-differential equation of an arbitrary order k, since the method for solving the equation is the same in both cases. Suppose we are given the equation

$$(2.2)\qquad Ax \equiv \sum_{j=0}^{k}\Big[\lambda_j x(t)+\mu_j \operatorname{sign} t x(t)+\frac{1}{\sqrt{2\pi}}\int_{-\infty}^{+\infty} a_j(t-s)x^{(j)}(s)\,ds+$$

$$+\frac{1}{\sqrt{2\pi}}\int_{-\infty}^{+\infty} b_j(t-s)\operatorname{sign} s\, x^{(j)}(s)\,ds\Big] = y(t), \quad y\in L^2(\boldsymbol{R}),$$

where $\lambda_1, \ldots, \lambda_k$, $\mu_1, \ldots, \mu_k$ are constant, $a_1, \ldots, a_k$, $b_1, \ldots, b_k \in L^2(\boldsymbol{R})$.

Using the notation introduced in Section 1, we start with the following

LEMMA 2.1. *Let k be a non-negative integer arbitrarily fixed and let*

$$(2.3)\qquad X_k = \{x\in L^2(\boldsymbol{R}):\ x^{(k)}\in L^2(\boldsymbol{R})\}.$$

Then for every $j = 0, 1, \ldots, k$ the operator

$$(2.4)\qquad T = \sum_{j=0}^{k}(S_\infty M^j - M^j S_\infty)$$

is compact in the space FX_k (in the norm $\|\ \|_2$).

PROOF. Let $x\in X_k$. Observe that $x^{(j)}\in L^2(\boldsymbol{R})$ for $j = 0, 1, \ldots, k$. Hence Formulae (1.22), (1.23) imply that $(1+|s|)^j Fx\in L^2(\boldsymbol{R})$ for $j = 0, 1, \ldots, k$. Since the operators

$$(T_j x)(t) = \frac{1}{\pi i}\int_{-\infty}^{+\infty}\frac{(is)^j-(it)^j}{s-t}x(s)\,ds = \frac{i^{j-1}}{\pi}\int_{-\infty}^{+\infty}\sum_{m=0}^{j-1} t^m s^{j-1-m}x(s)\,ds$$

are well defined on the space FX_k and are finite-dimensional, we conclude that they are compact in that space (in the norm $\|\ \|_2$). Thus the operator $T = \sum_{j=0}^{k} T_j$ is compact in the space FX_k.

We restrict ourselves to the space X_k, because only in that space we can find a solution of equation (2.2). We write for any $x\in X_k$

$$Ax = \sum_{j=0}^{k}[\lambda_j x^{(j)}+\mu_j S_H x^{(j)}+a_j * x^{(j)}+b_j * (S_H x^{(j)})]$$

and applying Formulae (1.8), (1.15), (1.21), (1.22) we obtain, if we write $\hat{a}_j = Fa_j$, $\hat{b}_j = Fb_j$,

$$\begin{aligned} FAx &= \sum_{j=0}^{k} [\lambda_j Fx^{(j)} + \mu_j FS_H x^{(j)} + F(a_j * x^{(j)}) + F(b_j * S_H x^{(j)})] \\ &= \sum_{j=0}^{k} [\lambda_j M^j Fx - \mu_j S_\infty Fx^{(j)} + (Fa_j)(Fx^{(j)}) + (Fb_j)(FS_H x^{(j)})] \\ &= \sum_{j=0}^{k} [\lambda_j M^j Fx - \mu_j S_\infty M^j Fx + \hat{a}_j M^j Fx - \hat{b}_j S_\infty Fx^{(j)}] \\ &= \sum_{j=0}^{k} [\lambda_j M^j Fx - \mu_j S_\infty M^j Fx + \hat{a}_j M^j Fx - \hat{b}_j S_\infty M^j Fx]. \end{aligned}$$

Hence if we write

(2.5) $$A_0 = \sum_{j=0}^{k} (\lambda_j + \hat{a}_j) M^j, \quad A_1 = -\sum_{j=0}^{k} (\mu_j + \hat{b}_j) M^j,$$

we have

$$FA = (A_0 + A_1 S_\infty) F + T \quad \text{on } X_k,$$

where T is defined in Lemma 2.1 by Formula (2.4), and

(2.6) $$A = F^{-1}(A_0 + A_1 S_\infty) F + F^{-1} T \quad \text{on } X_k.$$

We now write

(2.7) $$\tilde{a}_j = M^j \hat{a}_j, \quad \tilde{b}_j = M^j \hat{b}_j \quad (j = 0, 1, \ldots, k)$$

(M, $\hat{a}_j$, $\hat{b}_j$ commute with one another as operators of multiplication by the functions it, $\hat{a}_j$, $\hat{b}_j$, respectively). Formula (2.3) then implies that

$$\tilde{a}_j = M^j Fa_j = Fa_j^{(j)}, \quad \tilde{b}_j = M^j Fb_j = Fb_j^{(j)}.$$

We can therefore apply Theorem 2.12 of Chapter XII provided the functions

(2.8) $$\frac{b_j(s) - b_j(t)}{s - t} \quad \text{and} \quad \frac{a_j(s) - a_j(t)}{s - t} \quad (j = 0, 1, \ldots, k)$$

are square integrable on $\boldsymbol{R} \times \boldsymbol{R}$.

We observe that the Leibniz formula implies that $(xy)^{(k)} \in L^2(\boldsymbol{R})$ if xy, x, $x^{(k)}$, y, $y^{(k)} \in L^2(\boldsymbol{R})$. Therefore

(2.9) $$xy \in X_k \text{ if and only if } x \in X_k, \; y \in X_k \text{ and } xy \in L^2(\boldsymbol{R}).$$

This and together with assumptions (2.8) imply that $A_0, A_1 \in L^2(\boldsymbol{R})$ and that the operators $A_0 S_\infty - S_\infty A_0$, $A_1 S_\infty - S_\infty A_1$ are compact. Since the Fourier transform is invertible, $\varkappa_F = \varkappa_{F^{-1}} = 0$ on $L^2(\boldsymbol{R})$ and

$$\varkappa_A = \varkappa_{F^{-1}(A_0+N_1S_\infty)F+F^{-1}T} = \varkappa_{F^{-1}(A_0+A_1S_\infty)F} = \varkappa_{F^{-1}} + \varkappa_{A_0+A_1S_\infty} + \varkappa_F = \varkappa_{A_0+A_1S_\infty},$$

provided that this index is well-defined. Finally, Theorem 2.12 of Chapter XII implies

THEOREM 2.1. *If the functions a_j, b_j $(j = 0, 1, \dots, k)$ belonging to X_k satisfy x the following conditions:*

(i) $$\int_{-\infty}^{+\infty}\int_{-\infty}^{+\infty} \frac{a_j(s)-a_j(t)}{s-t}\, ds\, dt < +\infty; \qquad \int_{-\infty}^{+\infty}\int_{-\infty}^{+\infty} \frac{b_j(s)-b_j(t)}{s-t}\, ds\, dt < +\infty,$$

where $\tilde{a}_j = Fa_j^{(j)}$, $\tilde{b}_j = Fb_j^{(j)}$,

(ii) *the functions A_0, A_1 defined by Formulae* (2.5) *are continuous and bounded on $\boldsymbol{R}$ and, moreover,*

$$\inf_{\boldsymbol{R}} |A_0^2 - A_1^2| > 0, \tag{2.10}$$

then the operator A has finite d-characteristic on X_k and $\varkappa_A = \varkappa_{A_0+A_1S_\infty}$. Moreover, A is a Φ_H-operator with respect to the family

$$H = \{\xi\colon \xi(x) = \int_{-\infty}^{+\infty} \xi(t)\hat{x}(t)\, dt,\ \xi(t) \in X_k\}.$$

The last statement is clear if we write $H = F^{-1}\Xi F$, where Ξ is defined as in Theorem 2.12 of Chapter XII.

From Property 4.1 and Proposition 4.1 of Chapter XII we obtain

THEOREM 2.2. *Let all the conditions of Theorem* 2.1 *except* (2.10) *be satisfied. If one of the functions $A_0 + A_1$, $A_0 - A_1$ is equal to zero on a set $Z_0 \subset \boldsymbol{R}$ of measure* 0 *and the second one is invertible, then the operator A has the semi-finite d-characteristic* $(0, +\infty)$.

THEOREM 2.3. *If the functions $a_j, b_j \in X_k$ $(j = 0, 1, \dots, k)$ satisfy the following condition:*

(iii) *either $A_0 + A_1 \equiv 0$ and $A_0 - A_1 = 2A_0$ is invertible or $A_0 - A_1 \equiv 0$ and $A_0 + A_1 = 2A_0$ is invertible,* *then the operator A has an infinite d-characteristic. If* $A_0+A_1 \equiv 0$, *then the solution is* $x = \frac{1}{2}(1-\operatorname{sign} t)z+2A_0^{-1}Fy$, *where z is arbitrary, under the necessary and sufficient condition*

$$\tfrac{1}{2}[A_0(t)]^{-1}\hat{y}(t)+\frac{1}{2\pi i}\int_{-\infty}^{+\infty}\frac{[A_0(s)]^{-1}\hat{y}(s)}{s-t}\,ds = 0, \qquad \hat{y} = Fy.$$

If $A_0-A_1 \equiv 0$, *then* $x = \frac{1}{2}(1+\operatorname{sign} t)z+2A_0^{-1}Fy$, *where z is arbitrary, is a solution of equation* (2.2) *under the necessary and sufficient condition*

$$\tfrac{1}{2}[A_0(t)]^{-1}\hat{y}(t)-\frac{1}{2\pi i}\int_{-\infty}^{+\infty}\frac{[A_0(s)]^{-1}\hat{y}(s)}{s-t}\,ds = 0, \qquad \hat{y} = Fy.$$

The proof is immediate if we apply Theorem 5.2 of Chapter II to the equation $(A_0+A_1 S_\infty)Fx = Fy$, using Formulae (2.6) and (1.15), (1.16).

The last theorem could also be applied in the case of equations on a half-axis. We demonstrate this for $k = 1$, which is not an essential restriction. Let $x(t)$ be defined for $t \geqslant 0$ and write

$$\text{(2.11)} \qquad \tilde{x}(t) = \begin{cases} x(t) & \text{for } t \geqslant 0, \\ x(-t) & \text{for } t < 0. \end{cases}$$

We find

$$\int_{-\infty}^{+\infty}\tilde{x}(t)\,dt = \int_{-\infty}^{0}x(t)\,dt+\int_{0}^{+\infty}x(t)\,dt = -\int_{+\infty}^{0}x(u)\,du+\int_{0}^{+\infty}x(t)\,dt = 2\int_{0}^{+\infty}x(t)\,dt,$$

$$\text{(2.12)} \qquad \operatorname{sign} t\,\tilde{x}(t) = \begin{cases} x(t) & \text{for } t \geqslant 0, \\ -x(-t) & \text{for } t < 0, \end{cases}$$

and

$$\int_{-\infty}^{+\infty}\operatorname{sign} t\,\tilde{x}(t)\,dt = 0.$$

From Formulae (2.11) and (2.12) we obtain

$$\text{(2.13)} \qquad x(t) = \tfrac{1}{2}[\tilde{x}(t)+\operatorname{sign} t\,\tilde{x}(t)].$$

Now consider the equation

$$\text{(2.14)} \qquad \lambda x(t)+\frac{1}{\sqrt{2\pi}}\int_{0}^{+\infty}k(t-s)x(s)\,ds = y(t),$$

where $k(t)$ is defined on the whole of the real line. The substitution (2.13) transforms equation (2.14) into the equation

$$\frac{\lambda}{2}[\tilde{x}(t)+\operatorname{sign} t\,\tilde{x}(t)]+\frac{1}{\sqrt{2\pi}}\int_{-\infty}^{+\infty}\tilde{K}(t-s)\,[\tilde{x}(s)+\operatorname{sign} s\,\tilde{x}(s)]\,ds = \tilde{y}(t), \tag{2.15}$$

where $\tilde{y}(t)$ and $\tilde{K}(t)$ are defined in a manner similar to that in Formula (2.11).

Note that, arguing as before, we obtain $A_1 = -A_0$, and hence the operator in equation (2.14) has an infinite d-characteristic. We solve this equation by applying the first part of Theorem 2.3.

Other results concerning the equation in question can be found in the book of Gohberg and Feldman [1], see also Prössdorf [1].

3. GENERALIZED ABEL EQUATIONS

The classical *Abel equation* is of the form

$$\int_0^t \frac{x(s)\,ds}{(t-s)^\alpha} = y(t), \qquad 0 < \alpha < 1, \tag{3.1}$$

where $y' \in C[0, 1]$. We can solve this equation by acting on both sides with the operator $(Tx)(u) = \int_s^u \frac{x(t)\,dt}{(u-t)^{1-\alpha}}$. Indeed, since

$$\int_s^u \frac{dt}{(u-t)^{1-\alpha}(t-s)^\alpha} = \int_0^1 \frac{dt}{(1-t)^{1-\alpha}t^\alpha} = \frac{\pi}{\sin\alpha\pi},$$

if we write $y_1 = \sin(\alpha\pi)Ty/\pi$, we obtain the equation $\int_0^u x(s)\,ds = y_1(u)$ and

$$x(u) = y_1'(u) = \frac{\sin\alpha\pi}{\pi}\frac{d}{du}\int_s^u \frac{y(t)\,dt}{(u-t)^{1-\alpha}} = \frac{\sin\alpha\pi}{\pi}\left\{\left[\frac{y(t)}{(u-t)^{1-\alpha}}\right]_s^u + \right.$$

$$\left. + \frac{1}{\alpha-1}\int_s^u \frac{y(t)\,dt}{(u-t)^{2-\alpha}}\right\} = \frac{\sin\alpha\pi}{\pi}\left[\frac{y(0)}{u^{1-\alpha}} + \int_0^u \frac{y'(t)\,dt}{(u-t)^{1-\alpha}}\right],$$

after integration by parts (Abel [1], see also Pogorzelski [1]).

Now we will consider the generalized Abel equations[1] of the following types:

$$(3.2)\qquad A_\alpha x \equiv \frac{1}{\Gamma(\alpha)} \int_{-\infty}^{+\infty} \frac{a_0(t)+a_1(t)\operatorname{sign}(t-s)}{|t-s|^{1-\alpha}} x(s)\,ds = y(t),$$

$$(3.3)\qquad A^\alpha u \equiv \frac{1}{\Gamma(\alpha)} \int_{-\infty}^{+\infty} \frac{a_0(s)+a_1(s)\operatorname{sign}(t-s)}{|t-s|^{1-\alpha}} u(s)\,ds = w(t),$$

where $t \in \boldsymbol{R}, 0 < \alpha < 1$; $y, w \in L^p(\boldsymbol{R})$, $1 < p < 1/\alpha$ and the coefficients a_0, a_1 do not vanish simultaneously on $\boldsymbol{R}$, and $\Gamma(t)$ denotes the *Euler function.*

Equations (3.2) and (3.3) are called, respectively, the *Abel equation with external coefficients* and the *Abel equation with internal coefficients.* The method of solving them is to transform these equations into certain singular integral equations with the Hilbert transform, and then to solve them using the method of Samko [7] (in fact we slightly simplify his method).

We first introduce the so-called *operators of fractional integration*

$$(3.4)\qquad \begin{aligned} (I_+^\alpha x)(t) &= \frac{1}{\Gamma(\alpha)} \int_{-\infty}^{t} x(s)(t-s)^{\alpha-1} dt; \\ (I_-^\alpha x)(t) &= \frac{1}{\Gamma(\alpha)} \int_{-t}^{+\infty} x(s)(s-t)^{\alpha-1} dt. \end{aligned}$$

It is easy to check, that

$$(3.5)\qquad A_\alpha = (a_0+a_1)I_+^\alpha + (a_0-a_1)I_-^\alpha; \qquad A^\alpha = I_+^\alpha(a_0+a_1)+I_-^\alpha(a_0-a_1).$$

It is known (see: Hardy, Littlewood and Pólya [1]) that the operators A_α and A^α map $L^p(\boldsymbol{R})$ into $L^r(\boldsymbol{R})$, where $r = p(1-\alpha p)^{-1}$, provided that a_0 and a_1 are bounded. Moreover, if H is the Hilbert transform defined by Formula (1.14′) of Chapter XII,

$$(3.6)\qquad (Hx)(t) = \frac{1}{\pi} \int_{-\infty}^{+\infty} \frac{x(\tau)}{\tau - t} d\tau \qquad (H^2 = -I),$$

then the operators I_+^α and I_-^α commute with H on $L^p(\boldsymbol{R})$, $1 < p < 1/\alpha$, $0 < \alpha < 1$:

$$(3.7)\qquad HI_+^\alpha - I_+^\alpha H = 0; \qquad HI_-^\alpha - I_-^\alpha H = 0$$

[1] Generalized Abel equations have recently been studied in different settings by von Wolfersdorf [1] and Samko [1]–[4], [7].

and

$$I_-^\alpha = [\cos(\alpha\pi)I+\sin(\alpha\pi)H]I_+^\alpha, \tag{3.8}$$

$$I_+^\alpha = [\cos(\alpha\pi)I-\sin(\alpha\pi)H]I_-^\alpha. \tag{3.9}$$

The operators I_+^α and I_-^α are invertible and we have

$$\begin{aligned}(I_+^\alpha x)(t) &= \alpha\Gamma^{-1}(1-\alpha)\int_{-\infty}^{t}[x(t)-x(s)](t-s)^{-\alpha-1}ds,\\ (I_-^{-\alpha}x)(t) &= \alpha\Gamma^{-1}(1-\alpha)\int_{t}^{+\infty}[x(t)-x(s)](t-s)^{-\alpha-1}ds.\end{aligned} \tag{3.10}$$

Moreover, these operators have the same ranges.

Applying Formulae (3.8) and (3.9) we can rewrite the operators A_α, A^α in the following way

$$\begin{aligned}A_\alpha &= (a_0+a_1)I_+^\alpha+(a_0-a_1)I_-^\alpha = (a_0+a_1)I_+^\alpha+(a_0-a_1)[\cos(\alpha\pi)I+\sin(\alpha\pi)H]I_+^\alpha\\ &= \{[(a_0+a_1)+(a_0-a_1)\cos(\alpha\pi)]I+(a_0-a_1)(\sin(\alpha\pi))H\}I_+^\alpha\\ &= \{[a_0(1+\cos\alpha\pi)+a_1(1-\cos\alpha\pi)]I+(a_0-a_1)\sin(\alpha\pi)H\}I_+^\alpha.\end{aligned}$$

If we write

$$b_0 = a_0(1+\cos\alpha\pi)+a_1(1-\cos\alpha\pi), \quad b_1 = (a_0-a_1)\sin\alpha\pi, \tag{3.11}$$

we obtain

$$A_\alpha = (b_0I+b_1H)I_+^\alpha, \tag{3.12}$$

and similarly

$$A^\alpha = I_+^\alpha(b_0I+b_1H). \tag{3.13}$$

If we now write

$$A_\alpha^* = I_+^\alpha(b_0I-b_1H), \quad A_*^\alpha = (b_0I-b_1H)I_+^\alpha. \tag{3.14}$$

Hence $A_\alpha^*A_\alpha = I_+^\alpha(b_0I-b_1H)(b_0I+b_1H)I_+^\alpha$. Since $H^2 = -I$ and I_+^α, I_-^α commute with H for constant b_0, b_1 we obtain

$$\begin{aligned}A_\alpha^*A_\alpha &= A_\alpha A_\alpha^* = (b_0^2+b_1^2)(I_+^\alpha)^2,\\ A_*^\alpha A^\alpha &= A^\alpha A_*^\alpha = (b_0^2+b_1^2)(I_+^\alpha)^2.\end{aligned} \tag{3.15}$$

These identities hold if $p\alpha < 1/2$. Indeed, if for example A_α transforms $L^p(\boldsymbol{R})$ into $L^r(\boldsymbol{R})$, where $r = p(1-p\alpha)^{-1}$ for some p, then A_α^* transforms $L^r(\boldsymbol{R})$ into $L^q(\boldsymbol{R})$, where $q = r(1-r\alpha)^{-1} = p(1-2\alpha p)^{-1} > p > 1$ for $0 < \alpha < 1$.

Observe that

$$b_0^2+b_1^2 = [a_0(1+\cos\alpha\pi)+a_1(1-\cos\alpha\pi)]^2+(a_0-a_1)^2\sin^2\alpha\pi$$

$$= 2(a_0^2+a_1^2)+2(a_0^2-a_1^2)\cos\alpha\pi = 4\left[a_0^2\cos^2\frac{\alpha\pi}{2}+a_1^2\sin^2\frac{\alpha\pi}{2}\right] \neq 0$$

for $0<\alpha<1$. Hence all the operators A_α, A^α, A_α^*, A_*^α are invertible in the space $L^q(\boldsymbol{R})$, where $q = p(1-2\alpha p)^{-1} > p$. And acting on both sides of equation (3.2) (resp. (3.3)) with the operator A_α^* (resp. A_*^α) we obtain the equivalent equations

$$(3.16)\qquad (I_+^\alpha)^2 x = (b_0^2+b_1^2)^{-1}A_\alpha^* y,$$

$$(3.17)\qquad (I_+^\alpha)^2 u = (b_0^2+b_1^2)^{-1}A_*^\alpha w.$$

Applying the inversion Formulae (3.10), we conclude that

$$x = (I_+^{-\alpha})^2(b_0^2+b_0^2)^{-1}A_\alpha^* y, \qquad u = (I_+^{-\alpha})^2(b_0^2+b_1^2)^{-1}A_*^\alpha w.$$

Now consider the case where b_0 and b_1 are not constant. We assume that the operators b_0H-Hb_0 and b_1H-Hb_1 are compact in $L^p(\boldsymbol{R})$. Thus, since in this case also $b_0^2+b_1^2\neq 0$, by Theorem 4.3 of Chapter II we conclude that the operator b_0I+b_1H has finite d-characteristic, moreover, and it is a Φ_Ξ-operator with respect to the family

$$\Xi = \left\{\xi : \xi(x) = \int_{-\infty}^{+\infty} x(t)\xi(t)\,dt,\ \ \xi(t)\in L^{p'}(\boldsymbol{R}),\ 1/p+1/p' = 1\right\}$$

(Corollary 4.2 in Chapter II).

Hence applying Formula (3.12) we can write equation (3.2) in the form

$$(b_0I+b_1H)I_+^\alpha x = y.$$

Therefore, $x = I_+^{-\alpha}z$, where z is a solution of the equation $(b_0I+b_1H)z = y$. Applying Formula (3.13), we write equation (3.3) in the form $I_+^\alpha(b_0I+b_1H)x = y$. Hence u is a solution of the equation $(b_0I+b_1H)u = I_+^{-\alpha}w$. We recall that equations with the Hilbert transform were discussed in Section 2 of Chapter XII.

Equations (3.2) and (3.3) both have the same index:

$$\varkappa_{A_\alpha} = \varkappa_{A^\alpha} = \varkappa_{b_0I+b_1H} = \frac{1}{2\pi}\int_{-\infty}^{\infty} d_t\arg\frac{b_0(t)-b_1(t)}{b_0(t)+b_1(t)}$$

$$= \frac{1}{2\pi}\int_{-\infty}^{+\infty} d_t\arg\frac{a_0(1+\cos\alpha\pi)+a_1(1+\sin\alpha\pi)}{a_0(1+\cos\alpha\pi)+a_1(1-\sin\alpha\pi)}.$$

4. EQUATIONS WITH THE FOURIER TRANSFORM, THE SINE AND COSINE FOURIER TRANSFORM, AND THE HANKEL TRANSFORM

In order to investigate operator equations with the Fourier transform it is important to identify linear spaces X with the property that $FX \subset X$. We have studied two such spaces: $L^2(\boldsymbol{R}^n)$ and $X_k = \{x \in L^2(\boldsymbol{R}): x^{(k)}L^2(\boldsymbol{R})\}$. Since the Fourier transform is an involution of order 4 (Formula (1.12)), we have from Corollary 3.3 in Chapter II

THEOREM 4.1. *Let X be one of the spaces $L^2(\boldsymbol{R})$, X_k $(k = 1, 2, \dots)$. Let $A(F) = A_0+A_1F+A_2F^2+A_3F^3$, where the operators A_0, A_1, A_2, A_3 acting in X commute with one another and with the Fourier transform F. If the operators*

$$B_1 = A_0+A_1+A_2+A_3, \quad B_3 = A_0+iA_1-A_2-iA_3,$$
$$B_2 = A_0-A_1+A_2-A_3, \quad B_4 = A_0-iA_1-A_2+iA_3,$$

are invertible, then the equation $A(F)x = y$, $y \in X$ has the unique solution $x = [A(F)]^{-1}y = \sum_{j=0}^{3} B_j^{-1}P_j$, where the operators P_j are defined as follows

$$P_1 = \tfrac{1}{4}(I+iF)(I-F^2); \quad P_2 = \tfrac{1}{4}(I-F)(I+F^2),$$
$$P_3 = \tfrac{1}{4}(I-iF)(I-F^2); \quad P_4 = \tfrac{1}{4}(I+F)(I+F^2).$$

Observe that by Formula (1.25) the operators $M^{2k}+D^{2k}$, where $(Mx)(t) = itx(t)$; $Dx = dx/dt$, commute with the Fourier transform on the space X_{2k} $(k = 0, 1, \dots)$.

EXAMPLE 4.1. We apply Theorem 4.1 to the functional equation with reflection in $L^2(\boldsymbol{R}^n)$:

$$a_0x(t)+a_1x(-t) = y(t), \tag{4.1}$$

where a_0, a_1 are scalars. Since $(F^2x)(t) = x(-t)$, we have $(a_0I+F^2)x = y$. Therefore if $a_0^2-a_1^2 \neq 0$, the solution of equation (4.1) can be written in the form

$$\begin{aligned} x &= \tfrac{2}{4}[(a_0+a_1)^{-1}2I+(i-1)F-(i+1)F^3+(a_0-a_1)^{-1}2I-(i-1)F+(i+1)F^3]y \\ &= (a_0^2-a_1^2)^{-1}[a_0I-a_1(i-1)F+a_1(i+1)F^3]y, \end{aligned}$$

which implies

$$x(t) = (a_0^2-a_1^2)^{-1}\Big\{a_0y(t)+(2\pi)^{-n/2}a_1\int_{R^n}[\cos ts-\sin ts]y(s)ds\Big\}.$$

Consider the Hankel transform (1.30)

$$(H_0x)(t) = \int_0^\infty x(t)t^{n-1}V_{(n-1)/2}(t)dt, \tag{4.2}$$

where $V_k(t) = J_k(t)/t^k$ and J_k is the Bessel function of the first kind of order k, on the space X of all continuous functions defined for $t > 0$ and satisfying the condition $\int_0^\infty x(t)t^{n-1}dt < +\infty$ (Section 1). Since H_0 is an involution on the space X, we have (Section 1):

THEOREM 4.2. *If A_0, A_1 are operators which commute with the Hankel transform and if the operators $A_0 - A_1$ and $A_0 + A_1$ are invertible, then the equation $(A_0 + A_1 H_0)x = y$, $y \in X$, has the unique solution*

$$x = (A_0^2 - A_1^2)^{-1}(A_0 - A_1 H_0)y.$$

Let F_s and F_c be, respectively, the sine and the cosine Fourier transforms defined by Formulae (1.27). Since $F_c^3 - F_c = 0$ and $F_s^3 - F_s = 0$ on $L^2(\mathbf{R}^n)$, we have

THEOREM 4.3. *Let the operators A_0, A_1, A_2 commute with one another and with F_c (F_s respectively) on $L(\mathbf{R}^n)$. If the operators A_0, $A_0 + A_1 + A_2$, $A_0 - A_1 + A_2$ are invertible, then the equation $(A_0 + A_1 T + A_2 T^2)x = y$, where $y \in L^2(\mathbf{R}^n)$ and $T = F_c$ ($= F_s$ resp.) has the unique solution*

$$x = [A_0^{-1}P_0 + (A_0 + A_1 + A_2)^{-1}P_1 + (A_0 - A_1 + A_2)^{-1}P_2]y,$$

where

$$P_0 = I - T^2, \quad P_1 = \tfrac{1}{2}(T^2 + T), \quad P_2 = \tfrac{1}{2}(T^2 - T).$$

CHAPTER XVI

Cyclic Functional Equations

In the preceding Chapters we have studied certain functional equations:

In Chapter V we determined the periodic solutions of a difference equation (Example V.1.2).

In Chapter VII we examined functional equations with reflection (Theorem 1.2 and Example 1.2, also Example 1.4 in Chapter XV) and functional equations with rotation (Theorem 2.2).

Chapter VIII contains, in Theorem 1.1, the general form of a function satisfying the Carleman condition (i.e. the general form of the solutions of the so-called Babbage functional equation) on an arc homeomorphic with the real line.

In this Chapter, we start by solving the so-called *cyclic functional equations* [1], i.e. equations of the form

$$\sum_{k=0}^{n-1} a_k(t)\,x\big(g_k(t)\big) = y(t),$$

where $g(t) \not\equiv t$ is a Carleman function of order $n \geqslant 2$ on a set Ω [2] and $y(t)$, $a_0(t), \ldots, a_{n-1}(t)$ are defined on Ω.

We then generalize these results to a larger class of equations.

1. CYCLIC FUNCTIONAL EQUATIONS

Consider the functional equation

$$\sum_{k=0}^{n-1} a_k x\big(g_k(t)\big) = y(t), \tag{1.1}$$

where $g(t) \not\equiv t$ is a Carleman function of order $n \geqslant 2$ on a set Ω [3], $y(t)$ is defined

[1] For references see Kuczma [1], Chapter XIII.

[2] See Chapter VIII, introduction and Section 1.

[3] See Chapter VIII, Section 1.

on Ω and a_k are constant. Let X be a linear space of functions defined on Ω, and such that $y \in X$. We introduce a linear operator S by means of the equality

$$(1.2) \qquad (Sx)(t) = x(g(t)) \qquad (t \in \Omega, x \in X).$$

This is an involution of order n on X because

$$(S^n x)(t) = x(g_n(t)) = x(t) \quad \text{for } t \in \Omega, x \in X.$$

Applying Theorem 5.2 of Chapter II we obtain the following

THEOREM 1.1. *A necessary and sufficient condition for equation* (1.1) *to have a solution is*

$$(1.3) \qquad \sum_{k=0}^{n-1} \varepsilon^{-km} y(g_k(t)) = 0 \quad \textit{for all } m \textit{ such that } a(\varepsilon^m) = 0 \ (m = 1, 2, \ldots, n),$$

where

$$(1.4) \qquad a(t) = \sum_{k=0}^{n-1} a_k t^k, \qquad \varepsilon = e^{2\pi i/n}.$$

Under this condition the general form of the solutions is

$$(1.5) \qquad x(t) = \frac{1}{n} \sum_{m:a(\varepsilon^m)\neq 0} \frac{1}{a(\varepsilon^m)} \sum_{k=0}^{n-1} \varepsilon^{-km} y(g_k(t)) + \frac{1}{n} \sum_{m:a(\varepsilon^m)=0} \sum_{k=0}^{n-1} \varepsilon^{-km} z(g_k(t)),$$

where $z(t)$ is an arbitrary function belonging to X, the first sum runs over all $m, m = 1, 2, \ldots, n$, such that $a(\varepsilon^m) \neq 0$, and the second sum over all m, $m = 1, 2, \ldots, n$ such that $a(\varepsilon^m) = 0$.

If we put $a_k = a_k(t)$, where $a_k(t)$ $(k = 0, 1, \ldots, n)$ are functions belonging to X and such that

$$(1.6) \qquad a_k(g(t)) = a_k(t) \quad \text{on } \Omega \ (k = 0, 1, \ldots, n),$$

then Theorem 1.1, remains true. Indeed, it is easy to check that the operator of multiplication by a function satisfying condition (1.6) commutes with S.

Theorem 1.2 and Corollary 1.2 of Chapter VIII imply that, if Ω is an open arc and $g(t)$ is continuous on Ω, the order n cannot be greater than 2. Hence $n = 2$.

EXAMPLE 1.1. Consider the case where $a_k = 1$ for $k = 0, 1, \ldots, n-1$ and $y(t) = 0$ on Ω. Equation (1.1) is then of the form

$$\sum_{k=0}^{n-1} x(g_k(t)) = 0.$$

Since

$$a(\varepsilon^m) = \sum_{k=0}^{n-1} \varepsilon^{km} = \begin{cases} n & \text{for } m = n, \\ 0 & \text{for } m = 1, 2, \ldots, n-1, \end{cases}$$

we have

$$x(t) = \frac{1}{n}\sum_{m=1}^{n-1}\sum_{k=0}^{n-1} \varepsilon^{-km} z(g_k(t)) = \sum_{k=0}^{n-1}\left[\frac{1}{n}\sum_{m=0}^{n-1}\varepsilon^{-km}\right] z(g_k(t))$$

$$= \sum_{k=0}^{n-1}\left[\frac{1}{n}\sum_{m=1}^{n}\varepsilon^{-km} - \frac{1}{n}\right] z(g_k(t)) = z(t) - \frac{1}{n}\sum_{k=0}^{n-1} z(g_k(t)),$$

where $z(t)$ is an arbitrary function defined on Ω. For $n = 2$ we obtain $x(t) = z(t) - z(g(t))$.

EXAMPLE 1.2. Let $t = (t_1, \ldots, t_n) \in \boldsymbol{R}^n$. Let X be a linear space of functions defined on $\boldsymbol{R}^n$. Let

(1.7) $$g(t_1, \ldots, t_n) = (t_n, t_1, \ldots, t_{n-1}) \quad \text{for all } t \in \boldsymbol{R}^n.$$

The above is a Carleman function of order n on $\boldsymbol{R}^n$. Indeed, we have

$$g_2(t) = g(g(t)) = (t_{n-1}, t_n, \ldots, t_{n-3}, t_{n-2}),$$

and after k steps

$$g_k(t) = g(g_{k-1}(t)) = (t_{n+1-k}, \ldots, t_{n-k-1}, t_{n-k}).$$

Hence

$$g_n(t) = (t_1, t_2, \ldots, t_{n-1}, t_n) = t \quad \text{for all } t \in \boldsymbol{R}^n.$$

Now consider the following cyclic equation[1]:

(1.8) $$a_0(t)x(t_1, t_2, \ldots, t_{n-1}, t_n) + a_1(t)x(t_n, t_1, \ldots, t_{n-2}, t_{n-1}) + \ldots + \\ + a_{n-1}(t)x(t_2, t_3, \ldots, t_n, t_1) = y(t),$$

where $y, a_0, \ldots, a_{n-1} \in X$ and $a_k(g(t)) = a_k(t)$ (i.e. the a_k are symmetric with respect to all the variables) for $k = 0, 1, \ldots, n-1$. From Theorem 1.1 we infer that:

A necessary and sufficient condition for equation (1.8) to have a solution is

$$\frac{1}{n}\sum_{k=0}^{n-1} \varepsilon^{-km} y(t_{n+1-k}, \ldots, t_{n-k-1}, t_{n-k}) = 0$$

for all m such that $a(\varepsilon^m) = 0$ $(t_{n+1} = t_1)$, where $a(\varepsilon^m) = \sum_{k=0}^{n-1} \varepsilon^{km} a_k$, $\varepsilon = e^{2\pi i/n}$. Under this condition the solution is

$$x(t) = \frac{1}{n}\sum_{m:\,a(\varepsilon^m)\neq 0} [a(\varepsilon^m)]^{-1} \sum_{k=0}^{n-1} \varepsilon^{-km} y(t_{n+1-k}, \ldots, t_{n-k}) +$$

$$+ \sum_{m:\,a(\varepsilon^m)=0} \sum_{k=0}^{n-1} \varepsilon^{-km} z(t_{n+1-k}, \ldots, t_{n-k-1}, t_{n-k}),$$

[1] See Presič [1].

where $z \in X$ is arbitrary, the first sum runs over all m such that $a(\varepsilon^m) \neq 0$, and the second sum over all m such that $a(\varepsilon^m) = 0$ $(m = 1, 2, \ldots, n)$. For instance, if $n = 2$, we have equation (1.8) in the form

$$a_0(t_1, t_2)x(t_1, t_2) + a_1(t_1, t_2)x(t_2, t_1) = y(t_1, t_2).$$

If $a_0^2(t) - a_1^2(t) \neq 0$, then the solution is of the form

$$x(t) = (a_0 + a_1)^{-1} y(t_1, t_2) + (a_0 - a_1)^{-1} y(t_2, t_1).$$

2. MULTICYCLIC FUNCTIONAL EQUATIONS

In this section we consider two types of functional equations which will be called *multicyclic functional equations* for obvious reasons.

Let $\Omega \subset \mathbf{R}^q$ and consider p functions $g_1(t), \ldots, g_q(t) \not\equiv t$ defined on Ω and such that the pth function is a Carleman function of an order n_p with respect to the variable t_p, where $t = (t_1, \ldots, t_q)$. This means that

$$\text{(2.1)} \quad \begin{cases} g_p(s) \text{ is determined for } s = t_p, & \text{where } (t_1, \ldots, t_q) \in \Omega \subset \mathbf{R}^q, \\ g_{p,n_p}(t_p) = t_p & (p = 1, 2, \ldots, q), \end{cases}$$

where

$$g_{p,k} = g_p\big(g_{p,k-1}(t)\big) \quad \text{and} \quad g_{p,0} = g_p \quad (p = 1, 2, \ldots, q).$$

Now consider the equation

$$\text{(2.2)} \quad \sum_{(0)_q \leqslant k \leqslant n-(1)_q} a(t)x\big(g_{1,k}(t_1), \ldots, g_{q,k}(t_q)\big) = y(t),$$

where (as in Section 7 of Chapter II) we write

$$k = (k_1, \ldots, k_q), \quad n = (n_1, \ldots, n_q) \quad \text{and} \quad (0)_q = (0, \ldots, 0).$$

Let X be a linear space of functions defined on Ω and containing the functions $y(t)$, $a_k(t)$, $(0)_q \leqslant k \leqslant n-(1)_q$. Define the operators S_p by means of the equalities

$$\text{(2.3)} \quad (S_p x)(t) = x\big(t_1, \ldots, t_{p-1}, g_p(t_p), t_{p+1}, \ldots, t_q\big) \quad (p = 1, \ldots, q).$$

It is easy to check that the operators S_p are involutions of the orders n_p.

Now write $S^k = S_1^{k_1} \ldots S_q^{k_q}$, $(0)_q \leqslant k \leqslant n$. S is a multi-involution of order $n = (n_1, \ldots, n_q)$ (see Section 7 of Chapter II). We rewrite equation (2.2) in the form

$$\sum_{(0)_q \leqslant k \leqslant n-(1)_q} a_k S^k x = y,$$

and we apply Theorem 7.1 of Chapter II, to obtain the following

THEOREM 2.1. *If the functions $a_k(t)$, $y(t)$ are defined on a set $\Omega \subset \mathbf{R}^q$, $g_1, \ldots, g_q$ are Carleman functions of one variable of orders $n_1, \ldots, n_q$ respectively, such that*

$$\big(g_1(t_1), \ldots, g_q(t_q)\big) \in \Omega \quad \textit{if and only if } (t_1, \ldots, t_q) \in \Omega,$$

and, moreover,

$$a_k(g_1(t_1), \ldots, g_q(t_q)) = a_k(t_1, \ldots, t_q) \quad \textit{for} \ (0)_q \leqslant k \leqslant n-(1)_q, \ n = (n_1, \ldots, n_q),$$

then a necessary and sufficient condition for equation (2.2) *to have a solution is*

$$P_m y = 0 \quad \textit{for all } m \textit{ such that } a(\varepsilon^m) = 0,$$

where

$$a(\varepsilon^m) = \sum_{(0)_q \leqslant k \leqslant n-(1)_q} a_k \varepsilon^{km}, \quad \varepsilon^k = \varepsilon_1^{k_1} \ldots \varepsilon_q^{k_q}, \quad \varepsilon_p = e^{2\pi i/n_p},$$

$$P_m = P_{1,m_1} \ldots P_{q,m_q}, \quad (1)_q \leqslant m \leqslant n,$$

$$P_{p,r} = \frac{1}{n_p} \sum_{j=0}^{n_p-1} \varepsilon_p^{-jr} S_p^j \quad (p = 1, \ldots, q;\ r = 1, \ldots, n_p).$$

Under this condition the solution is

$$x = \sum_{m: a(\varepsilon^m) \neq 0} [a(\varepsilon^m)]^{-1} P_m y + \sum_{m: a(\varepsilon^m) = 0} P_m z,$$

where z is an arbitrary function (belonging to the space of functions under consideration).

The second equation which we examine, is of the form

$$\sum_{(0)_q \leqslant k \leqslant n-(1)_q} a_k(t) x(g_k(t)) = y(t), \tag{2.4}$$

where

$$g_k = g_{1,k_1} \circ g_{2,k_2} \circ \ldots \circ g_{q,k_q}{}^{1}, \tag{2.5}$$

g_p is a Carleman function of order n_p on a set $\Omega \subset \boldsymbol{R}^N$, i.e.

$$g_{p,n_p}(t) \equiv t \quad \text{on } \Omega, \quad \text{where } g_{p,j} = g \circ g_{p,j-1}, g_{p,0} = g_p \tag{2.6}$$

$$(p = 1, 2, \ldots, q) \ (j = 1, 2, \ldots).$$

The functions $a_k(t)$, $y(t)$ are defined on Ω and, moreover,

$$a_k(g_1 \circ g_2 \circ \ldots \circ g_q(t)) = a_k(t) \quad \text{for} \quad (0)_q \leqslant k \leqslant n-(1)_q = (n_1-1, \ldots, n_q-1). \tag{2.7}$$

Let X be a linear space of functions defined on Ω and such that $a_k, y \in X$. Let $(S_p x)(t) = x(g_p(t))$ on Ω $(p = 1, 2, \ldots, q)$ and let $S^k = S_1^{k_1} \ldots S_q^{k_q}$, $(0)_q \leqslant k \leqslant n-(1)_q$. Since all the operators S_p are involutions of orders n_p $(p = 1, \ldots$

---

[1] The symbol $g \circ f$ denotes the superposition of two functions: $(g \circ f)(t) = g(f(t))$.

..., q) respectively, S is a multi-involution of order $n = (n_1, ..., n_q)$. Thus from Theorem 7.1 of Chapter II we obtain the following

THEOREM 2.2. *If the functions* a_k, y *are defined on a set* $\Omega \subset \boldsymbol{R}^N$, $g_1, ..., g_q$ *are Carleman functions (not identically equal to t) of orders* $n_1, ..., n_q$, *respectively on* Ω *and, moreover, condition* (2.5) *is satisfied, then a necessary and sufficient condition for equation* (2.4) *to have a solution is*

$$P_m y = 0 \quad \textit{for all } m \textit{ such that } a(\varepsilon^m) = 0,$$

where

$$a(\varepsilon^m) = \sum_{(0)_q \leqslant k \leqslant n-(1)_q} a_k \varepsilon^{km}, \quad \varepsilon^k = \varepsilon_1^{k_1} \dots \varepsilon_q^{k_q}, \quad \varepsilon_p = e^{2\pi i/n_p},$$

$$P_m = P_{1,m} \dots P_{q,m_q},$$

$$(P_{p,j}x)(t) = \frac{1}{n_p} \sum_{j=0}^{n_p-1} \varepsilon_p^{-jr} x(g_{p,r}(t)) \quad (p = 1, 2, ..., q;\ r = 1, 2, ..., n_p).$$

Under this condition the solution is

$$x = \sum_{m: a(\varepsilon^m) \neq 0} [a(\varepsilon^m)]^{-1} P_m y + \sum_{m;\, a(\varepsilon^m)=0} P_m z,$$

where z is an arbitrary function from the space of functions under consideration, the first sum runs over all m such that $a(\varepsilon^m) \neq 0$, the second sum over all m such that $a(\varepsilon^m) = 0$.

It is obvious that, by combining Theorems 2.1 and 2.2, we can obtain a new result, and so on.

It is well known (Kuroš [1]), that every finite Abelian group G is a direct sum of cyclic groups G_p of finite orders, i.e.

$$G_p = \{e_p, a_p, ..., a_p^{n_p-1}\}, \quad \text{and} \quad a_p^{n_p} = e_p,$$

where n_p is the order of G_p and e_p is the unity in G_p. This implies that every finite Abelian group of transformations of a set Ω onto itself can be written in the form (2.5).

CHAPTER XVII

Some Miscellaneous Topics

In this chapter we present some results which are not strictly connected with the main field of our studies. However, since they are obtained by the method of algebraic operators, their proper place seems to be here.

1. MATRIX AND OPERATOR EQUATIONS

The matrix equation

$$\sum_{r=1}^{N} A_r X B_r = Y \tag{1.1}$$

was solved by Lumer and Rosenblum [1] in the case where the coefficients A_r and B_r are $n \times n$ matrices (see also Bellman [1]). Here we shall solve this equation in the case where A_r and B_r are algebraic operators with simple characteristic roots. In particular we shall obtain a simple proof of Lumer's and Rosenblum's result for matrices with simple characteristic roots.

Let $A_1, \ldots, A_N$ and $B_1, \ldots, B_N$ be algebraic operators acting in an infinite dimensional space X (over the complex field). Suppose that the characteristic polynomials of the operators $A_1, \ldots, A_N, B_1, \ldots, B_N$ have only simple roots. We denote by $a_{r1}, \ldots, a r_{n_r}$ the roots of the operator A_r and by $P_{r1}, \ldots, P_{rn_r}$ the projectors associated with A_r. The respective quantities for B_r will be denoted by $b_{r1}, \ldots, b_{rm_r}$ and $Q_{r1}, \ldots, Q_{rm_r}$. Hence for $r = 1, 2, \ldots, N$

$$P_{rk} P_{rj} = \begin{cases} P_{rj} & \text{if } k = j, \\ 0 & \text{if } k \neq j, \end{cases} \qquad \sum_{j=1}^{n_r} P_{rj} = I, \tag{1.2}$$

$$A_r P_{rj} = P_{rj} A_r = a_{rj} P_{rj} \qquad (j = 1, \ldots, n_r),$$

$$Q_{rk} Q_{rj} = \begin{cases} Q_{rj} & \text{if } k = j, \\ 0 & \text{if } k \neq j, \end{cases} \qquad \sum_{j=1}^{m_r} Q_{rj} = I, \tag{1.3}$$

$$B_r Q_{rj} = Q_{rj} B_r = b_{rj} Q_{rj} \qquad (j = 1, 2, \ldots, m_r)$$

and

(1.4) $$A_r = \sum_{j=1}^{n_r} a_{rj} P_{rj}, \quad B_r = \sum_{j=1}^{m_r} b_{rj} Q_{rj}.$$

Let $\boldsymbol{j}$ denote an N-dimensional *multi-index*, i.e. $\boldsymbol{j} = (j_1, \ldots, j_N)$, where $j_1, \ldots, j_N$ are non-negative integers, and let $\mathbf{1}$ denote the multi-index $\boldsymbol{j}$ such that $j_1 = \ldots \ldots = j_N = 1$. We recall that two multi-indices $\boldsymbol{j}$ and $\boldsymbol{k}$ are equal if and only if $j_r = k_r$ for $r = 1, 2, \ldots, N$, and $\boldsymbol{j} \leqslant \boldsymbol{k}$ if and only if $j_r \leqslant k_r$ for $r = 1, 2, \ldots, N$ (compare Section 7 of Chapter II).

Let

(1.5) $$P_{\boldsymbol{j}} = \prod_{i=1}^{N} P_{ij_i}, \quad Q_{\boldsymbol{k}} = \prod_{i=1}^{N} Q_{ik_i},$$

where $\mathbf{1} \leqslant \boldsymbol{j} \leqslant \boldsymbol{n}$, $\mathbf{1} \leqslant \boldsymbol{k} \leqslant \boldsymbol{m}$, $\boldsymbol{n} = (n_1, \ldots, n_N)$, $\boldsymbol{m} = (m_1, \ldots, m_N)$.

It easily follows from Formulae (1.2) and (1.3) that

(1.6) $$\sum_{\boldsymbol{j}=\mathbf{1}}^{\boldsymbol{n}} P_{\boldsymbol{j}} = I^{1}, \quad \sum_{\boldsymbol{k}=\mathbf{1}}^{\boldsymbol{m}} Q_{\boldsymbol{k}} = I$$

and

(1.7) $$A_r P_{\boldsymbol{j}} = a_{rj_r} P_{\boldsymbol{j}}, \quad B_r Q_{\boldsymbol{k}} = b_{rk_r} Q_{\boldsymbol{k}} \quad (r = 1, 2, \ldots, N).$$

We now write for $\mathbf{1} \leqslant \boldsymbol{j} \leqslant \boldsymbol{n}$, $\mathbf{1} \leqslant \boldsymbol{k} \leqslant \boldsymbol{m}$,

(1.8) $$d_{\boldsymbol{jk}} = \sum_{r=1}^{N} a_{rj_r} b_{rk_r},$$

(1.9) $$\alpha_{\boldsymbol{jk}} = \begin{cases} 0 & \text{if } d_{\boldsymbol{jk}} = 0, \\ d_{\boldsymbol{jk}}^{-1} & \text{otherwise,} \end{cases} \qquad \beta_{\boldsymbol{jk}} = \begin{cases} \lambda^{\boldsymbol{jk}} & \text{if } d_{\boldsymbol{jk}} = 0, \\ 0 & \text{otherwise,} \end{cases}$$

where $\lambda_{\boldsymbol{jk}}$ are arbitrary scalars.

This implies that

(1.10) $$\alpha_{\boldsymbol{jk}} d_{\boldsymbol{jk}} = \begin{cases} 0 & \text{if } d_{\boldsymbol{jk}} = 0, \\ 1 & \text{otherwise,} \end{cases} \qquad \beta_{\boldsymbol{jk}} d_{\boldsymbol{jk}} = 0,$$

$(\mathbf{1} \leqslant \boldsymbol{j} \leqslant \boldsymbol{n},\ \mathbf{1} \leqslant \boldsymbol{k} \leqslant \boldsymbol{m})$.

---

[1] By $\sum_{\boldsymbol{j}=\mathbf{1}}^{\boldsymbol{n}}$ we mean $\sum_{j_1=1}^{n_1} \sum_{j_2=1}^{n_2} \ldots \sum_{j_N=1}^{n_N}$.

THEOREM 1.1. *Let* $A_r A_q - A_q A_r = 0 = B_r B_q - B_q B_r$ *for* $r, q = 1, 2, \ldots, N$. *Then the solution of equation* (1.1) *is an operator of the form*

$$V = \sum_{j=1}^{n} \sum_{k=1}^{m} (\alpha_{jk} P_j Y Q_k + \beta_{jk} P_j Q_k) \tag{1.11}$$

if and only if

$$P_\mu Y Q_\nu = 0 \quad \text{for any } \mu \text{ and } \nu \text{ such that } d_{\mu\nu} = 0. \tag{1.12}$$

PROOF. Observe that the commutativity of the operators $A_1, \ldots, A_N$ (resp. $B_1, \ldots, B_N$) implies the commutativity of the respective projectors.

Necessity. Let X be a solution of equation (1.1) and let $d_{\mu\nu} = 0$. Acting on both sides of equation (1.1) by P_μ on the left and by Q_ν on the right we obtain the equality

$$\sum_{r=1}^{N} P_\mu A_r X B_r Q_\nu = P_\mu Y Q_\nu. \tag{1.13}$$

This and Properties (1.7) together imply that

$$P_\mu Y Q_\nu = \sum_{r=1}^{N} P_\mu A_r X B_r Q_\nu = \sum_{r=1}^{N} a_{r\mu_r} b_{r\nu_r} P_\mu X Q_\nu = d_{\mu\nu} P_\mu X Q_\nu = 0$$

which proves the necessity of condition (1.12).

Sufficiency. Let condition (1.12) be satisfied. We check that the operator V given by Formula (1.11) is a solution of equation (1.1). Indeed,

$$\sum_{r=1}^{N} A_r V B_r = \sum_{r=1}^{N} \sum_{j=1}^{n} \sum_{k=1}^{m} (\alpha_{jk} A_r P_j Y Q_k B_r + \beta_{jk} A_r P_j Q_k B_r)$$

$$= \sum_{j=1}^{n} \sum_{k=1}^{m} \Big(\sum_{r=1}^{N} a_{rj_r} b_{rk_r}\Big)\Big(\alpha_{jk} P_j Y Q_k + \beta_{jk} P_j Q_k\Big)$$

$$= \sum_{j=1}^{n} \sum_{k=1}^{m} d_{jk} (\alpha_{jk} P_j Y Q_k + \beta_{jk} P_j Q_k).$$

Since $P_\mu Y Q_\nu = 0$ when $d_{\mu\nu} = 0$, we obtain, by Formulae (1.10) and (1.6)

$$\sum_{r=1}^{N} A_r V B_r = \sum_{j=1}^{n} \sum_{k=1}^{m} P_j Y Q_k = \Big(\sum_{j=1}^{n} P_j\Big) Y \Big(\sum_{k=1}^{m} Q_k\Big) = Y,$$

which was to be proved.

The above-mentioned result of Lumer and Rosenblum [1] for matrices with simple characteristic roots follows from our theorem, since every $n \times n$ matrix is an algebraic operator in an n-dimensional vector space.

2. POLYHARMONIC SOLUTIONS OF THE EQUATION $\sum_{k=0}^{n-1} a_k \Delta^k u = v$

A function $u(t, s)$ is said to be *n-harmonic* in a simply connected domain D if it satisfies the equation $\Delta^n u = 0$ in D, where Δ is the *Laplace operator*: $\Delta = \frac{\partial^2}{\partial t^2} + \frac{\partial^2}{\partial s^2}$, and if u has continuous partial derivatives in this domain of the order $2n$. If $u(t, s)$ is n-harmonique for an $n \geqslant 2$, we say briefly that u is *polyharmonic.*

In this section we shall determine the unique polyharmonic function in the simply connected domain D on the plane of the complex variable $z = t + is$ which satisfy the equation

$$\sum_{k=0}^{n-1} a_k \Delta^k u(t, s) = v(t, s), \quad \text{in } D, \tag{2.1}$$

where $n \geqslant 2$, the coefficients a_k are complex constants and the given function $v(t, s)$ is polyharmonic in the domain D. For this purpose we need the following

THEOREM 2.1. *Let X be a linear space (over the complexes) and let S be an algebraic operator of the order n with the characteristic polynomial $P(t) = t^n$; i.e.*

$$S^n = 0 \quad \text{on } X. \tag{2.2}$$

Then the equation

$$\sum_{k=0}^{n-1} a_k S^k u = v \tag{2.3}$$

(where the a_k are scalars, $a_0 \neq 0$ and $v \in X$) has a unique solution

$$u = a_0^{-n} \sum_{k=0}^{n-1} (-1)^k d_{nk} S^k v, \tag{2.4}$$

where

$$d_n = \begin{vmatrix} a_0 & a_1 & a_2 & \dots & a_{n-1} \\ 0 & a_0 & a_1 & \dots & a_{n-2} \\ 0 & 0 & a_0 & \dots & a_{n-3} \\ \dots & \dots & \dots & \dots & \dots \\ \dots & \dots & \dots & \dots & \dots \\ 0 & 0 & 0 & \dots & a_0 \end{vmatrix} = a_0^n \neq 0 \tag{2.5}$$

and d_{nk} $(k = 0, 1, \ldots, n-1)$ *denotes the subdeterminant of* d_n *obtained by cancelling the first column and the* $(k+1)$*th row.*

PROOF. Acting on both sides of equation (2.3) with the operators $S, S^2, \ldots$ $\ldots, S^{n-1}$ and using identity (2.2), we obtain the following system of equations:

$$\begin{aligned} a_0 u + a_1 Su + a_2 S^2 u + \ldots + a_{n-2} S^{n-2} u + a_{n-1} S^{n-1} u &= v, \\ a_0 Su + a_1 S^2 u + \ldots + a_{n-3} S^{n-2} u + a_{n-2} S^{n-1} u &= Sv, \\ a_0 S^2 u + \ldots + a_{n-4} S^{n-2} u + a_{n-3} S^{n-1} u &= S^2 v, \\ \ldots\ldots\ldots\ldots\ldots\ldots\ldots\ldots\ldots\ldots\ldots\ldots \\ \ldots\ldots\ldots\ldots\ldots\ldots\ldots\ldots\ldots\ldots\ldots\ldots \\ a_0 S^{n-1} u &= S^{n-1} v. \end{aligned}$$

Since the determinant d_n of this system has only zeros under the diagonal, we conclude that $d_n = a_0^n$ and hence by our assumption $d_n \neq 0$. Therefore the solution is unique and it is determined by the Cramer Formulae, which yields the required Formula (2.4).

THEOREM 2.2. *Let* S *be a linear operator mapping a linear space* X *into itself. Let*

$$X_n = \{x \in X\colon S^n x = 0\} \quad (n = 0, 1, 2, \ldots) \text{ and } \tilde{X} = \bigcup_{n=0}^{\infty} X_n.$$

Then equation (2.3) *with* $a_0 \neq 0$ *and* $v \in \tilde{X}$ *has a unique solution in* $\tilde{X}$. *More precisely: if* $v \in X_m$, *then there is a unique* $u \in X_N$ *satysfying equation* (2.3) *and such that*

$$u = a_0^{-N} \sum_{k=0}^{N-1} (-1)^k d_{Nk} S^k v, \quad \text{where } N = \max(n, m) \tag{2.6}$$

and

$$a_n = a_{n+1} = \ldots = a_{N-1} = 0, \quad \text{if } n < N \tag{2.7}$$

(d_n, d_{nk} *are defined in Theorem* 2.1).

PROOF. Let $v \in X_m$ and suppose that $m > n$. Putting $a_n = a_{n+1} = \ldots = a_{m-1} = 0$, we can write $\sum_{k=0}^{n-1} a_k S^k u = \sum_{k=0}^{m-1} a_k S^k u$. This and Theorem 2.1 imply that

$$u = a_0^{-m} \sum_{k=0}^{m-1} (-1)^k d_{mk} S^k v \in X_m.$$

In the second case we have $m \leqslant n$ and $X_m \subset X_n$. Indeed, if $x \in X_m$, then $S^n x = S^{n-m}(S^m x) = 0$, which implies $x \in X_n$. We can therefore assume that $v \in X_n$ and apply Theorem 2.1 with $X = X_n$, to obtain the required result.

Vekua [1] has proved that every function n-harmonic in a simply connected domain is given by the following formula:

$$u(t, s) = \sum_{k=0}^{n-1} [U_m(z, \zeta, h_n; c_m) + U_m(\zeta, z; h_m^*, c_m)], \tag{2.8}$$

where

$$U_m(z, \zeta; h, c) = \frac{\zeta^m}{m!} \int_0^z \frac{(z-t)^m}{m!} h(t)\, dt + \tfrac{1}{2} c \frac{z^m \zeta^m}{m!m!}, \tag{2.9}$$

$z = t+is$, $\zeta = t-is$, the functions $h_m(z)$ are holomorphic in the domain D, the functions $h_m^*(\zeta)$ are holomorphic in the domain

$$\bar{D} = \{\zeta = t-is \colon t+is \in D\},$$

and the c_m are complex constants.

This representation is unique. Indeed, if we put

$$\frac{\partial}{\partial z} = \tfrac{1}{2}\left(\frac{\partial}{\partial t} - i\frac{\partial}{\partial s}\right); \quad \frac{\partial}{\partial \zeta} = \tfrac{1}{2}\left(\frac{\partial}{\partial t} + i\frac{\partial}{\partial s}\right), \tag{2.10}$$

we have $\Delta = \dfrac{\partial^2}{\partial t^2} + \dfrac{\partial^2}{\partial s^2} = 4\dfrac{\partial^2}{\partial z\, \partial \zeta}$ and

$$\begin{gathered} c_m = \left(\frac{\partial^m u}{\partial z^m \partial \zeta^m}\right)_{z=0,\zeta=0} \quad (m = 0, 1, \ldots, n-1), \\ h_m(z) = \left(\frac{\partial^{2m+1} u}{\partial z^{m+1} \partial \zeta^m}\right)_{\zeta=0}; \quad h_m^*(\zeta) = \left(\frac{\partial^{2m+1} u}{\partial z^m \partial^{m+1}}\right)_{z=0}. \end{gathered} \tag{2.11}$$

Hence the constants c_m and the functions h_m, h_m^* are determined by the values of the function u and of its derivatives on the respective characteristics ($z \in D$, $\zeta = 0$) and ($z = 0$, $\zeta \in \bar{D}$).

We now apply Theorem 2.1. Denote by X_n the set of all functions which are n-harmonic in D and let $S = \frac{1}{4}\Delta = \partial^2/\partial z\, \partial \zeta$. Observe that the operations $U_m(z, \zeta; h, c)$ are linear with respect to h and c. Indeed, we have for all complexes λ and μ

$$U_m(z, \zeta; \lambda h + \mu \tilde{h}, \lambda c + \mu \tilde{c}) = \lambda U_m(z, \zeta; h, c) + \mu U_m(z, \zeta; \tilde{h}, \tilde{c}). \tag{2.12}$$

This implies that X_n is a linear space. Moreover,

$$S^k U_m(z, \zeta; h, c) = \begin{cases} U_{m-k}(z, \zeta; h, c) & \text{if } k \leqslant m \\ 0 & \text{if } k > m \end{cases} \quad (m, k = 0, 1, \ldots, n-1). \tag{2.13}$$

Indeed,

$$\frac{\partial}{\partial \zeta} U_m(z, \zeta; h, c) = \frac{\zeta^{m-1}}{(m-1)!} \int_0^z \frac{(z-t)^m}{m!} h(t)\, dt + \frac{c}{2} \frac{\zeta^{m-1} z^m}{(m-1)!\, m!},$$

$$SU_m(z, \zeta; h, c) = \frac{\partial^2}{\partial z\, \partial \zeta} U_m(z, \zeta; h, c) = \frac{\zeta^{m-1}}{(m-1)!} \left[\frac{(z-t)^m}{m!}\right]_{t=z} +$$

$$+ \frac{\zeta^{m-1}}{(m-1)!} \int_0^z \frac{(z-t)^{m-1}}{(m-1)!} h(t)\, dt + \frac{c}{2} \frac{\zeta^{m-1} z^{m-1}}{(m-1)!(m-1)!}$$

$$= \frac{m-1}{(m-1)!} \int_0^z \frac{(z-t)^{m-1}}{(m-1)!} h(t)\, dt + \frac{c}{2} \frac{\zeta^{m-1} z^{m-1}}{(m-1)!(m-1)!}$$

$$= U_{m-1}(z, \zeta; h, c).$$

Hence $S^k U_m = U_{m-k}$ for $k = 0, 1, \ldots, m$. Since the functions h_m^* are holomorphic in the domain $\bar{D}$, we obtain a similar result for $U_m(\zeta, z; h^*, c)$. From definition (2.9) we find

$$S^{m+1} U_m(z, \zeta; h, c) = S^{m+1} U_m(\zeta, z; h^*, c) = 0.$$

This implies that $S^n u = 0$ for every $u \in X_n$. Now denote by $\tilde{X}$ the class of all functions polyharmonic in the domain D, so that we have $\tilde{X} = \bigcup_{n=1}^{\infty} X_n$.

THEOREM 2.3. *Equation (2.1) with $a_0 \neq 0$ has a unique solution in the space X. If the given function is n'-harmonic, then the solution is N-harmonic, where $N = \max(n, n')$, and is given by the formula*

$$u(t, s) = \sum_{m=0}^{N-1} [U_m(z, \zeta; h_m, c_m) + U_m(\zeta, z; h_m^*, c_m)], \tag{2.14}$$

where

$$c_m = a_0^{-N} \sum_{k=0}^{N-1-m} (-1)^k d'_{Nk} C_{m+k},$$

$$h_m(z) = a_0^{-N} \sum_{k=0}^{N-1-m} (-1)^k d'_{Nk} H_{m+k}(z), \qquad m = 0, 1, \ldots, N-1, \tag{2.15}$$

$$h_m^*(\zeta) = a_0^{-N} \sum_{k=0}^{N-1-m} (-1)^k d'_{Nk} H_{m+k}^*(\zeta),$$

the determinant d'_{Nk} *are obtained from the determinants* d_{Nk} *defined in Theorem* 2.1 *if we put* $4^k a_k$ *instead of* a_k, $a_n = a_{n+1} = \ldots = a_{N+1} = 0$ *when* $n < N$ *and*

$$(2.16)\quad \begin{aligned} C_m &= \left(\frac{\partial^{2m} v}{\partial z^m \partial^m}\right)_{z=0,\,\zeta=0} \qquad (m = 0, 1, \ldots, N-1), \\ H_m(z) &= \left(\frac{\partial^{2m+1} v}{\partial z^{m+1} \partial \zeta^m}\right)_{\zeta=0}; \quad H_m^*(\zeta) = \left(\frac{\partial^{2m+1} v}{\partial z^m \partial \zeta^{m+1}}\right)_{z=0}. \end{aligned}$$

PROOF. If we put $\Delta = 4S$ in equation (2.1), we obtain a new equation:

$$(2.17)\qquad \sum_{k=0}^{n-1} 4^k a_k S^k u = v.$$

Arguing as in the proof of Theorem 2.2 we can assume that v is N-harmonic, where $N = \max(n, n')$ and represent it (by (2.16)) in the form

$$v(t, s) = \sum_{k=0}^{N-1} U_m(z, \zeta; H_m, C_m) + U_m(\zeta, z; H_m^*, C_m)].$$

Now observe that for $m = n'+1, \ldots, N-1$ we have $n' < N$ and $H_m = H_m^* = C_m = 0$. Theorem 2.2 therefore implies that

$$u(t, s) = a_0^{-N} \sum_{k=0}^{N-1} (-1)^k d'_{Nk} S^k v.$$

Hence from Formulae (2.15), (2.16), (2.17)

$$\begin{aligned} u(t, s) &= a_0^{-N} \sum_{k=0}^{N-1} (-1)^k d'_{Nk} S^k \left\{ \sum_{m=0}^{N-1} [U_m(z, \zeta; H_m, C_m) + U_m(\zeta, z; H_m^*, C_m)] \right\} \\ &= a_0^{-N} \sum_{k=0}^{N-1} (-1)^k d'_{Nk} \sum_{m=k}^{N-1} [U_{m-k}(z, \zeta; H_m, C_m) + U_{m-k}(\zeta, z; H_m^*, C_m)] \\ &= a_0^{-N} \sum_{m=0}^{N-1} \sum_{k=0}^{m} (-1)^{m-k} d'_{N,m-k} [U_k(z, \zeta; H_m, C_m) + U_k(\zeta, z; H_m^*, C_m)] \\ &= a_0^{-N} \sum_{k=0}^{N-1} \sum_{m=0}^{N- \ -k} (-1)^m d'_{Nm} [U_k(z, H_{m+k}, C_{m+k}) + U_k(\zeta, z; H_{m+k}^*, C_{m+k})]. \end{aligned}$$

Applying Formulae (2.12) and notation of (2.15) and interchanging the roles of the indices k and m, we obtain Formula (2.14). Since representation (2.8) of an n-harmonic function is unique, we conclude from Theorem 2.2 that this solution is unique.

THEOREM 2.4. *Solution* (2.14) *of equation* (2.1) *satisfies the following initial conditions*

$$\left(\frac{\partial^k u}{\partial \zeta^k}\right)_{\zeta=0} = u_k(z); \quad \left(\frac{\partial^k u}{\partial z^k}\right)_{z=0} = u_k^*(\zeta) \quad (k = 0, 1, \dots, N-1) \tag{2.18}$$

with the coincidence conditions

$$u_k^{(m)}(0) = u_m^{*(k)}(0) \quad (k, m = 0, 1, \dots, N-1) \tag{2.19}$$

if and only if these initial functions and the given function v *satisfy the following equalities*;

$$u_{mk}^{(m)}(0) = a_0^{-N} \sum_{k=0}^{N-1-m} (-1)^k d'_{Nk} C_{m+k} \quad (m = 0, 1, \dots, N-1),$$

$$u_m^{(m+1)}(z) = a_0^{-N} \sum_{k=0}^{N-1-m} (-1)^k d'_{Nk} H_{m+k}(z),$$

$$u_m^{*(m+1)}(\zeta) = a_0^{-N} \sum_{k=0}^{N-1-m} (-1)^k d'_{Nk} H_{m+k}^*(\zeta).$$

If these conditions are satisfied, then the unique polyharmonic solution of equation (2.1) *satisfying the initial conditions* (2.18) *has the form*

$$u(t, s) = \sum_{m=0}^{N-1} [U_m(z, \zeta; u_m^{(m+1)}, u_m^{(m)}(0)) + U_m(\zeta, z; u_m^{*(m+1)}, u_m^{(m)}(0))]. \tag{2.21}$$

The proof follows immediately from Theorem 2.3 if we note (Vekua [1]) that the function u satisfies conditions (2.18) if and only if

$$c_m = u_m^{(m)}(0); \quad h_m(z) = u_m^{(m+1)}(z); \quad h_m^*(\zeta) = u_m^{*(m+1)}(\zeta) \tag{2.22}$$
$$(m = 0, 1, \dots, N-1).$$

Finally we remark that a real N-harmonic function has the following representation

$$u(t, s) = \sum_{k=0}^{N-1} c_k r^{2k} + \operatorname{re} \sum_{k=0}^{N-1} \frac{z^k}{k!} \int_0^z \frac{(z-t)^k}{k!} h_k(t)\, dt,$$

where $r = |z|$, the c_k are real constants and the functions $h_k(z)$ are holomorphic in the domain D. Hence a solution u of equation (2.1) is real if $h_m^*(\bar{z}) = \overline{h_m(z)}$ and if the c_m are real. This and Formulae (2.15) together imply that this solution is real if $a_0, \dots, a_{N-1}$ are real and $H_m^*(\bar{z}) = \overline{H_m(z)}$ and C_m are real. This means that u is real if $a_0, \dots, a_{N-1}$ and the given function v are real.

Note that Theorem 2.3 remains true if instead of the constant coefficients a_k, we consider coefficients $a_k(t, s)$ invariant with respect to the Laplacian, i.e. functions satisfying the equation

$$\Delta a_k - a_k = 0 \quad (k = 0, 1, \ldots, N-1).$$

Theorems 2.1 and 2.2 could also be applied to solving similar problems for equations with other operators.

3. CHARACTERIZATION OF COMMUTATORS WITH SINGULAR INTEGRAL OPERATORS

In this section we give necessary and sufficient conditions for the commutator of the Hilbert transform with an operator bounded in $L^2(\boldsymbol{R})$ to be compact. Similar results will be obtained for the cotangent Hilbert transform and for the Cauchy singular integral on a closed arc. The proofs are based on a simple property of commutators with an involution, which we present at the begining of the section.

Let $\mathscr{X}$ be an algebra (a linear ring) with unit e over the field of complex numbers. We recall that an $a \in \mathscr{X}$ is an involution if $a \neq e$ and $a^2 = e$, and that an $a \in \mathscr{X}$ is an almost involution with respect to a proper two-sided ideal $J \subset \mathscr{X}$ if the coset $[a]$ is an involution in the quotient ring $\mathscr{X}/J$, i.e. if there is a $b \in J$ such that $a^2 = e+b$ (see Section 2 of Chapter II). We denote the commutator and the anticommutator of two elements $a, b \in \mathscr{X}$ respectively by $[a, b] = ab-ba$ and $(a, b) = ab+ba$.

PROPOSITION 3.1. *Let a be an involution in the algebra $\mathscr{X}$ with unit e. An element $b \in \mathscr{X}$ commutes with a if and only if there is a $b_0 \in \mathscr{X}$ such that $b = (a, b_0)$.*

PROOF. Let $b = (a, b_0)$, where $b_0 \in \mathscr{X}$. Then

$$[a, b] = ab-ba = a(ab_0+b_0a)-(ab_0+b_0a)a = a^2b_0+ab_0a-ab_0a-b_0a^2$$
$$= eb_0-b_0e = b_0-b_0 = 0.$$

Conversely, let $[a, b] = 0$. We put $b_0 = ab/2$. Then

$$(a, b_0) = ab_0+b_0a = \tfrac{1}{2}a^2b+\tfrac{1}{2}aba$$
$$= \tfrac{1}{2}(eb+aba-ba^2+ba^2) = \tfrac{1}{2}b+(ab-ba)a+be = b,$$

which was to be proved.

We also have a dual statement:

PROPOSITION 3.2. *Let a be an involution in the algebra $\mathscr{X}$ with unit e. An element $b \in \mathscr{X}$ anticommutes with a if and only if there is a $b_0 \in \mathscr{X}$ such that $b = [a, b_0]$.*

The proof is just the same as the preceding one if we interchange the roles of the signs "+" and "−".

Corollary 3.1. *Let a be an involution in the algebra* $\mathscr{X}$ *with unit e. Let J be a proper two-ideal in* $\mathscr{X}$. *Then for all* $b \in \mathscr{X}$

$$[a, b] \in J \text{ if and only if } b = da+ad+g, \quad \text{where } d \in \mathscr{X} \text{ and } g \in J.$$

The proof is immediate if we apply Proposition 3.1 to the quotient algebra $\mathscr{X}/J$.

Corollary 3.2. *Let J be a proper two-sided ideal in an algebra* $\mathscr{X}$ *with unit e. Let* $a \in \mathscr{X}$ *be an almost involution with respect to J. Then for all* $b \in \mathscr{X}$

$$[a, b] \in J \text{ if and only if } b = da+ad+g, \quad \text{where } d \in \mathscr{X} \text{ and } g \in J.$$

The proof follows immediately from Proposition 3.1 applied to the quotient algebra $\mathscr{X}/J$.

It is plain that we can obtain results dual (in the sense of Proposition 3.2) to Corollaries 3.1 and 3.2.

Let $\mathscr{X} = B(L^2(\boldsymbol{R}))$ be the algebra of all bounded operators mapping $L^2(\boldsymbol{R})$ into itself. Here and in the sequel, all functions under consideration are complex-valued. In this algebra there is only one proper closed two-sided ideal, namely the ideal $T(L^2(\boldsymbol{R}))$ of compact operators (see: Gohberg, Markus, Feldman [1]). Consider the Hilbert transform (see Section 1 of Chapter XII):

$$(Hx)(t) = \frac{1}{\pi i} \int_{-\infty}^{+\infty} \frac{x(s)}{s-t} ds \qquad (x \in L^2(\boldsymbol{R})).$$

We recall that $H \in B(L^2(\boldsymbol{R}))$ and that $H^2 = I$ on $L^2(\boldsymbol{R})$, where I is the identity operator. From Proposition 3.1 we immediately obtain

Proposition 3.3. *If* $A \in B(L^2(\boldsymbol{R}))$, *then*

$$HA-AH = 0 \text{ if and only if } A = HA_0+A_0H, \quad \text{where } A_0 \in B(L^2(\boldsymbol{R})).$$

From Corollary 3.1 we obtain

Proposition 3.4. *If* $A \in B(L^2(\boldsymbol{R}))$, *then the commutator* $HA-AH$ *is compact if and only if* $A = HA_0+A_0H+T$, *where* $A_0 \in B(L^2(\boldsymbol{R}))$ *and T is compact.*

Let $\mathscr{X} = B(L^2(0, 2\pi))$ be the algebra of all bounded operators transforming $L^2(0, 2\pi)$ into itself. In this algebra as well the unique proper two-sided ideal is the ideal $T(L^2(0, 2\pi))$ of compact operators (see: Gohberg, Markus, Feldman [1]). Now consider the cotangent Hilbert transform:

$$(H_0 x)(t) = \frac{1}{2\pi i} \int_0^{2\pi} x(s) \cot \frac{s-t}{2} ds \qquad (x \in L^2(0, 2\pi)).$$

We recall (see Section 1 of Chapter XII) that $H_0 \in B(L^2(0, 2\pi))$ and that

$$H_0^2 = I - K, \quad \text{where} \quad (Kx)(t) = \frac{1}{2\pi} \int_0^{2\pi} x(s) ds. \tag{3.1}$$

Since K is one-dimensional, it is compact. Hence H_0 is an almost involution with respect to the ideal $T(L^2(0, 2\pi))$. Corollary 3.2 implies

PROPOSITION 3.5. *If $A \in B(L^2(0, 2\pi))$, then the commutator $H_0 A - AH_0$ is compact if and only if $A = H_0 A_0 + A_0 H_0 + T$, where $A_0 \in B(L^2(0, 2\pi))$ and T is compact.*

By straightforward calculation we obtain

COROLLARY 3.3. *If $A \in B(L^2(0, 2\pi))$, then $H_0 A - AH_0 = 0$ if and only if $A = H_0 A_0 + A_0 H_0 + T$, where $A_0 \in B(L^2(0, 2\pi))$, T is compact and $H_0 T - TH_0 = A_0 K - KA_0$, where K is defined by formula* (3.1).

Let $B(H^\mu(L))$, $\mu < 1$, be the algebra of all bounded operators mapping $H^\mu(L)$ into itself. Let us denote by $T(H^\mu(L))$ the ideal of compact operators acting in the space $H^\mu(L)$. Corollary 4.1 and Theorem 4.2 of Chapter I yield the following property: If T is a continuous map from $C(L)$ into $H^\mu(L)$, then the operator $\tilde{T} = T|_{H^\mu(L)}$ is compact in the space $H^\mu(L)$. Consider now the singular integral operator with the Cauchy kernel

$$(Sx)(t) = \frac{1}{\pi i} - \int_L \frac{x(s)}{s-t} ds \qquad (x \in H^\mu(L)), \tag{3.2}$$

Theorem 1.1 of Chapter XII asserts that $S \in B(H^\mu(L))$, $\mu < 1$. From Theorem 1.3 in Chapter XII it follows that $S^2 = I$ on $H^\mu(L)$. Hence from Proposition 3.1 and Corollary 3.1 we have

PROPOSITION 3.6. *If $A \in B(H^\mu(L))$, then $SA - AS = 0$ if and only if $A = SA_0 + A_0 S$, where $A_0 \in B(H^\mu(L))$.*

PROPOSITION 3.7. *If $A \in B(H^\mu(L))$, then $SA - AS$ is compact if and only if $A = SA_0 + A_0 S + T$, where $A_0 \in B(H^\mu(L))$, T is compact.*

Propositions 3.6 and 3.7 are true also when L is an oriented system (see Section 1 of Chapter XII).

The list of similar results can easily be extended.

Bibliography

Abel, N. S.

[1] Auflösung einer mechanischen Aufgabe, *Journal für reine und angewandte Math.* 1 (1826) 153–157.

Aĭzengendler, R. G., Vaĭnberg, M. M.

[1] The branching of periodic solutions of autonomous systems and of differential equations in Banach spaces (Russian). *Dokl. Akad. Nauk SSSR* 176 (1967), 9–12.

[2] The branching of periodic solutions of differential-difference equations (Russian), *ibidem* 186 (1969), 495–498.

Aliev, B. D., Aliev, R. M.

[1] Properties of the solutions of elliptic equations with deviating arguments (Russian), Special Problems of Functional Analysis, *Izd. Akad. Nauk Azerb. SSSR*, Baku 1968.

Antosik, P., Mikusiński, J.

[1] On Hermite expansions, *Bull. Acad. Polon. Sci.* 16 (1968), 787–791.

Arens, R. F., Eells, J. Jr.

[1] On embedding uniform and topological spaces, *Pacific J. Math.* 6 (1956), 397–403.

Arscott, M. F.

[1] *Periodic differential equations*, Pergamon Press 1964.

Artola, M.

[1] Sur une équation d'évolution du premier ordre à argument retardé, *C.R. Acad. Sci. Paris, Sér. A—B* 268 (1969), A1540–A1543.

Artjušenko, L. M.

[1] The existence of analytic almost periodic solutions of a class of linear integro-differential equations (Russian), *Proc. Seventh. Sci. Conf. Dept. Higher Math.* (Frunze 1962), 95–99. *Frunze Politeh. Inst. Frunze*, 1963

Arzelà, C.

[1] Sulle funzioni di linee, *Mem. Acad. Sci. Ist. Bologna, Cl. Sci. Fis. Mat.* 5 (1895), 55–74.

Aširov, S.

[1] Operator equation of the Volterra type with a retarded argument (Russian), *Izv. Akad. Nauk Turkmen. SSR*, 5 (1966), 3–9.

Athans, M., Falb, P. L.

[1] Optimal control. *An introduction to the theory and its applications*, McGraw-Hill Book Co. New York 1966.

Averbuh, V. I., Smoljanov, O. G.

[1] Different definitions of derivative in linear topological spaces (Russian), *Uspehi Mat. Nauk.* 23 (1968), 67–116.

[2] An addendum to the article "Different definitions of the derivative in linear topological spaces" (Russian), *ibidem*, 223–224.

Avramescu, C.

[1] Sur l'existence des solutions périodiques pour des équations intégrales, *An. Şti. Univ. "Al.I. Cuza", Iaşi*, 15 (1965), 59–69.

Azamatova, V. I.

[1] The exceptional case of a certain integral equation on the half axis (Russian), *Vesci Akad. Navuk BSSR* 1 (1969), 49–58.

Azbel'ev, N. V., Rachmatullina, L. F.

[1] On linear equations with delayed argument (Russian), *Differencial'nye Uravnenija* 6 (1970), 616–628.

Azbel'ev, N. V., Rahmatullina, L. F., Čigirev, A. I.

[1] Existence, uniqueness and convergence of successive approximations for non-linear integral equations with deviating argument (Russian), *ibidem* 6 (1970), 223–229.

Babbage, Ch.

[1] Essays towards the calculus of the functions, *Philos. Trans.* I, 1815, 389–429; II, 1816, 179–256.

[2] *Examples of the solution of the functional equations*, Cambridge 1820.

[3] Des équations functionelles, *Ann. Mat. Pura Appl.* 12 (1821/22), 73–103.

Banks, H. T., Jacobs, M. Q.

[1] The optimization of trajectories of linear functional differential equations, *SIAM J. Control* 8, 4 (1970).

Bellert, S.

On the continuation of the idea of Heaviside in the Operational Calculus, *J. Franklin Inst.* 1963 (411–440).

Bellman, R.

Introduction to matrix analysis, McGraw-Hill Book Co. New York–Toronto–London 1960.

Bellman, R., Cooke, K. L.

Differential-difference equations, Academic Press, New York–London 1963, Russian ed.: Izd. Mir, Moscow 1967.

Bicadze, A. V.

[1] Inversion of a system of singular integral equations (Russian), *Dokl. Akad. Nauk SSSR*, 43 (1953).

Bielecki, A.

[1] Une remarque sur la méthode de Banach–Cacciopoli–Tikhonov dans la théorie des équations différentielles ordinaires, *Bull. Acad. Polon. Sci.* 4 (1956), 261–264.

[2] Une remarque sur l'application de la méthode de Banach–Cacciopoli–Tikhonov dans la théorie de l'équation $s = f(x, y, z, p, q)$, *ibidem* 4 (1956), 265–268.

[3] Differential equations and their certain generalizations (Polish), *Biuro Kształcenia i Doskonalenia Kadr Naukowych Polskiej Akademii Nauk*, Warszawa 1961.

Bittner, R.

[1] *Algebraic and analytic properties of solutions of abstract differential equations*, Rozprawy Mat. 41, Warszawa 1961.

Bochner, S., Chandrasekharan, K.

[1] Fourier transforms, *Ann. of Math.* 19, Princeton University Press, Princeton 1949.

Borisovič, Ju. G., Subbotan, V. F.

[1] Shift operator on trajectories of evolution equations and periodic solutions (Russian), *Dokl. Akad. Nauk SSSR* 175 (1967), 9–12.

Borok, V. M.

[1] A boundary value problem in an infinite layer for linear differential-difference systems (Russian), *Dokl. Akad. Nauk SSSR* 189 (1969), 12–15.

[2] On a boundary value problem in an infinite slab for linear differential-difference systems (Russian), *Uspehi Matem. Nauk* 24 (1969), 229–230.

Borzdyko, V. I.

[1] Positive periodic solutions of differential equations with deviating argument in a Banach space (Russian), *Dokl. Akad. Nauk Tadžik. SSR* 10 (1967), 11–14.

Bracewell, R.

The Fourier transform and its applications, McGraw-Hill Book Co. New York 1965.

Brand, L.

[1] Periodic solutions of linear differential equations, *Arch. Rational. Mech. Anal.* 27 (1967), 365–372.

Browder, F. E.

[1] *Periodic solutions of non-linear equations of evaluation equation in infinite-dimensional space. Lectures in differential equations*, vol. I, Van Nostrand, New York–Toronto–London 1969.

Bykov, Ja. V.

[1] *On certain problems of the theory of integro-differential equations* (Russian), Frunze 1957.

[2] Certain problems in the qualitative theory of difference equations (Russian), *Proc. Sventh. Sci. Conf. Dept. Higher Math.* (Frunze 1962), 10–11, Frunze politehn. Inst. Frunze 1963.

Calderón, A. P., Zygmund, A.

[1] Singular integrals and periodic functions, *Studia Math.* 14 (1954), 249–271.

Carleman, T.

[1] Sur la résolution de certaines équations intégrales, *Arkiv. Mat. Astronom. Fys.* 16 (1922), 19.

[2] Sur la théorie des équations intégrales et les applications, *Verhandl. Internation. Math. Kongress.* I. Zürich 1932, 138–151.

Cartwright, M. L.

[1] Comparison theorems for almost periodic functions, *J. London Math. Soc.* 1 (1969), 11–19.

Cerneau, S.

[1] Solutions périodiques de systèmes différentiels singuliers à argument retardé, *C.R. Acad. Sci. Paris* 269 (1969), A770–A773.

Čerskiĭ, Ju. I.

[1] General singular equation and equations of convolution type (Russian), *Mat. Sb.* 41 (83) (1957), 277–296.

[2] An intero-differential Wiener–Hopf equation and its applications (Russian), *Izv. Vysš. Učebn. Zaved. Matematika* 2 (45) (1965), 188–200.

Charatašvili, G. L.

[1] The maximum principle in the theory of optimal processes with delay (Russian), *Dokl. Akad. Nauk SSSR* 136 (1961).

[2] The optimal in the sense of speed of response processes in linear systems with delay (Russian), *Soobšč. Akad. Nauk Gruz. SSR* 33 (1964).

Chyung, D. H., Lee, E. B.

[1] Optimal systems with time delay, *III-rd IFAC congress, 7F*, London 1966.

[2] Delayed Action Control Problems, *Automatika* 6 (1970), 396–400.

Coddington, E. A., Levinson, N.

[1] *Theory of ordinary differential equations*, McGraw-Hill Book Co. New York–Toronto–London 1955.

Coĭ, K. M.

[1] On the problem of stability of oscillations of quasi-linear autonomous systems with time lag (Russian), *Differencial'nye Uravnenija* 4 (1968), 868–874.

Coĭ, K. M., Šimanov, S. N.

[1] Periodic oscillations of quasi-linear autonomous systems with retardation (Russian), *Izv. Vysš. Učebn. Zaved Radiofizika* 10 (1967), 345–352.

Conner, P. E., Floyd, E. E.

[1] *Differentiable Periodic Maps*, Springer-Verlag, Berlin–Göttingen–Heidelberg 1964.

Cooke, K. L.

[1] Functional-differential systems, some models and perturbation problem, *Intern. Symp. Diff. Eqs. Dynamical systems*, Puerto-Rico 1965.

[2] Functional differential equations close to differential equations, *Bull. Amer. Math. Soc.* 72 (1966), 285–288.

[3] Some recent works on functional-differential equations, *Proc. US Japan Seminar on Diff. and Functional Equat.*, New York 1967, 27–47.

Courduneanu, C.

[1] On a class of functional-integral equations, *Bull. Math. Soc. Sci. Math. R.S. Roumanie* 12 (60) (1968), 43–53.

Devinatz, A., Shinbrot, M.

[1] General Wiener–Hopf operators, *Amer. Math. Soc. Transl.* 145 (1965), 467–494.

Dimitriev, Ju. A.

[1] Criteria for the dissipativity and existence of periodic solutions of impulsive automatic systems with one non-linear block (Russian), *Dokl. Akad. Nauk SSSR* 175 (1967), 989–992.

Dirac, A. M.

[1] *The principles of Quantum Mechanics*, 4th ed., Oxford 1958.

Driver, R.

[1] Existence theory for a delay-differential system, *Cont. Diff. Eqs.* 1 (1963).

Drobot, S., Mikusiński, J.

[1] Sur l'unicité de quelques équations différentielles dans les espaces abstraits II, *Studia Math.* 11 (1950), 38–40.

Dybin, V. B.

[1] Normalization of Wiener–Hopf operator (Russian), *Dokl. Akad. Nauk SSSR*, 191 (1970), 759–762.

[2] Integral Wiener–Hopf operator in classes of functions with polynomial character of behaviour at infinity (Russian), *Izv. Akad. Nauk Armjan. SSSR* 2 (1967), 250–270.

Dybin, V. B., Karapietianc, N. K.

[1] On integral equations of the convolution type in the class of generalized functions (Russian), *Sibirsk. Mat. Ž.* 7 (1966), 532–545.

[2] Application of the method of normalization to a class of infinite systems of linear algebraic equations (Russian), *Izv. Vysš. Učebn. Zaved. Matiematika* 10 (65) (1967), 39–49.

Džuraiev, A.

[1] The effect of a regular part in singular integral equations in a two-dimensional bounded domain (Russian), *Dokl. Akad. Nauk SSSR*, 198 (1971), 27–30.

Elsgolc, L. E.

[1] Some properties of periodic solutions of linear and quasi-linear differential equations with deviating argument (Russian), *Vestnik Moskov. Univ. Ser. I. Mat.* 5 (1959), 65–72.

[2] *Introduction to the theory of differential equations with deviating argument* (Russian), Nauka, Moscow 1964. Polish edition: *Równania różniczkowe z odchylonym argumentem*, PWN, Warszawa 1966.

Erugin, N. P.

[1] *Linear systems of ordinary differential equations*, Academic Press, New York–London 1966.

Euler, L.

[1] Nova methodus innumerabiles aequationes differentiales secundi gradus reducendi ad aequationes differentiales primi gradus, *Commentarii Academiae Scientiarum Imperialis Petropolitanae* 3 (1728), 124–137, ed. 1732.

Fetter, E. A.

[1] Some singular integral equations connected with groups of iterations of algebraic functions (Russian), *Trudy seminara po kraievym zadačem. Kazan. Gos. Univ.* 6 (1969), 208–213.

Findeisen, W., Pułaczewski, J., Manitius, A.

[1] Multilevel optimization and dynamic coordination of mass flows in a beet sugar plant, *Automatika* 6 (1970), 581–589.

Flaišer, N. M.

[1] New method of solving in a closed form of some classes of singular integral equations with a regular part, *Rev. Roumaine Math. Pures Appl.* 10 (1965), 615–620.

Gabasov, R., Čurakova, S. V.

[1] Concerning contrability of linear systems with delay (Russian), *Techničeskaja Kibernetika* 4 (1969).

[2] Necessary conditions of optimality in systems with time delay (Russian), *Avtomat. i Telemeh.* 1, 2 (1968).

Gahov, R. D., Čibrikova, L. I.

[1] On some types of singular integral equations solvable in a closed form (Russian), *Mat. Sb.* 35 (77) (1954).

Mc Garvey, D. C.

[1] Operators commuting with translation by one. I. Differential operators with periodic coefficients in $L_p(-\infty, \infty)$. III. Perturbation results for periodic differential operators, *J. Math. Anal. Appl.* 11 (1965), 564–596; 12 (1965), 187–234.

Gdykalykov, S.

[1] *A singular integral equation with shift in the space of generalized functions* (Russian), Izd. AN Kirg. SSR 1967, 26–31.

Gegelia, T. G.

[1] On the inversion formula of Bicadze operator (Russian), *Trudy Vyčisl. Centra Akad. Nauk Gruzin. SSSR*, 3 (1963).

Ginzburg, R. E.

[1] Oscillations of linear systems with autonomous self-regulating lag (Russian), *Differencial'nye Uravnenija* 6 (1970), 1257–1264.

Gnoenskiĭ, L. S., Kamenskiĭ, G. A., Elsgolc, E. L.

[1] *Mathematical foundations of the theory of control systems* (Russian), Izd. Nauka, Moscow 1969.

Gohberg, I. C.

[1] On systems of singular integral equations (Russian), *Učen. Zap. Kišin. Gos. Univ.* 11 (1954).

[2] Factorization problems in normed rings, functions of isometric and symmetric operators and singular integral equations (Russian), *Uspehi Mat. Nauk* 19 (1964), 71–124.

Gohberg, I. C., Feldman, I. A.

[1] *Projective methods of solving of Wiener–Hopf equations* (Russian), Kišinev 1967.

[2] Integro-difference Wiener–Hopf equations (Russian), *Acta Sci. Math.* (*Szeged*) 30 (1969), 199–224.

[3] *Equations with convolution and projective method of solving* (Russian), Izdat. Nauka, Moscow 1971.

Gohberg, I. C., Kreĭn, M. G.

[1] Fundamental theorems on defect numbers, root numbers and indices of linear operators (Russian), *Uspehi Mat. Nauk* 12 (1957), 43–118.

[2] *Introduction to the theory of linear non-selfadjoint operators* (Russian), Moscow 1965.

Gohberg, I. C., Markus, A. S., Feldman, I. A.

[1] On normally solvable operators and ideals connected with them (Russian), *Izd. Mold. Fil. Akad. Nauk SSSR* 10 (76) (1960), 51–70.

Gordadze, E. G.

[1] On multidimensional singular integrals (Russian), *Bull. of the Acad. of Sciences of the Georgian SSSR* 48 (1967), 513–518.

Grafton, R. B.

[1] A periodicity theorem for autonomous functional differential equations, *J. Differential Equations* 6 (1969), 87–109.

Grimm, L. J., Schmitt, K.

[1] Boundary value problems for differential equations with deviating arguments, *Aequationes Math.* 4 (1970), 176–189.

Gumowski, I.

[1] Sur le calcul des solutions périodiques de l'équation de Cherwell–Wright, *C.R. Acad. Sci. Paris* 268 (1969), A157–A159.

[2] Sur les solutions périodiques d'une équation différentielle-fonctionnelle autonome d'ordre 1, *ibidem* 270 (1970), A123–A125.

Guseĭnov, A. I., Mamiedov, Ja. D.

[1] Study of solutions of non-linear equations with retarded argument (Russian), *Učen. Zap. Azerb. Univ. Fiz.-Mat. i Chim. Nauk* 3 (1960), 3–9.

Halanay, A.

[1] Solutions périodiques des systèmes linéaires à argument retardé, *C.R. Acad. Sci. Paris* 249 (1959), 2708–2709.

[2] Sur les systèmes d'équations différentielles linéaires à argument retardé, *ibidem* 250 (1960), 797–798.

[3] Periodic and almost periodic solutions of systems of differential equations with retarded argument (Russian), *Rev. Roumaine Math. Pures Appl. Acad. RPR* 4 (1959), 685–691.

[4] On some properties of periodic and almost periodic systems with retardation (Russian), *ibidem* 9 (1964), 667–75.

[5] Periodic invariant magnifolds for a certain class of systems with retardation (Russian), *ibidem* 10 (1965), 251–259.

[6] *Differential equations, stability, oscillations, time lag*, Academic Press 1966.

[7] Almost periodic solutions for a class of non-linear systems with time lag, *Rev. Roumaine Math. Pures Appl.* 14 (1969), 1249–1276.

Halanay, A., Kurzweil, J.

[1] A theory of invariant magnifolds for flows, *Rev. Roumaine Math. Pures Appl.* 13 (1968), 1079–1087.

Hale, J. K.

[1] Linear Functional Differential Equations with constant coefficients, *Contr. Diff. Equetions* 2 (1963), 291.

[2] Periodic and almost periodic solutions of functional-difference equations, *Archiv. Rational. Mech. Anal.* 15 (1964), 289–304.

[3] Geometric theory of functional-differential equations, Differential Equations and Dynamical Systems (*Proc. Intern. Sympos. Mayaguez P.R. 1965*) Academic Press, New York 1967, 247–266.

[4] Solutions near simple periodic orbits of functional differential equations, *J. Differential Equations* 7 (1970), 126–138.

[5] *Functional Differential Equations*, Springer Verlag, Berlin–Heidelberg–New York 1971.

Hale, J. K., Cruz, M. A.

[1] Asymptotic behavior of neutral functional differential equations, *Arch. Rational. Mech. Anal.* 34 (1969), 331–353.

[2] Existence, uniqueness and continuous dependence for hereditary systems, *Ann. Math. Pura Appl.* 85 (1970), 63–81.

Hale, J. K., Meyer, K. R.

[1] A class of functional equations of neutral type, *Mem. Amer. Math. Soc.* 76 (1967).

Hardy, G. H., Littlewood, J. E., Pólya, G.

[1] *Inequalities*, Cambridge 1934.

Heard, M. L.

[1] On asymptotic behavior and periodic solutions of a certain Volterra equation, *J. Differential Equations* 6 (1969), 172–186.

Hermite, Ch.

[1] Sur la formule d'interpolation de Lagrange, *J. Reine Angew. Math.* 84 (1878), 70–79.

Hirschman, I. I., Widder, D. V.

[1] *The convolution transform*, Princeton Univ. Press, Princeton, 1955 (Russian edition: Moscow 1958).

Hvedelidze, B. V.

[1] Singular integral equations with Cauchy–Lebesgue integral (Russian), *Soobšč. Akad. Nauk Gruz. SSR* 7 (1947).

Iftimie, V.

[1] Solutions périodiques des équations de convolution, *C.R. Acad. Sci. Paris* 269 (1969), A454–A456.

Ince, E. L.

[1] *Ordinary differential equations* (Longmans, Green, 1926), London 1927.

Jacobson, N.

[1] Structure of the rings, *Amer. Math. Soc. Transl.* Providence 1956.

Jakovleva, G. F.

[1] The existence of bounded and periodic solutions of integro-differential equations with retarded argument (Russian), *Differencialnyje Uravnenija* 3 (1967), 912–925.

Johnson, B. E., Sinclair, A. M.

[1] Continuity of derivation and a problem of Kaplansky, *Amer. J. Math.* 90 (1968), 1067–1073.

Jones, G. S. Jr.

[1] Asymptotic behavior and periodic solutions of a non-linear differential-difference equation, *Proc. Nat. Acad. Sci. USA* 47 (1961), 879–882.

[2] Periodic functions generated as solutions of non-linear differential-difference equations, *Intern. Sympos. Nonlinear Differ. Equations and Nonlinear Mechanics. Colorado Springs 1961*, New York–London 1963, 105–112.

Jumarie, G.

[1] Contribution à l'étude des solutions périodiques des équations aux différences non linéaires, *Rev. CETHEDEC*, 19 (1969), 69–89.

Judaev, G. S.

[1] On the theory of infinite systems with retardations (Russian), *Izv. Vysš. Učebn. Zaved. Matematika* 7 (1969), 101–107.

Kahane, Ch.

[1] On operators commuting with differentiation, *Amer. Math. Monthly* 76 (1969), 171–173.

Kakičev, V. A.

[1] Boundary problems of linear conjugation for functions holomorphic in bicylindric domains (Russian), *Teoria Funkcii̇, Funkc. Anal. i ih Prilož.* 5 (1967), 35–58.
[2] On regularization of singular integral equations with Cauchy kernel for bicylindric domains (Russian), *Izv. Vyzov MVO* 7 (62), 1967.
[3] Boundary value problems of linear conjugation for functions holomorphic in bicylindrical regions (Russian), *Soviet Math. Dokl.* 9 (1968), 222–226.
[4] Degenerated two-dimensional integral equations with Cauchy kernels for bicylindrical domains (Russian), *Teoria Funkcii̇, Funkc. Analiz i ih Prilož.* 7 (1968), 13–19.
[5] On a problem of linear conjugation for bicylindric domains and its applications (Russian), *Materialy konferencii po kraievym zadačam.* Izd. Kazansk. Univer. 1970.
[6] Multiplicative boundary problems for functions holomorphic in bicylindric domains (Russian), *Teoria Funkcii̇, Funkc. Analiz i ih Prilož.* 12 (1970), 12–20.

Kakutani, S., Markus, L.

[1] *On the non-linear difference-differential equation* $y'(t) = A - By(t-\tau)y(t)$. *Contribution to the non linear oscillations*, Vol. 4, Princeton University Press, Princeton–New York 1958, 1–18.

Kaplansky, I.

[1] The structure of certain operator algebras, *Trans. Amer. Math. Soc.* 70 (1951), 219–255.
[2] *Infinite abelian groups*, Ann. Arbor, Michigan, 1954.
[3] *An introduction to differential algebra*, Paris 1957.

Karapietianc, N. K., Samko, S. G.

[1] On the index of certain classes of integral operators (Russian), *Dokl. Akad. Nauk SSSR*, 194 (1970), 504–508.
[2] On a class of integral equations of convolution type and its applications (Russian), *ibidem*, 193 (1970), 981–984.

Karimov, S.

[1] On periodic solutions of systems of difference equations with two variables (Russian), *Izv. Vysš. Učebn. Zaved. Matematika* 4 (71) (1968), 33–39.

Kemper, G. A.

[1] Almost periodic functional differential equations, *SIAM J. Appl. Math.* 16 (1968) 155–161.

Kirillova, F. M.

[1] Applications of functional analysis to the theory of optimal processes, *SIAM J. Control* 5 (1967).

Kisielewicz, M.

[1] On the existence of solutions of differential-integral equations with lagging argument, *Comment. Math.* 13 (1970), 255–266.

[2] *Some properties of partial functional-differential equations of the hyperbolic type* (Polish), Lubuskie Towarzystwo Naukowe, Poznań–Zielona Góra 1971.

Kleĭmenov, A. F.

[1] The existence and stability of periodic solutions of systems with lag which are close to a Liapunov system (Russian), *Differencialnyje Urawnenija* 4 (1968), 1433–1440.

[2] Oscillation in nearly-Liapounov self-contained systems with lag (Russian), *Prikl. Mat. Meh.* 32 (1968), 567–574 (translated as *J. Appl. Mat. Mech.* 32 (1968), 589–596).

Kleĭmenov, A. F., Šimanov, S. N.

[1] Periodic solutions of systems with lag closely related to Liapounov systems (Russian), *ibidem* 33 (1969), 403–412; 33 (1969), 392–401.

Kolesov, Ju. S.

[1] Periodic solutions of a class of differential equations with hysteric non-linearity (Russian), *Dokl. Akad. Nauk SSSR*, 176 (1967), 1240–1243.

Komjak, I. I.

[1] A certain integral equation on the half-axis (Russian), *ibidem* 13 (1969), 197–201.

Konovalov, Ju. P., Šimanov, S. N.

[1] Periodic solutions of quasi-harmonic systems with retardation (Russian), *Izv. Vysš. Učebn. Zaved. Matematika* 6 (61) (1967), 59–67.

Kordzadze, R. A.

[1] A class of singular integral equations with shift, *Dokl. Akad. Nauk SSSR* 168 (1966), 1245–1247.

[2] The index of singular integro-differential operators (Russian), *ibidem* 185 (1969), 753–756.

Kordylewski, J., Kuczma, M.

[1] On some linear functional equations, *Ann. Polon. Math.* I. 9 (1960), 119–136; II. 11 (1962), 203–207.

Korenevskiĭ, D. G., Feĭčenko, S. F.

[1] The Cauchy problem for a hyperbolic equation with functionally perturbed argument (Russian), *Ukrain. Mat. Ž.* 21 (1969), 108–110.

Krasnosel'skiĭ, M. A.

[1] Alternative principle of the existence of periodic solutions for differential equations with time lag (Russian), *Dokl. Akad. Nauk SSSR* 152 (1963), 801–805.

[2] *Shift operator on trajectories of differential equations* (Russian), Izd. Nauka, Moscow 1966.

Krasnosel'skiĭ, M. A., Burd, V. S., Kolesov, Ju. S.

[1] *Non-linear almost periodic oscillations* (Russian), Izd. Nauka, Moscow 1970.

Krasnosel'skiĭ, M. A., Kreĭn, S. G.

[1] Operator equations in function spaces (Russian), *Proceeding IV Math Congress, Leningrad 1964*. Izd. Nauka, 1969, 292–299.

Krasnosel'skiĭ, M. A., Lifšic, E. A., Strugin, V. V.

[1] A new method in the problem of periodic solutions of equations with deviating argument (Russian), *Trudy Sem. Teor. Differencial. Uravnieniĭ s Otkoln. Arg. Univ. P. Lumumby* 5 (1967), 116–120.

Krassowskiĭ, N. N.

[1] *Theory of motion control (linear systems)* (Russian), Izd. Nauka, Moscow 1968.

Krylov, N. M., Bogoljubov, N. V.

[1] *Introduction to non-linear mechanics* (Russian), Izd. AN Ukrain. SSR, Kiev 1937; English translation: *Ann. of Math. Studies 11*, Princeton Univ. Press, Princeton–New York 1943.

Kschwendt, H.

[1] Legendre expansion and integral equations of dispacement type, *J. Computational Phys.* 5 (1970), 84–102.

Kuczma, M.

[1] *Functional equations in a single variable*, PWN—Polish Scientific Publishers, Warszawa 1968.

Kudrewicz, J.

[1] Periodic Oscillations in non-linear systems, *Arch. Automat. i Telemech.* 11 (1966), 373–389.

[2] Stability of linear systems with periodic time-varying parameters (Polish), *ibidem* 15 (1970), 259–278.

Kulesko, N. A.

On the theory of linear differential equations with periodic right-hand side and with distributed deviations of the argument (Russian), *Teoria Funkciĭ, Funkc. Analiz i ih Priložen.* 4 (1967), 67–78.

Kuller, R. G.

[1] On the differential equation $f' = f \circ g$, where $g \circ g = I$, *Math. Mag.* 42 (1969), 195–200.

Kuroš, A. G.

[1] *Theory of Groups* (Russian), Ed. III. Izd. Nauka, Moscow 1967.

Kuržanskiĭ, A. B.

[1] On the approximation of linear differential equations with lag (Russian), *Differentialnyje Uravnenija* 3 (1967), 2094–2107.

Kurzweil, J.

[1] Invariant magnifolds of a class of linear functional differential equations, *Rev. Roumaine Math. Pures Appl.* 13 (1968), 1113–1120.

Kwapisz, M.

[1] On quasilinear differential-functional equations with quasilinear conditions, *Math. Nachr.* 43 (1970), 215–222.

Lacroix, S. F.

[1] *Traité du calcul différentiel et du calcul intégral*, Vol. III. 2-me éd. Paris 1819 (Chapter 8).

Laĭterer, Ju.

[1] Criteria of normal solvability of systems of singular integral equations and Wiener–Hopf equations (Russian), *Mat. Sb.* 83 (125) (1970), 388–406.

Lasota, A., Olech, C.

[1] An optimal solution of Nicoletti's boundary value problem, *Ann. Polon. Math.* 18 (1966), 131–139.

Lasota, A., Opial, Z.

[1] Fixed-point theorems for multi-valued mappings and optimal control problems, *Bull. Acad. Polon. Sci.* 16 (1968), 645–649.

[2] Sur les solutions périodiques des équations différentielles ordinaires, *Ann. Polon. Math.* 16 (1964), 69–94.

Lasota, A., Yorke, J. A.

[1] Bounds for periodic solutions of differential equations in Banach spaces, *The Institute for Fluid Dynamics. Univ. of Maryland, Technical Note BN*-633, *January 1970.*

Lazer, A. C., Sánchez, D. A.

[1] On periodically perturbed systems, *Michigan Math. J.* 16 (1969), 193–200.

Lee, E. B., Markus, L.

[1] *Foundations of Optimal Control Theory*, J. Wiley and Sons Inc. New York–London–Sydney 1968.

Levin, J. J.

[1] Boundness and oscillation of some Volterra and delay equations, *J. Differential Equations* 5 (1969), 369–398.

Levitan, B. M.

[1] *Almost periodic functions* (Russian), Moscow 1953.

Lifšic, E. A.

[1] Fredholm's alternative for the problem of periodic solutions of differential equations with distributed argument (Russian), *Problemy Mat. Anal. Slož. Sistem.* 1 (1967), 53–57.

Lillo, J. C.

[1] Oscillatory solutions of the equations $y(x) = m(x)y(x-n(x))$, *J. Differential Equations* 6 (1969), 1–35.

[2] Periodic differential-difference equations of order *n*, *Amer. J. Math.* 91 (1969), 368–384.

Litvinčuk, G. S.

[1] Some systems of singular integral equations (Russian), *Uspehi Matem. Nauk* 18 (1963), 139–144.

[2] Noether theory of systems of singular integral equations with Carleman deviation and complex conjugate unknowns (Russian), *Izv. Akad. Nauk SSSR* 31 (1967), 563–586.

[3] Corrigenda to my paper "Noether theory..." (Russian), *ibidem* 32 (1968), 1414–1417.

Litwinčuk, G. S., Habasov, E. G.

[1] On theory of singular integral equations satysfying Fredholm alternative (Russian), *Dokl. Akad. Nauk SSSR* 140 (1961), 48–51.

[2] On Hilbert boundary problem with shift (Russian), *ibidem* 142 (1962), 274–277.

[3] On a class of singular integral equations with shift (Russian), *ibidem* 145 (1962), 731–734.

[4] A class of singular integral equations and boundary problem of Carleman type (Russian), *Izv. Vysš. Učebn. Zaved. Matematika* I.4 (1964), 99–110; II.5 (1964), 41–53.

[5] A class of singular integral equations and generalized boundary problem of Carleman type (Russian), *Sibirsk. Mat. Ž.* 5 (1964 , 608–625.

Litvinčuk, G. S., Nečaev, A. P.

[1] A contribution to the theory of Carleman's generalized boundary value problem, *Soviet Math. Dokl.* 10 (1969), 1341–1345.

Litvinčuk, G. S., Zverovič, Z. I.

[1] Boundary problems with shift and singular functional equations (Russian), *Uspehi Mat. Nauk* 23 (1968), 67–121.

Ljubič, Ju. I., Tkačenko, V. A.

[1] On a Floquet theory for equations with retarded argument (Russian), *Diferencjal'nye Uravneniïa* 5 (1969), 648–656.

Lord, M. E.

[1] A discrete algebraic derivative, *J. Math. Anal. Appl.* 25 (1969), 701–709.

Lumer, G., Rosenblum, M.

[1] Linear operator equations, *Proc. Amer. Math. Soc.* 10 (1959), 32–41.

Manitius, A.

[1] *Optimal control of systems with delays in state* (Polish), Ph. D. dissertation, Politechnika Warszawska, Warszawa 1968.

[2] *Optimum control of linear time lag systems*, IV IFAC Congress, session 13, Warszawa 1969.

[3] Optimal control of processes with delays—A review and some new results (Polish), *Arch. Automat. i Telemech.* 15 (1970), 2J5–221.

Martynjuk, D. I.

Periodic solutions of non-linear second order differential equations (Ukrainian), *Ukrain. Mat. Ž.* 19 (1967), 125–132.

Martynjuk, D. I., Šamoĭlenko, A. M.

Periodic solutions of non-linear systems with delay (Russian), *Math. Phys.* 3 (1967), 128–148.

Mażbic-Kulma, B.

[1] On an equation with reflection of order *n*, *Studia Math.* 35 (1970), 69–76.

[2] Differential equations in a linear ring, *ibidem* 39 (1971), 157–161.

[3] *Algebraic derivative and functional differential equations* (Polish), Ph. D. dissertation, Institute of Mathematics, Polish Academy of Sciences, Warszawa 1971.

de Medrano, S. L.

[1] *Involutions on Manifolds*, Springer Verlag, Berlin–Heidelberg–New York 1971.

Meyer, K. R.

On the existence of solutions of linear differential-difference equations. Differential Equations and Dynamical Systems, Acad. Press. New York 1967, 239–245.

Mihlin, S. G. (Mikhlin, S. G.)

[1] Singular Integral Equations (Russian), *Uspehi Mat. Nauk.* 3 (1948), 31–112.

[2] *Integral Equations and their applications to certain problems of mechanics, mathematical physics and technology*, Pergamon Press, New York 1964 (I-st Russian edition: Moscow–Leningrad 1949).

Mikołajska, Z.

[1] Une remarque sur les solutions bornées d'une équation différo-différentielle non-linéaire, *Ann. Polon. Math.* 15 (1964), 23–32.

[2] Remarques sur l'allure des équations différentielles à paramètre retardé, *ibidem* 16 (1964), 59–68.

[3] Sur l'allure asymptotique des solutions d'une équation à paramètre retardé, *ibidem* 16 (1965), 214–219.

[4] Une remarque sur l'existence d'une solution périodique d'une équation différo-différentielle au deuxième membre croissant, *ibidem* 18 (1966), 53–58.

[5] Une modification de la condition de Liapounov pour les équations à paramètre retardé, *ibidem* 21 (1969), 103–111.

[6] Une remarque sur des notes de Razumichin et Krasovskij sur la stabilité asymptotique, *ibidem* 22 (1969), 1–4.

[7] Sur l'existence d'une solution périodique d'une équation différentielle du premier ordre avec paramètre retardé, *ibidem* 23 (1970), 25–36.

[8] Une remarque sur la méthode des approximations successives dans la recherche des solutions périodiques des équations différentielles à paramètre retardé, *ibidem* 24 (1970), 55–64.

Mikusiński, J.

[1] Un théorème d'unicité pour quelques systèmes d'équations différentielles considérées dans les espaces abstraits, *Studia Math.* 12 (1951), 80–83.

[2] Sur les solutions linéairement indépendantes des équations différentielles à coefficients constants, *ibidem* 16 (1958), 41–47.

[3] Sur les théorèmes d'unicité et le nombre de solutions linéairement indépendantes, *ibidem* 16 (1958), 95–98.

[4] Sur les espaces linéaires avec dérivation, *ibidem* 16 (1958), 113–123.

[5] Extension de l'espace linéaire avec dérivation, *ibidem* 16 (1958), 156–172.

[6] *Operational calculus*, Pergamon Press 1959 (I-st Polish edition: Warszawa 1953).

Miller, R. K., Sell, G. R.

[1] Volterra integral equations and topological dynamics, *Mem. Amer. Math. Soc.* 102.

Minorsky, N.

[1] Self-excited mechanical oscillations, *J. Appl. Phys.* 19 (1948), 332–338.

Misnik, V. P.

[1] On periodic solutions of integro-differential difference equations (Russian), *Izv. Vysš. Učebn. Zaved. Matematika* 2 (45) (1965), 110–117.

Mitropol'skiĭ, Ju. A., Tkač, B. P.

[1] Periodic solutions of non-linear systems of partial differential equations of neutral type (Russian), *Ukrain. Math. Ž.* 21 (1969), 475–486.

Monari, C.

[1] Soluzioni periodiche di una equazione parabolica con un termine di ritardo non-lineare, *Ist. Lombardo Accad. Sci. Lett. Rend. A* 103 (1969), 688–703.

Montgomery, D., Zippin, L.

[1] *Topological Transformation groups*, Interscience Publ. London 1955.

Motkin, A. S.

[1] Singular integral equations on an infinite contour (Russian), *Vesci Akad. Nauk Belarusk. SSR* 4 (1967), 36–46.

Muchamadiev, E.

[1] On the theory of periodic completely continuous vector field (Russian), *Uspehi Mat. Nauk* 22 (1967), 127–128.

Munteanu, M.

[1] On a linear functional equation. (Roum.), *Bul. Sti. Inst. Politehn. Cluj* 10 (1967), 33–37.

Muskhelishvili, N. I.

[1] *Singular integral equations*, Groningen 1953 (III-rd Russian complemented and corrected edition: Moscow 1968).

[2] *Certain problems of mathematical theory of elasticity* (Russian), Moscow 1954.

Myškis, A. D.

[1] General theory of differential equations with retarded argument (Russian), *Uspehi Mat. Nauk* 4 (1949), 99–141.

[2] *Linear differential equations with retarded argument* (Russian), Moscow–Leningrad 1951.

Myškis, A. D., Elsgolc, L. E.

[1] The status and problems of the theory of differential equations with deviating argument (Russian), *Uspehi Mat. Nauk* 22 (1967), 21–57.

Neĭmark, Ju. I., Fišman, L. Z.

[1] Behavior in the large of phase trajectories of quasi-linear differential equations with retarded argument (Russian), *Dokl. Akad. Nauk SSSR* 171 (1966), 44–47.

Neresjan, A. B.

[1] Application of certain transformation operators to boundary value problem for equations with time lag (Russian), *Mat. Sbornik* 63 (1964), 341–355.

Neuman, F.

[1] Criterion of periodicity of solutions of a certain differential equation with a periodic coefficient, *Ann. Mat. Pura. Appl.* 75 (1967), 385–396.

Noether, F.

[1] Über eine Klasse singulärer Integralgleichungen, *Mat. Ann.* 82 (1921), 21–52.

Nohel, J. A.

[1] La théorie géometrique d'une classe d'équations non linéaires différentielles avec arguments retardés, *Enseignement Mat.* 12 (1966), 173–182.

Nordgren, E. A.

[1] Invariant subspaces of a direct sum of weighted shifts, *Pacific J. Math.* 27 (1968), 587–598.

Norkin, S. B.

[1] *Differential equation of second order with delay*, Izd. Nauka, Moscow 1965.

Oğuztörelli, N. N.

[1] *Time lag control systems*, Academic Press. New York–London 1966.

[2] A class of non-linear differential equations of parabolic type with a delayed argument, *Bull. Inst. Politehn. Iaşi* (N.S.), 14 (1968), 43–49.

[3] Un problema misto concernente un'equazione integro-differenziale di tipo parabolico con argomento ritardato, *Rend. Mat.* 2 (1969), 245–294.

Orth, D.

[1] Singular integral equations with shifts. I, *J. Math. Mech.* 18 (1968), 491–516. II. *Math. J. Indiana Univ.* 20 (1971), 603–621.

Paradoksova, I. A.

[1] On a singular integral equation with a cyclic group of rational transformations, *Naučn. Dokl. Vysš. Školy. Fiz. Mat. Nauki* 6 (1958), 36–40.

Perello, C.

[1] *A note on periodic solutions of non linear differential equations with time lags*, Differential Equations and Dynamical Systems (Proc. Internal. Sympos. Mayaguez, P.R. 1965, 185–187), Academic Press, New York 1967.

[2] Periodic solutions of differential equations with time lag containing a small parameter, *J. Differential Equations* 4 (1968), 160–175.

Pinney, E.

[1] *Ordinary difference-differential equations*, Univ. of California Press, Berkeley and Los Angeles 1958.

Plaat, O.

[1] Linear difference operators on periodic functions, *Proc. Amer. Math. Soc.* 18 (1967), 257–262.

Plemelj, J.

[1] Ein Ergänzungsatz zur Cauchy'schen Integraldarstellung analytischer Funktionen Randwerte betreffend, *Monatsch. Math. u. Physik* 19 (1908), 205–210.

Pogorzelski, W.

[1] *Integral Equations and their applications.* I, Pergamon Press, PWN—Polish Scientific Publishers, Warszawa 1966 (First Polish ed. by PWN: Part I—1953, Part II—1958, Part III—1960).

[2] *Integral equations and their applications*, Part IV (Polish), PWN—Polish Scientific Publishers, Warszawa 1970.

[3] Sur les problèmes aux limites discontinues dans la théorie des fonctions analytiques, *J. Math. Mech.* 9 (1960), 583–606.

Poincaré, H.

[1] *Leçons de méchanique céleste*, III. Paris 1910.

Poisson, S. D.

[1] Mémoire sur les équations aux différence mélées, *J. de l'École Polytechnique* 6 (1806), 126–147.

Pontryagin, L. S., Boltyanskii, V. G., Gamkrelidze, R. V., Miscenko, E. F.

[1] *The mathematical theory of optimal processes*, Interscience Publishers Inc., New York 1962.

Presič, S. B.

[1] A method for solving a class of cyclic functional equations, *Mat. Vestnik* 5 (1968), 375–377.

Privalov, I. I.

[1] Sur les fonctions conjugées, *Bull. Soc. Math. France* 44 (1916), 100–103.

[2] Cauchy's Integral (Russian), *Izv. Saratovsk. Univ.* 1 (1918), Saratov 1919.

[3] *Boundary properties of univalent analytic functions* (Russian), II-nd ed. Moscow 1950.

Prössdorf, S.

[1] Zur Theorie der Faltungsgleichungen nicht normalen Typs, *Math. Nachr.* 42 (1969), 103–131.

[2] Über eine Klasse singulärer Integralgleichungen nicht normalen Typs, *Math. Ann.* 183 (1969), 130–150.

[3] Zur Lösung eines systems singulärer Integralgleichungen mit entartetem Symbol, Elliptische Differentialgleichungen, 1 (1970), 111–118.

Prouse, G.

Periodic or almost periodic solutions of a non linear functional equation, *Atti Accad. Naz. Lincei. Rend.* I. 43 (1967), 161–167; II. *ibidem* 281–287; III. *ibidem* 448–452; IV. *ibidem* 44 (1968), 1–8.

Przeworska-Rolewicz, D.

[1] Sur les équations involutives et leurs applications, *Studia Math.* 20 (1961), 95–117,

[2] Sur les involutions d'ordre *n*, *Bull. Acad. Polon. Sci.* 8 (1960), 735–739.

[3] Sur les équations involutives, *ibidem* 741–746.

[4] Sur les opérations satisfaisantes à l'identité polynomiale, *Studia Math.* 22 (1962). 43–58.

[5] Équations avec opérations algébriques, *ibidem* 22 (1963), 337–367.

[6] Sur l'unique solution polyharmonique d'équation $\sum_{k=0}^{n-1} \Delta^k u = v$, *Atti Accad. Naz. Lincei, Sér.* VIII, 5, 34 (1963), 34–39.

[7] Sur les équations avec opérations presque algébrique, *Studia Math.* 25 (1965), 163–180.

[8] On periodic solutions of linear differential-difference equations with constant coefficients, *ibidem* 31 (1968), 69–73.

[9] On equations with several involutions of different orders and its applications to partial differential-difference equations, *ibidem* 32 (1969), 102–112.

[10] On equations with reflection, *ibidem* 33 (1969), 191–200.

[11] On equations with rotation, *ibidem* 35 (1970), 51–68.

[12] A characterization of algebraic derivative, *Bull. Acad. Polon. Sci.* 9 (1969), 11–13.

[13] Algebraic derivative and abstract differential equations, *An. Acad. Brasil. Ci.* 42 (1970), 403–409.

[14] Sur l'intégrale de Cauchy pour un arc fermé à l'infini, *Ann. Polon. Math.* 8 (1960), 155–171.

[15] A characterization of commutators with Hilbert transforms, *Studia Math.* 44 (1972), 27–30.

[16] Generalized linear equations of Carleman type, *Bull. Acad. Polon. Sci.* 20 (1972), 635–639

[17] Right invertible operators and functional-differential equations, *Demonstratio Math.* (to appear).

Przeworska-Rolewicz, D., Rolewicz, S.

[1] Remarks on Φ-operators in linear topological spaces, *Prace Mat.* 9 (1965), 91–94.

[2] On d and d_Ξ-characteristic of linear operators, *Ann. Polon. Math.* 19 (1967), 117–121.

[3] On operators with finite d-characteristic, *Studia Math.* 24 (1964), 257–270.

[4] On operators preserving a conjugate space, *ibidem* 25 (1964), 251–255.

[5] On quasi-Fredholm ideals, *ibidem* 26 (1965), 67–71.

[6] *Equations in linear spaces*, PWN—Polish Scientific Publishers, Warszawa 1968.

[7] On periodic solutions of non-linear differential-difference equations, *Bull. Acad. Polon. Sci.* 16 (1968), 577–580.

[8] On control of linear periodic time lag systems, *Studia Math.* 32 (1969), 142–152.

Reissig, R., Sansone, G., Conti, R.

[1] *Qualitative Theorie nichtlinearer Differentialgleichungen*, Edizioni Cremonese, Roma 1963.

Riesz, F.

[1] Über lineare Funktionengleichungen, *Acta Math.* 41 (1918), 71–98.

Riesz, M.

[1] Sur les fonctions conjugées, *Math. Z.* 27 (1927), 218–244.

Rjabov, Ju. A.

[1] On the analysis of non-linear oscillations of systems with small retardation (Russian), *Abh. Deutsch. Akad. Wiss. Berlin* 1 (1965), 94–95.

Rochberg, R.

[1] The equation $(I-S)g = f$ for shift operator in Hilbert space, *Proc. Amer. Math. Soc.* 19 (1968), 123–129.

Röhl, H.

[1] Ω-degenerate singular internal equations and holomorfic affine boundles over compact Riemann surfaces I, *Comm. Math. Helv.* 38 (1963), 84–120.

Rolewicz, S.

[1] On perturbations of deviations of periodic differential-difference equations in Banach spaces, *Studia Math.* 47 (1973), 31–35.

Rožkov, V. I.

[1] Equations of neutral type with small delay dependent on unknown function (Russian), *Izv. Vysš. Učebn. Zaved. Matematika* 6 (73) (1968), 90–97.

Ryan, J. P.

[1] *The shift and commutativity*, Preprint from the author's Ph.D. thesis at the Yale University. University of Kentucky, Lexington, 1970.

Sadyhov, M. M., Ahmedov, K. I., Gasanov, K. K.

[1] Differential-extremal equations with retarded argument (Russian), *Azerbaĭdžan Gos. Univ. Učen. Zap.* 4 (1967), 63–71.

Samko, S. G.

[1] Solution of generalized Abel equation by means of an equation with Cauchy kernel (Russian), *Soviet Math. Dokl.* 8 (1967), 1259–1262.

[2] A general singular equation on an open curve and the generalized Abel equation (Russian), *ibidem* 8 (1967), 1377–1381.

[3] On a generalized Abel equation and operators of fractional integration (Russian), *Differencial'nye Uravnenya* 4 (1968), 298–314.

[4] Abel's generalized equation, Fourier transform and convolution-type equations (Russian), *Soviet Math. Dokl.* 10 (1969), 942–946.

[5] General singular operator and integral operator with automorphic kernel (Russian), *Izv. Vysš. Učebn. Zaved. Matematika* 1 (8) (1969), 67–77.
[6] On solvability in closed form of singular integral equations (Russian), *Soviet Math. Dokl.* 10 (1969), 1445–1448.
[7] Generalized Abel integral equation on straight line (Russian), *Izv. Vysš. Učebn. Zaved. Matematika* 8 (99) (1970), 83–93.

Šamoĭlenko, A. M.
[1] A numerical-analytic method for analyzing countable systems of periodic differential equations (Russian), Math. Phys. (1966), 115–132, Naukova Dumka, Kiev 1966.

Schatte, P.
[1] Funktionen-theoretische Untersuchungen im Mikusińskischen Operatorkörper, *Math. Nachr.* 35 (1967), 19–56.

Schmeidler, W.
[1] *Integralgleichungen mit Anwendungen in Physik und Technik.* I, Lineare Integralgleichungen. Akademische Verlagsgesellschaft Geese & Portig K.-G., Leipzig 1955.

Schmitt, K.
[1] Periodic solutions of linear second order differential equations with deviating argument, *Proc. Amer. Math. Soc.* 26 (1970), 282–285.

McShane, N.
[1] On the periodicity of homeomorphism on real line, *Amer. Math. Monthly* 68 (1961).

Sikorski, R.
[1] On Mikusiński's algebraic theory of differential equations, *Studia Math.* 16 (1958), 230–236.

Silberstein, L.
[1] Solution of the equation $f'(x) = f(1/x)$, *Phil. Mag.* 30 (1940), 185–186.

Šimanov, S. N.
[1] On almost periodic oscillations in non-linear systems with delay (Russian), *Dokl. Akad. Nauk SSSR* 125 (1959), 1203–1206.
[2] On almost periodic oscillations of quasilinear systems with delay in a degenerate case (Russian), *ibidem* 133 (1960), 36–39.
[3] Concerning theory of linear differential equations with periodic coefficient and with time lag (Russian), *Prikl. Mat. Mech.* 27 (1963), 450–458.

Simonienko, I. B.
[1] Boundary Riemann problem with measurable coefficients (Russian), *Dokl. Akad. Nauk SSSR*, 135 (1960), 538–541.
[2] Boundary Riemann problem for n pairs of functions with measurable coefficients and its application to studies of singular integrals in L_p spaces with weights (Russian), *ibidem* 141 (1961), 36–40.
[3] On a maximal extrapolation property of functions possessing integral representation (Russian), *Mat. Sb.* 65 (1964), 390–397.

[4] Operators of convolution type (Russian), *ibidem* 74 (1967), 298–313.
[5] Concerning the problem of solvability of bisingular and polysingular equations (Russian), *Funkcionalnyĭ Analiz i Prilož.* 5 (1971), 93–94.

Singer, I.
[1] Sur une classe d'applications linéaires continues dans les espaces L_F^p $(1 \leqslant p < \infty)$, *Ann. Sci. École Norm. Sup.* 77 (1960), 235–256.

Singer, I. M., Wermer, J.
[1] Derivation on commutative normed algebras, *Math. Ann.* 129 (1955), 260–264.

Sosunov, A. S.
[1] Concerning the problem of regularization of singular integral equations with deviation satisfying the generalized Carleman condition (Russian), *Izv. Vysš. Učebn. Zaved. Matematika* 4 (59) (1967), 103–111.

Stephan, B. H.
[1] On the existence of periodic solutions of $z'(t) = -az(t-r+\mu k(t, z(t))+F(t)))$, *J. Differential equations* 6 (1969), 408–419.

Tažimuratov, I., Harasahal, V. V.
[1] *The existence of periodic solutions of systems of linear homogeneous partial differential equations* (Russian), Certain Problems in Differential Equations (Russian). Izd. Nauka, Kazah. SSR, Alma-Ata 1969, 57–62.

Titchmarsh, E. C.
[1] Reciprocal formulae involving series and integrals, *Math. Z.* 25 (1926), 321–347.
[2] *Introduction to the theory of Fourier integrals*, Oxford Univ. Press, Oxford 1937.

Tkač, B. P.
[1] Periodic solutions of a countable system of differential equations with deviating argument of neutral type (Russian), *Ukrain. Mat. Žurn.* 21 (1969), 73–85.
[2] Periodic solution of systems of partial differential equations of neutral type (Russian), *Differencial'nye Uravnenya* 5 (1969), 735–748.

Tricomi, F. G.
[1] Sulle equazioni lineari alle derivate parziali di 2° ordine di tipo misto, *Mem. Acad. Naz. Lincei*, 14 (1923).
[2] On the finite Hilbert transformation, *Quart. J. Math.* 2 (1951), 199–211.
[3] Sulle equazioni integrali del tipo di Carleman, *Ann. Mat. Pura Appl.* 39 (1955), 229–244.
[4] *Integral equations*, Interscience Publishers, London–New York 1957.

Turdiev, T.
[1] Solution of certain types of differential equations with retardation (Russian), *Trudy Sem. Teor. Diff. Uravn. s Otklon. Argum. Univ. P. Lumumby* 4 (1967), 1967–2004.

Ulashev, B.

[1] On an extremal-integro-differential equation with retarded argument. (Russian), *Izv. Akad. Nauk Uzb. SSR*, 4 (1969), 34–41.

Vekua, I. N.

[1] *New methods of solving of elliptic equations* (Russian), Moscow–Leningrad 1948.

Vekua, N. P.

[1] *Systems of singular integral equations and certain boundary problems* (Russian), Ed. II corrected and complemented. Izd. Nauka, Moscow 1970.

Viner, I. Ja.

[1] A method for the approximate determination of the solutions of quasi-linear differential-difference equations (Russian), *Ryazansk. Gos. Ped. Inst. Učen. Zap.* 35 (1963), 25–30.

[2] Approximate solution of certain classes of quasi-linear differential-difference equations (Russian), *ibidem* 31–37.

[3] Differential equations with involutions (Russian), *Differencial'nye Uravnenya* 5 (1969), 1131–1137.

Vinokurov, V. R.

[1] Volterra integral equations with infinite interval of integration (Russian), *Differencial'nye Uravnenya* 5 (1969).

Volterra, V.

[1] Sur la théorie mathématique des phénomènes héréditaires, *J. Math. Pures Appl.* 7 (1928).

Volterra, E.

[1] On elastic continua with hereditary characteristic, *J. Appl. Mech.* 18 (1951), 273–279.

Wexler, D.

[1] Solution périodique de systèmes linéaires à argument retardé, *J. Differential Equations* 3 (1967), 236–247.

[2] Solution périodique des systèmes linéaires stationnaires de type neutre, *C. R. Acad. Sci. Paris* 265 (1967), A56–A58.

[3] Periodic solutions of some stationary linear systems, *J. Differential Equations* 5 (1969), 12–31.

[4] Solutions presque périodiques des systèmes linéaires à perturbation-distribution, *Rev. Roumaine Math. Pures Appl.* 13 (1968), 111–129.

Widom, H.

[1] Singular integral equations in L_p, *Trans. Amer. Math. Soc.* 97 (1960).

Wiener, N., Hopf, E.

[1] Über eine Klasse singulärer Integralgleichungen, *BMG* 1931, 696–706.

Wierzbicki, A.

[1] The maximum principle for systems with non-trivial delays of control (Russian), *Avtomat. i Telemeh.* (1971).

Wilkowski, J.

[1] On periodic solutions of differential-difference equations with stochastic coefficients, *Bull. Acad. Polon. Sci.* 21 (1973), 253–256.

Włodarska-Dymitruk, A.

[1] On a class of solutions of differential-difference equations, *Bull. Acad. Polon. Sci.* 19 (1971), 29–35.

[2] On the control of certain linear almost periodic time lag systems, *ibidem* 17 (1971), 587–591.

[3] On the control of certain non-linear almost periodic time lag systems, *ibidem* 19 (1971), 997-1001.

[4] *Exponential-periodic solutions of differential-difference equations*, Ph. D. dissertation, Institute of Mathematics, Polish Academy of Sciences, Warszawa 1971.

von Wolfersdorf, L.

[1] Abelsche Integralgleichung und Randwertprobleme für die verallgemeinerte Tricomi-Gleichung, *Math. Nachr.* 27 (1965), 161–178.

Zamanov, T. A.

[1] Differential equations with retarded argument in abstract spaces (Russian), *Izv. Akad. Nauk Azerb. SSR*, 1967, 3–7.

Zariski, O., Samuel, P.

[1] *Commutative algebra*, vol. I, Princeton 1959.

Zverkin, A. M.

[1] Concerning linear differential equations with retarded argument and periodic coefficients (Russian), *Dokl. Akad. Nauk SSSR* 128 (1959), 882–885.

[2] Appendix in Russian edition of Bellman and Cooke [1]: Differential-difference equations with periodic coefficients (Russian).

Zygmund, A.

[1] *Hilbert transform in E^n*, Proceedings of the International Congress of Mathematicians 1954, vol. III, 140–151.

[2] On singular integrals, *Rendiconti di Matematica* 16 (1957), 468–505.

[3] *Trigonometric Series*, vol. II. University Press. Cambridge 1959 (I-st Polish ed. Warszawa 1935).

[4] *Intégrales Singulières*, Lectures Notes in Mathematics 204. Springer Verlag. New York–Heidelberg–Berlin.

Žernak, A. N.

[1] The reduction of a system of difference equations with periodic coefficients (Russian), *Izv. Vysš. Učebn. Zaved. Matematika* 7 (1969), 48–53.

Name index

Subject index